STATA TIME-SERIES
REFERENCE MANUAL
RELEASE 9

A Stata Press Publication
StataCorp LP
College Station, Texas

Stata Press, 4905 Lakeway Drive, College Station, Texas 77845

The suggested citation for this software is

StataCorp. 2005. *Stata Statistical Software: Release 9*. College Station, TX: StataCorp LP.

Table of Contents

Cross-Referencing the Documentation

When reading this manual, you will find references to other Stata manuals. For example,

[U] **26 Overview of Stata estimation commands**

[R] **regress**

[D] **reshape**

The first is a reference to chapter 26, *Overview of Stata estimation commands* in the *Stata User's Guide*, the second is a reference to the `regress` entry in the *Base Reference Manual*, and the third is a reference to the `reshape` entry in the *Data Management Reference Manual*.

All the manuals in the Stata Documentation have a shorthand notation, such as [U] for the *User's Guide* and [R] for the *Base Reference Manual*.

The complete list of shorthand notations and manuals is as follows:

[GSM]	*Getting Started with Stata for Macintosh*
[GSU]	*Getting Started with Stata for Unix*
[GSW]	*Getting Started with Stata for Windows*
[U]	*Stata User's Guide*
[R]	*Stata Base Reference Manual*
[D]	*Stata Data Management Reference Manual*
[G]	*Stata Graphics Reference Manual*
[P]	*Stata Programming Reference Manual*
[XT]	*Stata Longitudinal/Panel Data Reference Manual*
[MV]	*Stata Multivariate Statistics Reference Manual*
[SVY]	*Stata Survey Data Reference Manual*
[ST]	*Stata Survival Analysis and Epidemiological Tables Reference Manual*
[TS]	*Stata Time-Series Reference Manual*
[I]	*Stata Quick Reference and Index*
[M]	*Mata Reference Manual*

Detailed information about each of these manuals may be found online at

http://www.stata-press.com/manuals/

Title

> **intro** — Introduction to time-series manual

Description

This entry describes this manual and what has changed since Stata 8.

Remarks

This manual documents Stata's time-series commands and is referred to as [TS] in references.

Following this entry, [TS] **time series** provides an overview of the ts commands. The other parts of this manual are arranged alphabetically. If you are new to Stata's time-series features, we recommend that you read the following sections first:

[TS] **time series**   Introduction to time-series commands
[TS] **tsset**   Declare a dataset to be time-series data

Stata is continually being updated, and Stata users are always writing new commands. To ensure that you have the latest features, you should install the most recent official update; see [R] **update**.

What's new

1. arima can now fit multiplicative seasonal ARIMA (SARIMA) models; see new options sarima(), mar(), and mma() in [TS] **arima**.

2. New command rolling performs rolling window or recursive estimations, including regressions, and collects statistics from the estimation on each window.

3. This manual now has a glossary that defines commonly used terms in time-series analysis and often explains how these terms are used in the manual; see the glossary of [TS].

4. Many existing commands that previously did not allow time-series operators in their varlists now fully support them. These commands include areg, binreg, biprobit, boxcox, cloglog, cnsreg, glm, heckman, heckprob, hetprob, impute, intreg, logistic, logit, lowess, mvreg, nbreg, orthog, pcorr, poisson, probit, pwcorr, rreg, testparm, treatreg, truncreg, xtcloglog, xtgls, xtintreg, xtlogit, xtpoisson, xtprobit, xtreg, xtsum, and xttobit.

5. Many commands requiring time-series data will now work on a single panel from a panel dataset when that panel is selected using an if expression or an in qualifier. Those commands include ac, corrgram, cumsp, dfgls, dfuller, pac, pergram, pperron, wntestb, wntestq, and xcorr.

6. The dialogs for analyzing IRF results now populate lists of models and variables from the current IRF results that may be chosen to produce tables and graphs. The improved dialogs include those for irf cgraph, irf ctable, irf graph, irf ograph, and irf table.

7. dfuller has a new option, drift, for testing the null hypothesis of a random walk with drift. The algorithm for calculating MacKinnon's approximate p-values is also now more accurate in cases where the p-value is relatively large; see [TS] **dfuller**.

8. corrgram and pac have a new option, yw, which computes partial autocorrelations using the Yule–Walker equations instead of the default regression-based method; see [TS] **corrgram**.

1

9. Time-series operators are now displayed differently in estimation and other result tables.

10. The `regress` postestimation commands `dwstat`, `durbina`, `bgodfrey`, and `archlm` have been brought together under the postestimation statistics command `estat` as `estat dwatson`, `estat durbina`, `estat bgodfrey`, and `estat archlm`. All of these commands will now work on a single panel from a panel dataset. The original commands are now undocumented but continue to work. See [R] **regress postestimation time series**.

11. The ability of `arima` and `arch` to estimate standard errors using either the observed information matrix (OIM) or the outer product of gradients (OPG) has been consolidated under the new `vce()` option.

12. New commands `tsline` and `tsrline` produce line graphs using the currently `tsset` date as the x-axis. See [TS] **tsline**.

    All graphs automatically produce better label and tick values for variables having date or time-series formats.

    `tsline` and `tsrline` support the seven new options for plotting data with time-formatted variables: `tscale()`, `tlabel()`, `tmlabel()`, `ttick()`, `tmtick()`, `tline()`, and `ttext()`; see [G] *axis_options*, [G] *added_line_options*, and [G] *added_text_options*. With these options you may directly specify date literals, such as `12sep2003` or `1990q2`, to identify positions.

What's new in Stata 8.2

The July 23, 2004, update of Stata 8.2 also included significant enhancements to Stata's time-series capabilities, including new commands for fitting and analyzing cointegrated vector error-correction models (VECMs); see [TS] **vec intro**. These new features were initially documented in the second edition of the *Stata Time-Series Reference Manual* for Stata 8. Here is the *What's New* from that manual.

1. New command `vec` fits cointegrated vector error-correction models, also known as VECMs.

2. New command `vecrank` produces statistics used to determine the number of cointegrating equations in a VECM.

3. New command `fcast` replaces the old command `varfcast` and produces dynamic forecasts of the dependent variables after fitting a VAR, SVAR, or VECM.

4. New command `irf` replaces the old command `varirf` and does everything the old command did and more. `irf` estimates the impulse–response functions, cumulative impulse–response functions, orthogonalized impulse–response functions, structural impulse–response functions, and forecast-error variance decompositions (FEVDs) after fitting a VAR, SVAR, or VECM. Results can be graphed and presented in tables.

    Old command `varirf` continues to work but is not documented. If you have old `.vrf` files, they will work both with the old `varirf` command and the new `irf` command.

5. `varsoc` can be used to obtain lag-order selection statistics for VECMs, as well as VARs.

6. New command `veclmar` computes Lagrange-multiplier test statistics for residual autocorrelation after fitting a VECM.

7. New command `vecnorm` computes a series of test statistics against the null hypothesis that the disturbances are normally distributed after fitting a VECM. For each equation, and for all equations jointly, three statistics are computed: a skewness statistic, a kurtosis statistic, and the Jarque–Bera statistic.

8. New command `vecstable` checks the eigenvalue stability condition after fitting a VECM.

9. New command `vecstable` and existing command `varstable` now have a `graph` option that produces publication-quality graphs for presenting the stability results.

10. New command `haver` makes it easy to load and to analyze the economic and financial databases available from Haver Analytics.

For a complete list of all the new features in Stata 9, see [U] **1.3 What's new**.

Also See

Complementary:	[U] **1.3 What's new**
Background:	[R] **intro**

Title

> **time series** — Introduction to time-series commands

Description

The *Time-Series Reference Manual* organizes the commands alphabetically, which makes it easy to find individual command entries if you know the name of the command. This overview organizes and presents the commands conceptually, that is, according to the similarities in the functions that they perform.

The commands listed under the heading **Data management tools and time-series operators** help you prepare your data for further analysis. The commands listed under the heading **Univariate time series** are grouped together because they are either estimators or filters designed for univariate time series or pre-estimation or postestimation commands that are conceptually related to one or more univariate time-series estimators. The commands listed under the heading **Multivariate time series** are similarly grouped together because they are either estimators designed for use with multivariate time series or pre-estimation or postestimation commands conceptually related to one or more multivariate time-series estimators. Within these three broad categories, similar commands have been grouped together.

(Continued on next page)

Data management tools and time-series operators

tsset	Declare a dataset to be time-series data
tsfill	Fill in missing times with missing observations in time-series data
tsappend	Add observations to a time-series dataset
tsreport	Report time-series aspects of a dataset or estimation sample
tsrevar	Time-series operator programming command
haver	Load data from Haver Analytics database
rolling	Rolling window and recursive estimation

Univariate time series

Estimators

arima	ARIMA, ARMAX, and other dynamic regression models
arima postestimation	Postestimation tools for arima
arch	Autoregressive conditional heteroskedasticity (ARCH) family of estimators
arch postestimation	Postestimation tools for arch
newey	Regression with Newey–West standard errors
newey postestimation	Postestimation tools for newey
prais	Prais–Winsten and Cochrane–Orcutt regression
prais postestimation	Postestimation tools for prais

Time-series smoothers and filters

tssmooth ma	Moving-average filter
tssmooth dexponential	Double-exponential smoothing
tssmooth exponential	Single-exponential smoothing
tssmooth hwinters	Holt–Winters nonseasonal smoothing
tssmooth shwinters	Holt–Winters seasonal smoothing
tssmooth nl	Nonlinear filter

Diagnostic tools

corrgram	Tabulate and graph autocorrelations
xcorr	Cross-correlogram for bivariate time series
cumsp	Cumulative spectral distribution
pergram	Periodogram
dfgls	DF-GLS unit-root test
dfuller	Augmented Dickey–Fuller unit-root test
pperron	Phillips–Perron unit-root test
estat dwatson	Durbin–Watson d statistic
estat durbinalt	Durbin's alternative test for serial correlation
estat bgodfrey	Breusch–Godfrey test for higher-order serial correlation
estat archlm	Engle's LM test for the presence of autoregressive conditional heteroskedasticity
wntestb	Bartlett's periodogram-based test for white noise
wntestq	Portmanteau (Q) test for white noise

Multivariate time series

Estimators

var	Vector autoregression models
var postestimation	Postestimation tools for var
svar	Structural vector autoregression models
svar postestimation	Postestimation tools for svar
varbasic	Fit a simple VAR and graph impulse–response functions
vec	Vector error-correction models

Diagnostic tools

varlmar	Obtain LM statistics for residual autocorrelation after var or svar
varnorm	Test for normally distributed disturbances after var or svar
varsoc	Obtain lag-order selection statistics for VARs and VECMs
varstable	Check the stability condition of VAR or SVAR estimates
varwle	Obtain Wald lag-exclusion statistics after var or svar
veclmar	Obtain LM statistics for residual autocorrelation after vec
vecnorm	Test for normally distributed disturbances after vec
vecrank	Estimate the cointegrating rank using Johansen's framework
vecstable	Check the stability condition of VECM estimates

Forecasting, inference, and interpretation

irf create	Obtain impulse–response functions and FEVDs
fcast compute	Compute dynamic forecasts of dependent variables after var, svar, or vec
vargranger	Perform pairwise Granger causality tests after var or svar

Graphs and tables

corrgram	Tabulate and graph autocorrelations
xcorr	Cross-correlogram for bivariate time series
pergram	Periodogram
irf graph	Graph impulse–response functions and FEVDs
irf cgraph	Combine graphs of impulse–response functions and FEVDs
irf ograph	Graph overlaid impulse–response functions and FEVDs
irf table	Create tables of impulse–response functions and FEVDs
irf ctable	Combine tables of impulse–response functions and FEVDs
fcast graph	Graph forecasts of variables computed by fcast compute
tsline	Time-series line plot
tsrline	Time-series range plot with lines
varstable	Check the stability condition of VAR or SVAR estimates
vecstable	Check the stability condition of VECM estimates
wntestb	Bartlett's periodogram-based test for white noise

Results management tools

irf add	Add results from an IRF file to the active IRF file
irf describe	Describe an IRF file
irf drop	Drop IRF results from the active IRF file
irf rename	Rename an IRF result in an IRF file
irf set	Set the active IRF file

Remarks

Remarks are presented under the headings

> *Data management tools and time-series operators*
> *Univariate time series*
> > *Estimators*
> > *Time-series smoothers and filters*
> > *Diagnostic tools*
> *Multivariate time series*
> > *Estimators*
> > *Diagnostic tools*

Data management tools and time-series operators

Since time-series estimators are, by definition, a function of the temporal ordering of the observations in the estimation sample, Stata's time-series commands require the data to be sorted and indexed by time, using the `tsset` command, before they can be used. `tsset` is simply a way for you to tell Stata which variable in your dataset represents time; `tsset` then sorts and indexes the data appropriately for use with the time-series commands. Once your dataset has been `tsset`, you can use Stata's time-series operators in data manipulation or programming using that dataset and when specifying the syntax for most time-series commands. Stata has time-series operators for representing the lags, leads, differences, and seasonal differences of a variable. The time-series operators are documented in [TS] **tsset**.

`tsset` can also be used to declare that your dataset contains cross-sectional time-series data, often referred to as panel data. When you use `tsset` to declare your dataset to contain panel data, you specify a variable that identifies the panels, as well as identifying the time variable. Once your dataset has been `tsset` as panel data, the time-series operators work appropriately for the data.

`tsfill`, which is documented in [TS] **tsfill**, can be used after `tsset` to fill in missing times with missing observations. `tsset` will report any gaps in your data, and `tsreport` will provide additional details about the gaps. `tsappend` adds observations to a time-series dataset using the information set by `tsset`. This can be particularly useful when you wish to predict out of sample after fitting a model with a time-series estimator. `tsrevar` is a programmer's command that provides a way to use *varlists* that contain time-series operators with commands that do not otherwise support time-series operators.

The `haver` commands documented in [TS] **haver** allow you to load and describe the contents of a Haver Analytics (*www.haver.com*) file.

`rolling` performs rolling regressions, recursive regressions, and reverse recursive regressions. Any command that saves results in `e()` or `r()` can be used with `rolling`.

Univariate time series

Estimators

The four univariate time-series estimators currently available in Stata are `arima`, `arch`, `newey`, and `prais`. The latter two, `prais` and `newey`, are really just extensions to ordinary linear regression. When you fit a linear regression on time-series data via ordinary least squares, if the disturbances are autocorrelated, the parameter estimates are usually consistent, but the estimated standard errors tend to be biased downward. A number of estimators have been developed to deal with this problem. One strategy is to use OLS for estimating the regression parameters and use a different estimator for the variances, one that is consistent in the presence of autocorrelated disturbances, such as the

Newey–West estimator that is implemented in newey. An alternative strategy is to attempt to model the dynamics of the disturbances. The estimators found in prais, arima, and arch are based on such a strategy.

prais implements two such estimators: the Prais–Winsten and the Cochrane–Orcutt GLS estimators. These estimators are generalized least-squares estimators, but they are fairly restrictive in that they permit only first-order autocorrelation in the disturbances. While they have certain pedagogical and historical value, they are somewhat obsolete. Faster computers with more memory have made it possible to implement Full Information Maximum Likelihood (FIML) estimators, such as Stata's arima command. These estimators permit much greater flexibility when modeling the disturbances and are more efficient estimators.

arima provides the means to fit linear models with autoregressive moving-average (ARMA) disturbances, or in the absence of linear predictors, autoregressive integrated moving-average (ARIMA) models. This means that, whether you think that your data are best represented as a distributed-lag model, a transfer-function model, or a stochastic difference equation, or you simply wish to apply a Box–Jenkins type filter to your data, the model can be fitted using arima. arch, a conditional maximum likelihood estimator, has similar modeling capabilities for the mean of the time series but can also model autoregressive conditional heteroskedasticity in the disturbances with a wide variety of specifications for the variance equation.

Time-series smoothers and filters

In addition to the estimators mentioned above, Stata also provides six time-series filters or smoothers. Included are a simple, uniformly weighted, moving-average filter with unit weights; a weighted moving-average filter in which you can specify the weights; single- and double-exponential smoothers; Holt–Winters seasonal and nonseasonal smoothers; and a nonlinear smoother.

Most of these smoothers were originally developed as *ad hoc* procedures and are used for reducing the noise in a time series (smoothing) or forecasting. While they have limited application for signal extraction, these smoothers have all been found to be optimal for some underlying modern time-series model.

Diagnostic tools

Stata's time-series commands also include a number of pre-estimation and postestimation diagnostic commands. corrgram estimates the autocorrelation function and partial autocorrelation function of a univariate time series, as well as Q statistics. These functions and statistics are often used to determine the appropriate model specification before fitting ARIMA models. corrgram can also be used with wntestb and wntestq to examine the residuals after fitting a model for evidence of model misspecification. Stata's time-series commands also include the commands pergram and cumsp, which provide the log-standardized periodogram and the cumulative sample spectral distribution, respectively, for time-series analysts who prefer to estimate in the frequency domain rather than the time domain.

xcorr estimates the cross-correlogram for bivariate time series and can similarly be used both for pre-estimation and postestimation. For example, the cross-correlogram can be used before fitting a transfer-function model to produce initial estimates of the impulse–response function. This estimate can then be used to determine the optimal lag length of the input series to include in the model specification. It can also be used as a postestimation tool after fitting a transfer function. The cross-correlogram between the residual from a transfer-function model and the pre-whitened input series of the model can be examined for evidence of model misspecification.

When you fit ARMA or ARIMA models, the dependent variable being modeled must be covariance-stationary (ARMA models), or the order of integration must be known (ARIMA models). Stata has three commands that can test for the presence of a unit root in a time-series variable: `dfuller` performs the augmented Dickey–Fuller test, `pperron` performs the Phillips–Perron test, and `dfgls` performs a modified Dickey–Fuller test.

The remaining diagnostic tools for univariate time-series are for use after fitting a linear model via OLS with Stata's `regress` command. They are documented collectively in [R] **regress postestimation time series**. They include `estat dwatson`, `estat durbinalt`, `estat bgodfrey`, and `estat archlm`. `estat dwatson` computes the Durbin–Watson d statistic to test for the presence of first-order autocorrelation in the OLS residuals. `estat durbinalt` likewise tests for the presence of autocorrelation in the residuals. By comparison, however, Durbin's alternative test is more general and easier to use than the Durbin–Watson test. With `estat durbinalt`, you can test for higher orders of autocorrelation, the assumption that the covariates in the model are strictly exogenous is relaxed, and there is no need to consult tables to compute rejection regions, as you must with the Durbin–Watson test. `estat bgodfrey` computes the Breusch–Godfrey test for autocorrelation in the residuals, and while the computations are different, the test in `estat bgodfrey` is asymptotically equivalent to the test in `estat durbinalt`. Finally, `estat archlm` performs Engle's LM test for the presence of autoregressive conditional heteroskedasticity.

Multivariate time series

Estimators

Stata provides commands for fitting the most widely applied multivariate time-series models. `var` and `svar` fit vector autoregressions and structural vector autoregressions to stationary data. `vec` fits cointegrating vector error-correction models.

Diagnostic tools

Before fitting a multivariate time-series model, you must specify the number of lags to include. `varsoc` produces statistics for determining the order of a VAR, SVAR, or VECM.

Several postestimation commands perform the most common specification analysis on a previously fitted VAR or SVAR. You can use `varlmar` to check for serial correlation in the residuals, `varnorm` to test the null hypothesis that the disturbances come from a multivariate normal distribution, and `varstable` to see if the fitted VAR or SVAR is stable. Two common types of inference about VAR models are whether one variable Granger-causes another and whether a set of lags can be excluded from the model. `vargranger` reports Wald tests of Granger causation, and `varwle` reports Wald lag exclusion tests.

Similarly, several postestimation commands perform the most common specification analysis on a previously fitted VECM. You can use `veclmar` to check for serial correlation in the residuals, `vecnorm` to test the null hypothesis that the disturbances come from a multivariate normal distribution, and `vecstable` to analyze the stability of the previously fitted VECM.

VARs and VECMs are frequently fitted to produce baseline forecasts. `fcast` produces dynamic forecasts from previously fitted VARs and VECMs.

Many researchers fit VARs, SVARs, and VECMs because they want to analyze how unexpected shocks affect the dynamic paths of the variables. Stata has a suite of `irf` commands for estimating IRF functions and interpreting, presenting, and managing these estimates; see [TS] **irf**.

References

Baum, C. F. 2005. Stata: The language of choice for time-series analysis? *Stata Journal* 5: 46–63.

Hamilton, J. D. 1994. *Time Series Analysis.* Princeton: Princeton University Press.

Lütkepohl, H. 1993. *Introduction to Multiple Time Series Analysis.* 2nd ed. New York: Springer.

Pisati, M. 2001. sg162: Tools for spatial data analysis. *Stata Technical Bulletin* 60: 21–37. Reprinted in *Stata Technical Bulletin Reprints*, vol. 10, pp. 277–298. College Station, TX: Stata Press.

Stock, J. H. and M. W. Watson. 2001. Vector autoregressions. *Journal of Economic Perspectives* 15(4): 101–115.

Also See

Complementary:	[U] **1.3 What's new**
Background:	[R] **intro**

Title

> **arch** — Autoregressive conditional heteroskedasticity (ARCH) family of estimators

Syntax

arch *depvar* [*indepvars*] [*if*] [*in*] [*weight*] [, *options*]

options	description
Model	
<u>noc</u>onstant	suppress constant term
<u>arch</u>(*numlist*)	ARCH terms
<u>g</u>arch(*numlist*)	GARCH terms
<u>saa</u>rch(*numlist*)	simple asymmetric ARCH terms
<u>ta</u>rch(*numlist*)	threshold ARCH terms
<u>aa</u>rch(*numlist*)	asymmetric ARCH terms
<u>na</u>rch(*numlist*)	nonlinear ARCH terms
narchk(*numlist*)	nonlinear ARCH terms with single shift
<u>ab</u>arch(*numlist*)	absolute value ARCH terms
<u>at</u>arch(*numlist*)	absolute threshold ARCH terms
<u>sd</u>garch(*numlist*)	lags of σ_t
<u>ea</u>rch(*numlist*)	news terms in Nelson's (1991) EGARCH model
<u>eg</u>arch(*numlist*)	lags of $\ln(\sigma_t^2)$
<u>p</u>arch(*numlist*)	power ARCH terms
<u>tp</u>arch(*numlist*)	threshold power ARCH terms
<u>ap</u>arch(*numlist*)	asymmetric power ARCH terms
<u>np</u>arch(*numlist*)	nonlinear power ARCH terms
nparchk(*numlist*)	nonlinear power ARCH terms with single shift
<u>pg</u>arch(*numlist*)	power GARCH terms
<u>c</u>onstraints(*constraints*)	apply specified linear constraints
Model 2	
archm	include ARCH-in-mean term in the mean-equation specification
<u>archml</u>ags(*numlist*)	include specified lags of conditional variance in mean equation
<u>archme</u>xp(*exp*)	apply transformation in *exp* to any ARCH-in-mean terms
arima($\#_p$,$\#_d$,$\#_q$)	specify ARIMA(p, d, q) model for dependent variable
ar(*numlist*)	autoregressive terms of the structural model disturbance
ma(*numlist*)	moving-average terms of the structural model disturbances
Model 3	
het(*varlist*)	include *varlist* in the specification of the conditional variance
<u>saves</u>pace	conserve memory during estimation

Priming

arch0(xb)	compute priming values based on the expected unconditional variance; the default
arch0(xb0)	compute priming values based on the estimated variance of the residuals from OLS
arch0(xbwt)	compute priming values based on the weighted sum of squares from OLS residuals
arch0(xb0wt)	compute priming values based on the weighted sum of squares from OLS residuals, with more weight at earlier times
arch0(zero)	set priming values of ARCH terms to zero
arch0(#)	set priming values of ARCH terms to #
arma0(zero)	set all priming values of ARMA terms to zero; the default
arma0(p)	begin estimation after observation p, where p is the maximum AR lag in model
arma0(q)	begin estimation after observation q, where q is the maximum MA lag in model
arma0(pq)	begin estimation after observation $(p+q)$
arma0(#)	set priming values of ARMA terms to #
condobs(#)	set conditioning observations at the start of the sample to #

SE/Robust

vce(*vcetype*)	*vcetype* may be robust, oim, or opg
robust	synonym for vce(robust)

Reporting

level(#)	set confidence level; default is level(95)
detail	report list of gaps in time series

Max options

maximize_options	control the maximization process; seldom used

You must tsset your data before using arch; see [TS] **tsset**.

depvar and *varlist* may contain time-series operators; see [U] **11.4.3 Time-series varlists**.

by, rolling, statsby, and xi may be used with arch; see [U] **11.1.10 Prefix commands**.

iweights are allowed; see [U] **11.1.6 weight**.

See [U] **20 Estimation and postestimation commands** for additional capabilities of estimation commands.

To fit an ARCH($\#_m$) model, type

. arch *depvar* ... , arch(1/$\#_m$)

To fit a GARCH($\#_m, \#_k$) model, type

. arch *depvar* ... , arch(1/$\#_m$) garch(1/$\#_k$)

You can also fit many other models.

Details of syntax

The basic model `arch` fits is

$$y_t = \mathbf{x}_t \boldsymbol{\beta} + \epsilon_t$$
$$\text{Var}(\epsilon_t) = \sigma_t^2 = \gamma_0 + A(\boldsymbol{\sigma}, \boldsymbol{\epsilon}) + B(\boldsymbol{\sigma}, \boldsymbol{\epsilon})^2 \tag{1}$$

The y_t equation may optionally include ARCH-in-mean and ARMA terms:

$$y_t = \mathbf{x}_t \boldsymbol{\beta} + \sum_i \psi_i g(\sigma_{t-i}^2) + \text{ARMA}(p, q) + \epsilon_t$$

If no options are specified, $A() = B() = 0$, and the model collapses to linear regression. The following options add to $A()$ (α, γ, and κ represent parameters to be estimated):

Option	Terms added to $A()$
`arch()`	$A() = A() + \alpha_{1,1}\epsilon_{t-1}^2 + \alpha_{1,2}\epsilon_{t-2}^2 + \cdots$
`garch()`	$A() = A() + \alpha_{2,1}\sigma_{t-1}^2 + \alpha_{2,2}\sigma_{t-2}^2 + \cdots$
`saarch()`	$A() = A() + \alpha_{3,1}\epsilon_{t-1} + \alpha_{3,2}\epsilon_{t-2} + \cdots$
`tarch()`	$A() = A() + \alpha_{4,1}\epsilon_{t-1}^2(\epsilon_{t-1} > 0) + \alpha_{4,2}\epsilon_{t-2}^2(\epsilon_{t-2} > 0) + \cdots$
`aarch()`	$A() = A() + \alpha_{5,1}(|\epsilon_{t-1}| + \gamma_{5,1}\epsilon_{t-1})^2 + \alpha_{5,2}(|\epsilon_{t-2}| + \gamma_{5,2}\epsilon_{t-2})^2 + \cdots$
`narch()`	$A() = A() + \alpha_{6,1}(\epsilon_{t-1} - \kappa_{6,1})^2 + \alpha_{6,2}(\epsilon_{t-2} - \kappa_{6,2})^2 + \cdots$
`narchk()`	$A() = A() + \alpha_{7,1}(\epsilon_{t-1} - \kappa_7)^2 + \alpha_{7,2}(\epsilon_{t-2} - \kappa_7)^2 + \cdots$

The following options add to $B()$:

Option	Terms added to $B()$
`abarch()`	$B() = B() + \alpha_{8,1}|\epsilon_{t-1}| + \alpha_{8,2}|\epsilon_{t-2}| + \cdots$
`atarch()`	$B() = B() + \alpha_{9,1}|\epsilon_{t-1}|(\epsilon_{t-1} > 0) + \alpha_{9,2}|\epsilon_{t-2}|(\epsilon_{t-2} > 0) + \cdots$
`sdgarch()`	$B() = B() + \alpha_{10,1}\sigma_{t-1} + \alpha_{10,2}\sigma_{t-2} + \cdots$

Each option requires a *numlist* argument (see [U] **11.1.8 numlist**), which determines the lagged terms included. For instance, `arch(1)` specifies $\alpha_{1,1}\epsilon_{t-1}^2$, `arch(2)` specifies $\alpha_{1,2}\epsilon_{t-2}^2$, `arch(1,2)` specifies $\alpha_{1,1}\epsilon_{t-1}^2 + \alpha_{1,2}\epsilon_{t-2}^2$, `arch(1/3)` specifies $\alpha_{1,1}\epsilon_{t-1}^2 + \alpha_{1,2}\epsilon_{t-2}^2 + \alpha_{1,3}\epsilon_{t-3}^2$, etc.

If options `earch()` or `egarch()` are specified, the basic model fitted is

$$y_t = \mathbf{x}_t \boldsymbol{\beta} + \sum_i \psi_i g(\sigma_{t-i}^2) + \text{ARMA}(p, q) + \epsilon_t$$
$$\ln \text{Var}(\epsilon_t) = \ln \sigma_t^2 = \gamma_0 + C(\ln\boldsymbol{\sigma}, \mathbf{z}) + A(\boldsymbol{\sigma}, \boldsymbol{\epsilon}) + B(\boldsymbol{\sigma}, \boldsymbol{\epsilon})^2 \tag{2}$$

where $z_t = \epsilon_t / \sigma_t$. $A()$ and $B()$ are given as above, but $A()$ and $B()$ now add to $\ln \sigma_t^2$ rather than σ_t^2. (The options corresponding to $A()$ and $B()$ are rarely specified in this case.) $C()$ is given by

Option	Terms added to $C()$
`earch()`	$C() = C() + \alpha_{11,1}z_{t-1} + \gamma_{11,1}(\lvert z_{t-1}\rvert - \sqrt{2/\pi})$ $\qquad\qquad + \alpha_{11,2}z_{t-2} + \gamma_{11,2}(\lvert z_{t-2}\rvert - \sqrt{2/\pi}) + \cdots$
`egarch()`	$C() = C() + \alpha_{12,1}\ln\sigma_{t-1}^2 + \alpha_{12,2}\ln\sigma_{t-2}^2 + \cdots$

Alternatively, if options `parch()`, `tparch()`, `aparch()`, `nparch()`, `nparchk()`, or `pgarch()` are specified, the basic model fitted is

$$y_t = \mathbf{x}_t\boldsymbol{\beta} + \sum_i \psi_i g(\sigma_{t-i}^2) + \mathrm{ARMA}(p,q) + \epsilon_t$$

$$\{\mathrm{Var}(\epsilon_t)\}^{\varphi/2} = \sigma_t^{\varphi} = \gamma_0 + D(\boldsymbol{\sigma},\boldsymbol{\epsilon}) + A(\boldsymbol{\sigma},\boldsymbol{\epsilon}) + B(\boldsymbol{\sigma},\boldsymbol{\epsilon})^2 \tag{3}$$

where φ is a parameter to be estimated. $A()$ and $B()$ are given as above, but $A()$ and $B()$ now add to σ_t^{φ}. (The options corresponding to $A()$ and $B()$ are rarely specified in this case.) $D()$ is given by

Option	Terms added to $D()$
`parch()`	$D() = D() + \alpha_{13,1}\epsilon_{t-1}^{\varphi} + \alpha_{13,2}\epsilon_{t-2}^{\varphi} + \cdots$
`tparch()`	$D() = D() + \alpha_{14,1}\epsilon_{t-1}^{\varphi}(\epsilon_{t-1} > 0) + \alpha_{14,2}\epsilon_{t-2}^{\varphi}(\epsilon_{t-2} > 0) + \cdots$
`aparch()`	$D() = D() + \alpha_{15,1}(\lvert\epsilon_{t-1}\rvert + \gamma_{15,1}\epsilon_{t-1})^{\varphi} + \alpha_{15,2}(\lvert\epsilon_{t-2}\rvert + \gamma_{15,2}\epsilon_{t-2})^{\varphi} + \cdots$
`nparch()`	$D() = D() + \alpha_{16,1}\lvert\epsilon_{t-1} - \kappa_{16,1}\rvert^{\varphi} + \alpha_{16,2}\lvert\epsilon_{t-2} - \kappa_{16,2}\rvert^{\varphi} + \cdots$
`nparchk()`	$D() = D() + \alpha_{17,1}\lvert\epsilon_{t-1} - \kappa_{17}\rvert^{\varphi} + \alpha_{17,2}\lvert\epsilon_{t-2} - \kappa_{17}\rvert^{\varphi} + \cdots$
`pgarch()`	$D() = D() + \alpha_{18,1}\sigma_{t-1}^{\varphi} + \alpha_{18,2}\sigma_{t-2}^{\varphi} + \cdots$

Common models

Common term	Options to specify
ARCH (Engle 1982)	`arch()`
GARCH (Bollerslev 1986)	`arch() garch()`
ARCH-in-mean (Engle, Lilien, and Robins 1987)	`archm arch()` $\big[$`garch()`$\big]$
GARCH with ARMA terms	`arch() garch() ar() ma()`
EGARCH (Nelson 1991)	`earch() egarch()`
TARCH, threshold ARCH (Zakoian 1990)	`abarch() atarch() sdgarch()`
GJR, form of threshold ARCH (Glosten, Jagannathan, and Runkle 1993)	`arch() tparch()` $\big[$`garch()`$\big]$
SAARCH, simple asymmetric ARCH (Engle 1990)	`arch() saarch()` $\big[$`garch()`$\big]$
PARCH, power ARCH (Higgins and Bera 1992)	`parch()` $\big[$`pgarch()`$\big]$
NARCH, nonlinear ARCH	`narch()` $\big[$`garch()`$\big]$
NARCHK, nonlinear ARCH with a single shift	`narchk()` $\big[$`garch()`$\big]$
A-PARCH, asymmetric power ARCH (Ding, Granger, and Engle 1993)	`aparch()` $\big[$`pgarch()`$\big]$
NPARCH, nonlinear power ARCH	`nparch()` $\big[$`pgarch()`$\big]$

In all cases, you type

$$\texttt{arch } depvar\ [\,indepvars\,]\,,\ options$$

where *options* are chosen from the table above. Each option requires that you specify as its argument a *numlist* that specifies the lags to be included. For most ARCH models, that value will be 1. For instance, to fit the classic first-order GARCH model on `cpi`, you would type

    . arch cpi, arch(1) garch(1)

If you wanted to fit a first-order GARCH model of `cpi` on `wage`, you would type

    . arch cpi wage, arch(1) garch(1)

If, for any of the options, you want first- and second-order terms, specify *optionname*(1/2). Specifying `garch(1) arch(1/2)` would fit a GARCH model with first- and second-order ARCH terms. If you specified `arch(2)`, only the lag 2 term would be included.

Reading arch output

The regression table reported by `arch` will appear as

op.depvar	Coef.	Std. Err.	z	P>|z|	[95% Conf. Interval]
depvar					
x1	# ...				
x2					
L1.	# ...				
L2.	# ...				
_cons	# ...				
ARCHM					
sigma2	# ...				
ARMA					
ar					
L1.	# ...				
ma					
L1.	# ...				
HET					
z1	# ...				
z2					
L1.	# ...				
L2.	# ...				
ARCH					
arch					
L1.	# ...				
garch					
L1.	# ...				
aparch					
L1.	# ...				
etc.					
_cons	# ...				
POWER					
power	# ...				

Dividing lines separate "equations".

The first one, two, or three equations report the mean model:

$$y_t = \mathbf{x}_t \boldsymbol{\beta} + \sum_i \psi_i g(\sigma^2_{t-i}) + \mathrm{ARMA}(p, q) + \epsilon_t$$

The first equation reports $\boldsymbol{\beta}$, and the equation will be named [*depvar*]; for examples, if you fitted a model on d.cpi, the first equation would be named [cpi]. In Stata, the coefficient on x1 in the above example could be referred to as [*depvar*]_b[x1]. The coefficient on the lag 2 value of x2 would be referred to as [*depvar*]_b[L2.x2]. Such notation would be used, for instance, in a subsequent test command; see [R] **test**.

The [ARCHM] equation reports the ψ coefficients if your model includes ARCH-in-mean terms; see options discussed under the **Model 2** tab above. Most ARCH-in-mean models include only a contemporaneous variance term, so the term $\sum_i \psi_i g(\sigma^2_{t-i})$ becomes $\psi\sigma^2_t$. The coefficient ψ will be [ARCHM]_b[sigma2]. If your model includes lags of σ^2_t, the additional coefficients will be [ARCHM]_b[L1.sigma2], and so on. If you specify a transformation $g()$ (option archmexp()), the coefficients will be [ARCHM]_b[sigma2ex], [ARCHM]_b[L1.sigma2ex], and so on. sigma2ex refers to $g(\sigma^2_t)$, the transformed value of the conditional variance.

The [ARMA] equation reports the ARMA coefficients if your model includes them; see options discussed under the **Model 2** tab above. This equation includes one or two "variables" named ar and ma. In subsequent test statements, you could refer to the coefficient on the first lag of the autoregressive term by typing [ARMA]_b[L1.ar] or simply [ARMA]_b[L.ar] (the L operator is assumed to be lag 1 if you do not specify otherwise). The second lag on the moving-average term, if there were one, could be referred to by typing [ARMA]_b[L2.ma].

The last one, two, or three equations report the variance model.

The [HET] equation reports the multiplicative heteroskedasticity if the model includes it; see *Other options affecting specification of variance*. When you fit such a model, you specify the variables (and their lags), determining the multiplicative heteroskedasticity; after estimation, their coefficients are simply [HET]_b[*op.varname*].

The [ARCH] equation reports the ARCH, GARCH, etc., terms by referring to "variables" arch, garch, and so on. For instance, if you specified arch(1) garch(1) when you fitted the model, the conditional variance is given by $\sigma^2_t = \gamma_0 + \alpha_{1,1}\epsilon^2_{t-1} + \alpha_{2,1}\sigma^2_{t-1}$. The coefficients would be named [ARCH]_b[_cons] (γ_0), [ARCH]_b[L.arch] ($\alpha_{1,1}$), and [ARCH]_b[L.garch] ($\alpha_{2,1}$).

The [POWER] equation appears only if you are fitting a variance model in the form of (3) above; the estimated φ is the coefficient [POWER]_b[power].

The naming convention for estimated ARCH, GARCH, etc., parameters is as follows (definitions for parameters α_i, γ_i, and κ_i can be found in the tables for $A()$, $B()$, $C()$, and $D()$ above):

Option	1st parameter	2nd parameter	Common parameter
arch()	α_1 = [ARCH]_b[arch]		
garch()	α_2 = [ARCH]_b[garch]		
saarch()	α_3 = [ARCH]_b[saarch]		
tarch()	α_4 = [ARCH]_b[tarch]		
aarch()	α_5 = [ARCH]_b[aarch]	γ_5 = [ARCH]_b[aarch_e]	
narch()	α_6 = [ARCH]_b[narch]	κ_6 = [ARCH]_b[narch_k]	
narchk()	α_7 = [ARCH]_b[narch]	κ_7 = [ARCH]_b[narch_k]	
abarch()	α_8 = [ARCH]_b[abarch]		
atarch()	α_9 = [ARCH]_b[atarch]		
sdgarch()	α_{10} = [ARCH]_b[sdgarch]		
earch()	α_{11} = [ARCH]_b[earch]	γ_{11} = [ARCH]_b[earch_a]	
egarch()	α_{12} = [ARCH]_b[egarch]		
parch()	α_{13} = [ARCH]_b[parch]		φ = [POWER]_b[power]
tparch()	α_{14} = [ARCH]_b[tparch]		φ = [POWER]_b[power]
aparch()	α_{15} = [ARCH]_b[aparch]	γ_{15} = [ARCH]_b[aparch_e]	φ = [POWER]_b[power]
nparch()	α_{16} = [ARCH]_b[nparch]	κ_{16} = [ARCH]_b[nparch_k]	φ = [POWER]_b[power]
nparchk()	α_{17} = [ARCH]_b[nparch]	κ_{17} = [ARCH]_b[nparch_k]	φ = [POWER]_b[power]
pgarch()	α_{18} = [ARCH]_b[pgarch]		φ = [POWER]_b[power]

Description

arch fits regression models in which the volatility of a series varies through time. Usually, periods of high and low volatility are grouped together. ARCH models estimate future volatility as a function of prior volatility. To accomplish this, arch fits models of autoregressive conditional heteroskedasticity (ARCH) using conditional maximum likelihood. In addition to ARCH terms, models may include multiplicative heteroskedasticity.

Concerning the regression equation itself, models may also contain ARCH-in-mean and ARMA terms.

Options

```
                Model
```

noconstant; see [TS] **estimation options**.

arch(*numlist*) specifies the ARCH terms (lags of ϵ_t^2).

Specify arch(1) to include first-order terms, arch(1/2) to specify first- and second-order terms, arch(1/3) to specify first-, second-, and third-order terms, etc. Terms may be omitted. Specify arch(1/3 5) to specify terms with lags 1, 2, 3, and 5. All the options work this way.

arch() may not be specified with aarch(), narch(), narchk(), nparchk(), or nparch(), as this would result in collinear terms.

garch(*numlist*) specifies the GARCH terms (lags of σ_t^2).

saarch(*numlist*) specifies the simple asymmetric ARCH terms. Adding these terms is one way to make the standard ARCH and GARCH models respond asymmetrically to positive and negative innovations. Specifying saarch() with arch() and garch() corresponds to the SAARCH model of Engle (1990).

saarch() may not be specified with narch(), narchk(), nparchk(), or nparch(), as this would result in collinear terms.

tarch(*numlist*) specifies the threshold ARCH terms. Adding these is another way to make the standard ARCH and GARCH models respond asymmetrically to positive and negative innovations. Specifying tarch() with arch() and garch() corresponds to one form of the GJR model (Glosten, Jagannathan, and Runkle 1993).

tarch() may not be specified with tparch() or aarch(), as this would result in collinear terms.

aarch(*numlist*) specifies the lags of the two-parameter term $\alpha_i(|\epsilon_t| + \gamma_i\epsilon_t)^2$. This term provides the same underlying form of asymmetry as including arch() and tarch() but is expressed in a different way.

aarch() may not be specified with arch() or tarch(), as this would result in collinear terms.

narch(*numlist*) specifies the lags of the two-parameter term $\alpha_i(\epsilon_t - \kappa_i)^2$. This term allows the minimum conditional variance to occur at a value of lagged innovations other than zero. For any term specified at lag L, the minimum contribution to conditional variance of that lag occurs when $\epsilon_{t-L}^2 = \kappa_L$—the squared innovations at that lag are equal to the estimated constant κ_L.

narch() may not be specified with arch(), saarch(), narchk(), nparchk(), or nparch(), as this would result in collinear terms.

narchk(*numlist*) specifies the lags of the two-parameter term $\alpha_i(\epsilon_t - \kappa)^2$; note that this is a variation of narch() with κ held constant for all lags.

narchk() may not be specified with arch(), saarch(), narch(), nparchk(), or nparch(), as this would result in collinear terms.

abarch(*numlist*) specifies lags of the term $|\epsilon_t|$.

atarch(*numlist*) specifies lags of $|\epsilon_t|(\epsilon_t > 0)$, where $(\epsilon_t > 0)$ represents the indicator function returning 1 when true and 0 when false. Like the TARCH terms, these ATARCH terms allow the effect of unanticipated innovations to be asymmetric about zero.

sdgarch(*numlist*) specifies lags of σ_t. Combining atarch(), abarch(), and sdgarch() produces the model by Zakoian (1990) that the author called the TARCH model. The acronym TARCH, however, refers to any model using thresholding to obtain asymmetry.

earch(*numlist*) specifies lags of the two-parameter term $\alpha z_t + \gamma(|z_t| - \sqrt{2/\pi})$. These terms represent the influence of news—lagged innovations—in Nelson's (1991) EGARCH model. For these terms, $z_t = \epsilon_t/\sigma_t$, and arch assumes $z_t \sim N(0, 1)$. Nelson derived the general form of an EGARCH model for any assumed distribution and performed estimation assuming a Generalized Error Distribution (GED). See Hamilton (1994) for a derivation where z_t is assumed normal. The z_t terms can be parameterized in either of these two equivalent ways. arch uses Nelson's original parameterization; see Hamilton (1994) for an equivalent alternative.

egarch(*numlist*) specifies lags of $\ln(\sigma_t^2)$.

For the following options, note that the model is parameterized in terms of $h(\epsilon_t)^\varphi$ and σ_t^φ. A single φ is estimated, even when more than one option is specified.

parch(*numlist*) specifies lags of $|\epsilon_t|^\varphi$. parch() combined with pgarch() corresponds to the class of nonlinear models of conditional variance suggested by Higgins and Bera (1992).

tparch(*numlist*) specifies lags of $(\epsilon_t > 0)|\epsilon_t|^\varphi$, where $(\epsilon_t > 0)$ represents the indicator function returning 1 when true and 0 when false. As with tarch(), tparch() specifies terms that allow for a differential impact of "good" (positive innovations) and "bad" (negative innovations) news for lags specified by *numlist*.

tparch() may not be specified with tarch(), as this would result in collinear terms.

aparch(*numlist*) specifies lags of the two-parameter term $\alpha(|\epsilon_t| + \gamma\epsilon_t)^\varphi$. This asymmetric power ARCH model, A-PARCH, was proposed by Ding, Granger, and Engle (1993) and corresponds to a Box–Cox function in the lagged innovations. The authors fitted the original A-PARCH model on over 16,000 daily observations of the Standard and Poor's 500, and not without good reason. As the number of parameters and the flexibility of the specification increase, larger amounts of data are required to estimate the parameters of the conditional heteroskedasticity. See Ding, Granger, and Engle (1993) for a discussion of how 7 popular ARCH models nest within the A-PARCH model.

Note that when γ goes to 1, the full term goes to zero for many observations and, at this point, can be numerically unstable.

nparch(*numlist*) specifies lags of the two-parameter term $\alpha|\epsilon_t - \kappa_i|^\varphi$.

nparch() may not be specified with arch(), saarch(), narch(), narchk(), or nparchk(), as this would result in collinear terms.

nparchk(*numlist*) specifies lags of the two-parameter term $\alpha|\epsilon_t - \kappa|^\varphi$; note that this is a variation of nparch() with κ held constant for all lags. This is the direct analog of narchk(), except for the power of φ. nparchk() corresponds to an extended form of the model of Higgins and Bera (1992) as presented by Bollerslev, Engle, and Nelson (1994). nparchk() would typically be combined with the option pgarch().

nparchk() may not be specified with arch(), saarch(), narch(), narchk(), or nparch(), as this would result in collinear terms.

pgarch(*numlist*) specifies lags of σ_t^φ.

constraints(*constraints*); see [TS] **estimation options**.

        | Model 2 |

archm specifies that an ARCH-in-mean term be included in the specification of the mean equation. This term allows the expected value of *depvar* to depend on the conditional variance. ARCH-in-mean is most commonly used in evaluating financial time series when a theory supports a trade-off between asset risk and return. By default, no ARCH-in-mean terms are included in the model.

archm specifies that the contemporaneous expected conditional variance be included in the mean equation. For example, typing

    . arch y x, archm arch(1)

specifies the model

$$\mathbf{y}_t = \beta_0 + \beta_1\mathbf{x}_t + \psi\sigma_t^2 + \epsilon_t$$
$$\sigma_t^2 = \gamma_0 + \gamma\epsilon_{t-1}^2$$

archmlags(*numlist*) is an expansion of archm that includes lags of the conditional variance σ_t^2 in the mean equation. To specify a contemporaneous and once-lagged variance, specify either archm archmlags(1) or archmlags(0/1).

archmexp(*exp*) applies the transformation in *exp* to any ARCH-in-mean terms in the model. The expression should contain an X wherever a value of the conditional variance is to enter the expression. This option can be used to produce the commonly used ARCH-in-mean of the conditional standard deviation. Using the example from archm, typing

    . arch y x, archm arch(1) archmexp(sqrt(X))

specifies the mean equation $y_t = \beta_0 + \beta_1 x_t + \psi \sigma_t + \epsilon_t$. Alternatively, typing

> . arch y x, archm arch(1) archmexp(1/sqrt(X))

specifies $y_t = \beta_0 + \beta_1 x_t + \psi / \sigma_t + \epsilon_t$.

arima(#$_p$,#$_d$,#$_q$) is an alternative, shorthand notation for specifying autoregressive models in the dependent variable. The dependent variable and any independent variables are differenced #$_d$ times, 1 through #$_p$ lags of autocorrelations are included, and 1 through #$_q$ lags of moving averages are included. For example, the specification

> . arch y, arima(2,1,3)

is equivalent to

> . arch D.y, ar(1/2) ma(1/3)

The former is easier to write for "classic" ARIMA models of the mean equation, but it is not nearly as expressive as the latter. If gaps in the AR or MA lags are to be modeled, or if different operators are to be applied to independent variables, the latter syntax is required.

ar(*numlist*) specifies the autoregressive terms of the structural model disturbance to be included in the model. For example, ar(1/3) specifies that lags 1, 2, and 3 of the structural disturbance be included in the model. ar(1,4) specifies that lags 1 and 4 be included, possibly to account for quarterly effects.

If the model does not contain regressors, these terms can also be considered autoregressive terms for the dependent variable; see [TS] **arima**.

ma(*numlist*) specifies the moving-average terms to be included in the model. These are the terms for the lagged innovations or white-noise disturbances.

Model 3

het(*varlist*) specifies that *varlist* be included in the specification of the conditional variance. *varlist* may contain time-series operators. This varlist enters the variance specification collectively as multiplicative heteroskedasticity; see Judge et al. (1985). If het() is not specified, the model will not contain multiplicative heteroskedasticity.

Assume that the conditional variance depends on variables x and w and also has an ARCH(1) component. We request this specification by using the options het(x w) arch(1), and this corresponds to the conditional-variance model

$$\sigma_t^2 = \exp(\lambda_0 + \lambda_1 x_t + \lambda_2 w_t) + \alpha \epsilon_{t-1}^2$$

Multiplicative heteroskedasticity enters differently with an EGARCH model because the variance is already specified in logs. For the options het(x w) earch(1) egarch(1), the variance model is

$$\ln(\sigma_t^2) = \lambda_0 + \lambda_1 x_t + \lambda_2 w_t + \alpha z_{t-1} + \gamma(|z_{t-1}| - \sqrt{2/\pi}) + \delta \ln(\sigma_{t-1}^2)$$

savespace conserves memory by retaining only those variables required for estimation. The original dataset is restored after estimation. This option is rarely used and should be specified only if there is insufficient memory to fit a model without the option. Note that arch requires considerably more temporary storage during estimation than most estimation commands in Stata.

⎡ Priming ⎤

arch0(*cond_method*) is a rarely used option that specifies how to compute the conditioning (presample or priming) values for σ_t^2 and ϵ_t^2. In the presample period, it is assumed that $\sigma_t^2 = \epsilon_t^2$ and that this value is constant. If arch0() is not specified, the priming values are computed as the expected unconditional variance given the current estimates of the β coefficients and any ARMA parameters.

arch0(xb), the default, specifies that the priming values are the expected unconditional variance of the model, which is $\sum_1^T \widehat{\epsilon}_t^2 / T$, where $\widehat{\epsilon}_t$ is computed from the mean equation and any ARMA terms.

arch0(xb0) specifies that the priming values are the estimated variance of the residuals from an OLS estimate of the mean equation.

arch0(xbwt) specifies that the priming values are the weighted sum of the $\widehat{\epsilon}_t^2$ from the current conditional mean equation (and ARMA terms) that places more weight on estimates of ϵ_t^2 at the beginning of the sample.

arch0(xb0wt) specifies that the priming values are the weighted sum of the $\widehat{\epsilon}_t^2$ from an OLS estimate of the mean equation (and ARMA terms) that places more weight on estimates of ϵ_t^2 at the beginning of the sample.

arch0(zero) specifies that the priming values are 0. Unlike the priming values for ARIMA models, 0 is generally not a consistent estimate of the presample conditional variance or squared innovations.

arch0(#) specifies that $\sigma_t^2 = \epsilon_t^2 = \#$ for any specified non-negative #. Thus arch0(0) is equivalent to arch0(zero).

arma0(*cond_method*) is a rarely used option that specifies how the ϵ_t values are initialized at the beginning of the sample for the ARMA component, if the model has one. This option has an effect only when AR or MA terms are included in the model (options ar(), ma(), or arima() specified).

arma0(zero), the default, specifies that all priming values of ϵ_t be taken as 0. This fits the model over the entire requested sample and takes ϵ_t as its expected value of 0 for all lags required by the ARMA terms; see Judge et al. (1985).

arma0(p), arma0(q), and arma0(pq) specify that estimation begin after priming the recursions for a certain number of observations. p specifies that estimation begin after the pth observation in the sample, where p is the maximum AR lag in the model; q specifies that estimation begin after the qth observation in the sample, where q is the maximum MA lag in the model; and pq specifies that estimation begin after the $(p + q)$th observation in the sample.

During the priming period, the recursions necessary to generate predicted disturbances are performed, but results are used only to initialize pre-estimation values of ϵ_t. To understand the definition of pre-estimation, say that you fit a model in 10/100. If the model is specified with ar(1,2), pre-estimation refers to observations 10 and 11.

The ARCH terms σ_t^2 and ϵ_t^2 are also updated over these observations. Any required lags of ϵ_t before the priming period are taken to be their expected value of 0, and ϵ_t^2 and σ_t^2 take the values specified in arch0().

arma0(#) specifies that the presample values of ϵ_t are to be taken as # for all lags required by the ARMA terms. Thus arma0(0) is equivalent to arma0(zero).

condobs(#) is a rarely used option that specifies a fixed number of conditioning observations at the start of the sample. Over these priming observations, the recursions necessary to generate predicted disturbances are performed, but only to initialize pre-estimation values of ϵ_t, ϵ_t^2, and σ_t^2. Any required lags of ϵ_t before the initialization period are taken to be their expected value of 0 (or the value specified in arma0()), and required values of ϵ_t^2 and σ_t^2 assume the values specified by arch0(). condobs() can be used if conditioning observations are desired for the lags in the ARCH terms of the model. If arma() is also specified, the maximum number of conditioning observations required by arma() and condobs(#) is used.

_____ SE/Robust _____

vce(*vcetype*); see [R] *vce_option*.

robust; see [TS] **estimation options**.

For ARCH models, the robust or quasi-maximum likelihood estimates (QMLE) of variance are robust to symmetric non-normality in the disturbances. The robust variance estimates generally are not robust to functional misspecification of the mean equation; see Bollerslev and Wooldridge (1992).

Note that the robust variance estimates computed by arch are based on the full Huber/White/sandwich formulation, as discussed in [P] **_robust**. In fact, many software packages report robust estimates that set some terms to their expectations of zero (Bollerslev and Wooldridge 1992), which saves them from calculating second derivatives of the log-likelihood function.

_____ Reporting _____

level(#); see [TS] **estimation options**.

detail specifies that a detailed list of any gaps in the series be reported, including gaps due to missing observations or missing data for the dependent variable or independent variables.

_____ Max options _____

maximize_options: difficult, technique(*algorithm_spec*), iterate(#), [no]log, trace, gradient, showstep, hessian, shownrtolerance, tolerance(#), ltolerance(#), gtolerance(#), nrtolerance(#), nonrtolerance, from(*init_specs*); see [R] **maximize**.

These options are often more important for ARCH models than for other maximum likelihood models because of convergence problems associated with ARCH models—ARCH model likelihoods are notoriously difficult to maximize.

The following options are all related to maximization and are either particularly important in fitting ARCH models or not available for most other estimators.

gtolerance(#) specifies the threshold for the relative size of the gradient; see [R] **maximize**. The default for arch is gtolerance(.05).

gtolerance(999) may be specified to disable the gradient criterion. If the optimizer becomes stuck with repeated "(backed up)" messages, it is likely that the gradient still contains substantial values, but an uphill direction cannot be found for the likelihood. With this option, results can often be obtained, but it is unclear whether the global maximum likelihood has been found.

When the maximization is not going well, it is also possible to set the maximum number of iterations (see [R] **maximize**) to the point where the optimizer appears to be stuck and to inspect the estimation results at that point.

from(*init_specs*) specifies the initial values of the coefficients. ARCH models may be sensitive to initial values and may have coefficient values that correspond to local maxima. The default starting values are obtained via a series of regressions, producing results that, based on asymptotic theory, are consistent for the β and ARMA parameters and generally reasonable for the rest. Nevertheless, these values may not always be feasible in that the likelihood function cannot be evaluated at the initial values arch first chooses. In such cases, the estimation is restarted with ARCH and ARMA parameters initialized to zero. It is possible, but unlikely, that even these values will be infeasible and that you will have to supply initial values yourself.

The standard syntax for from() accepts a matrix, a list of values, or coefficient name value pairs; see [R] **maximize**. In addition, arch allows the following:

from(archb0) sets the starting value for all the ARCH/GARCH/... parameters in the conditional-variance equation to 0.

from(armab0) sets the starting value for all ARMA parameters in the model to 0.

from(archb0 armab0) sets the starting value for all ARCH/GARCH/... and ARMA parameters to 0.

Remarks

The volatility of a series is not constant through time, periods of relatively low volatility and periods of relatively high volatility tend to be grouped together. This is a commonly observed characteristic of economic time series and is even more pronounced in many frequently sampled financial series. ARCH models seek to estimate this time-dependent volatility as a function of observed prior volatility. In some cases, the model of volatility is of more interest than the model of the conditional mean. As implemented in arch, the volatility model may also include regressors to account for a structural component in the volatility—usually referred to as multiplicative heteroskedasticity.

ARCH models were introduced by Engle (1982) in a study of inflation rates, and there has since been a barrage of proposed parametric and nonparametric specifications of autoregressive conditional heteroskedasticity. Overviews of the literature can found in Bollerslev, Engle, and Nelson (1994) and Bollerslev, Chou, and Kroner (1992). Introductions to basic ARCH models appear in many general econometrics texts, including Davidson and MacKinnon (1993, 2004), Greene (2003), Kmenta (1997), Johnston and DiNardo (1997), and Wooldridge (2002). Harvey (1989) and Enders (1995) provide introductions to ARCH in the larger context of econometric time-series modeling, and Hamilton (1994) gives considerably more detail in the same context.

arch fits models of autoregressive conditional heteroskedasticity (ARCH, GARCH, etc.) using conditional maximum likelihood. By "conditional", we mean that the likelihood is computed based on an assumed or estimated set of priming values for the squared innovations ϵ_t^2 and variances σ_t^2 prior to the estimation sample; see, for example, Hamilton (1994) or Bollerslev (1986). Sometimes additional conditioning is done on the first a, g, or $a + g$ observations in the sample, where a is the maximum ARCH term lag and g is the maximum GARCH term lag (or the maximum lags from the other ARCH family terms).

The original ARCH model proposed by Engle (1982) modeled the variance of a regression model's disturbances as a linear function of lagged values of the squared regression disturbances. We can write an ARCH(m) model as

$$y_t = \mathbf{x}_t \boldsymbol{\beta} + \epsilon_t \qquad \text{(conditional mean)}$$
$$\sigma_t^2 = \gamma_0 + \gamma_1 \epsilon_{t-1}^2 + \gamma_2 \epsilon_{t-2}^2 + \cdots + \gamma_m \epsilon_{t-m}^2 \qquad \text{(conditional variance)}$$

where

$$\epsilon_t \sim N(0, \sigma_t^2)$$

ϵ_t^2 is the squared residuals (or innovations)

γ_i are the ARCH parameters

The ARCH model has a specification for both the conditional mean and the conditional variance, and the variance is a function of the size of prior unanticipated innovations—ϵ_t^2. This model was generalized by Bollerslev (1986) to include lagged values of the conditional variance—a GARCH model. The GARCH(m, k) model is written as

$$y_t = \mathbf{x}_t\boldsymbol{\beta} + \epsilon_t$$
$$\sigma_t^2 = \gamma_0 + \gamma_1\epsilon_{t-1}^2 + \gamma_2\epsilon_{t-2}^2 + \cdots + \gamma_m\epsilon_{t-m}^2 + \delta_1\sigma_{t-1}^2 + \delta_2\sigma_{t-2}^2 + \cdots + \delta_k\sigma_{t-k}^2$$

where

$$\gamma_i \text{ are the ARCH parameters}$$

$$\delta_i \text{ are the GARCH parameters}$$

The GARCH model of conditional variance can be considered an ARMA process in the squared innovations, although not in the variances as the equations might seem to suggest; see, for example, Hamilton (1994). Specifically, the standard GARCH model implies that the squared innovations result from

$$\epsilon_t^2 = \gamma_0 + (\gamma_1 + \delta_1)\epsilon_{t-1}^2 + (\gamma_2 + \delta_2)\epsilon_{t-2}^2 + \cdots + (\gamma_k + \delta_k)\epsilon_{t-k}^2 + w_t - \delta_1 w_{t-1} - \delta_2 w_{t-2} - \delta_3 w_{t-3}$$

where

$$w_t = \epsilon_t^2 - \sigma_t^2$$
$$w_t \text{ is a white-noise process that is fundamental for } \epsilon_t^2$$

One of the primary benefits of the GARCH specification is its parsimony in identifying the conditional variance. As with ARIMA models, the ARMA specification in GARCH allows the conditional variance to be modeled with fewer parameters than with an ARCH specification alone. Empirically, many series with a conditionally heteroskedastic disturbance have been adequately modeled with a GARCH(1,1) specification.

An ARMA process in the disturbances can be added easily to the mean equation. For example, the mean equation can be written with an ARMA(1, 1) disturbance as

$$y_t = \mathbf{x}_t\boldsymbol{\beta} + \rho(y_{t-1} - \mathbf{x}_{t-1}\boldsymbol{\beta}) + \theta\epsilon_{t-1} + \epsilon_t$$

with an obvious generalization to ARMA(p, q) by adding terms; see [TS] **arima** for more discussion of this specification. This change affects only the conditional-variance specification in that ϵ_t^2 now results from a different specification of the conditional mean.

Much of the literature on ARCH models focuses on alternative specifications of the variance equation. `arch` allows many of these specifications to be requested using the options `saarch()` through `pgarch()`, which imply that one or more terms may be changed or added to the specification of the variance equation.

These alternative specifications also address asymmetry. Both the ARCH and GARCH specifications imply a symmetric impact of innovations. Whether an innovation ϵ_t^2 is positive or negative makes no difference to the expected variance σ_t^2 in the ensuing periods; only the size of the innovation matters—good news and bad news have the same effect. Many theories, however, suggest that positive and negative innovations should vary in their impact. For risk-averse investors, a large unanticipated drop in the market is more likely to lead to higher volatility than a large unanticipated increase (see Black 1976, Nelson 1991). `saarch()`, `tarch()`, `aarch()`, `abarch()`, `earch()`, `aparch()`, and `tparch()` allow various specifications of asymmetric effects.

`narch()`, `narchk()`, `nparch()`, and `nparchk()` imply an asymmetric impact of a very specific form. All the models considered so far have a minimum conditional variance when the lagged innovations are all zero. "No news is good news" when it comes to keeping the conditional variance small. `narch()`, `narchk()`, `nparch()`, and `nparchk()` also have a symmetric response to innovations, but they are not centered at zero. The entire news-response function (response to innovations) is shifted horizontally so that minimum variance lies at some specific positive or negative value for prior innovations.

ARCH-in-mean models allow the conditional variance of the series to influence the conditional mean. This is particularly convenient for modeling the risk/return relationship in financial series; the riskier an investment, with all else equal, the lower its expected return. ARCH-in-mean models modify the specification of the conditional mean equation to be

$$y_t = \mathbf{x_t}\beta + \psi\sigma_t^2 + \epsilon_t \qquad \text{(ARCH-in-mean)}$$

While this linear form in the current conditional variance has dominated the literature, `arch` allows the conditional variance to enter the mean equation through a nonlinear transformation $g()$ and for this transformed term to be included contemporaneously or lagged.

$$y_t = \mathbf{x_t}\beta + \psi_0 g(\sigma_t^2) + \psi_1 g(\sigma_{t-1}^2) + \psi_2 g(\sigma_{t-2}^2) + \cdots + \epsilon_t$$

Square root is the most commonly used $g()$ transformation because researchers want to include a linear term for the conditional standard deviation, but any transform $g()$ is allowed.

▷ Example 1: ARCH model

Consider a simple model of the US Wholesale Price Index (WPI) (Enders 1995, 106–110), which we also consider in [TS] **arima**. The data are quarterly over the period 1960q1 through 1990q4.

In [TS] **arima**, we fit a model of the continuously compounded rate of change in the WPI, $\ln(\text{WPI}_t) - \ln(\text{WPI}_{t-1})$. The graph of the differenced series—see [TS] **arima**—clearly shows periods of high volatility and other periods of relative tranquility. This makes the series a good candidate for ARCH modeling. Indeed, price indices have been a common target of ARCH models. Engle (1982) presented the original ARCH formulation in an analysis of UK inflation rates.

First, we fit a constant-only model by OLS and test ARCH effects using Engle's Lagrange-multiplier test (`estat archlm`).

```
. use http://www.stata-press.com/data/r9/wpi1
. regress D.ln_wpi
```

Source	SS	df	MS
Model	0	0	.
Residual	.02521709	122	.000206697
Total	.02521709	122	.000206697

Number of obs =	123
F(0, 122) =	0.00
Prob > F =	.
R-squared =	0.0000
Adj R-squared =	0.0000
Root MSE =	.01438

D.ln_wpi	Coef.	Std. Err.	t	P>|t|	[95% Conf. Interval]
_cons	.0108215	.0012963	8.35	0.000	.0082553 .0133878

```
. estat archlm, lags(1)
LM test for autoregressive conditional heteroskedasticity (ARCH)
```

lags(p)	chi2	df	Prob > chi2
1	8.366	1	0.0038

HO: no ARCH effects   vs.   H1: ARCH(p) disturbance

Because the LM test shows a p-value of 0.0038, which is well below 0.05, we reject the null hypothesis of no ARCH(1) effects. Thus we can further estimate the ARCH(1) parameter by specifying `arch(1)`. See [R] **regress postestimation time series** for more information on Engle's LM test.

The first-order generalized ARCH model (GARCH, Bollerslev 1986) is the most commonly used specification for the conditional variance in empirical work and is typically written GARCH(1, 1). We can estimate a GARCH(1, 1) process for the log-differenced series by typing

```
. arch D.ln_wpi, arch(1) garch(1)

(setting optimization to BHHH)
Iteration 0:   log likelihood =   355.2346
Iteration 1:   log likelihood =  365.64589
 (output omitted )
Iteration 10:  log likelihood =  373.23397

ARCH family regression

Sample:  1960q2 to 1990q4                         Number of obs    =        123
                                                  Wald chi2(.)     =          .
Log likelihood =   373.234                        Prob > chi2      =          .
```

D.ln_wpi	Coef.	OPG Std. Err.	z	P>|z|	[95% Conf. Interval]	
ln_wpi						
_cons	.0061167	.0010616	5.76	0.000	.0040361	.0081974
ARCH						
arch						
L1.	.4364123	.2437428	1.79	0.073	−.0413147	.9141394
garch						
L1.	.4544606	.1866605	2.43	0.015	.0886126	.8203085
_cons	.0000269	.0000122	2.20	0.028	2.97e-06	.0000508

We have estimated the ARCH(1) parameter to be .436 and the GARCH(1) parameter to be .454, so our fitted GARCH(1, 1) model is

$$y_t = .0061 + \epsilon_t$$
$$\sigma_t^2 = .436\,\epsilon_{t-1}^2 + .454\,\sigma_{t-1}^2$$

where $y_t = \ln(\mathtt{wpi}_t) - \ln(\mathtt{wpi}_{t-1})$.

Note that the model Wald test and probability are both reported as missing (.). By convention, Stata reports the model test for the mean equation. In this case and fairly often for ARCH models, the mean equation consists only of a constant, and there is nothing to test.   ◁

▷ Example 2: ARCH model with ARMA process

We can retain the GARCH(1, 1) specification for the conditional variance and model the mean as an ARMA process with AR(1) and MA(1) terms as well as a fourth-lag MA term to control for quarterly seasonal effects by typing

```
. arch D.ln_wpi, ar(1) ma(1 4) arch(1) garch(1)
(setting optimization to BHHH)
Iteration 0:    log likelihood =   380.99952
Iteration 1:    log likelihood =   388.57801
Iteration 2:    log likelihood =   391.34179
Iteration 3:    log likelihood =   396.37029
Iteration 4:    log likelihood =   398.01112
(switching optimization to BFGS)
Iteration 5:    log likelihood =   398.23657
BFGS stepping has contracted, resetting BFGS Hessian (0)
Iteration 6:    log likelihood =   399.21491
Iteration 7:    log likelihood =   399.21531  (backed up)
 (output omitted )
Iteration 12:   log likelihood =    399.4934  (backed up)
Iteration 13:   log likelihood =   399.49607
Iteration 14:   log likelihood =   399.51241
(switching optimization to BHHH)
Iteration 15:   log likelihood =   399.51441
Iteration 16:   log likelihood =   399.51443
Iteration 17:   log likelihood =   399.51443

ARCH family regression -- ARMA disturbances

Sample:  1960q2 to 1990q4                      Number of obs    =         123
                                               Wald chi2(3)     =      153.56
Log likelihood =   399.5144                    Prob > chi2      =      0.0000
```

D.ln_wpi	Coef.	OPG Std. Err.	z	P>|z|	[95% Conf. Interval]	
ln_wpi						
_cons	.0069541	.0039517	1.76	0.078	-.000791	.0146992
ARMA						
ar						
L1.	.7922673	.1072225	7.39	0.000	.582115	1.002419
ma						
L1.	-.3417738	.1499944	-2.28	0.023	-.6357574	-.0477902
L4.	.2451725	.1251131	1.96	0.050	-.0000446	.4903896
ARCH						
arch						
L1.	.2040451	.1244992	1.64	0.101	-.039969	.4480591
garch						
L1.	.694968	.189218	3.67	0.000	.3241075	1.065829
_cons	.0000119	.0000104	1.14	0.253	-8.52e-06	.0000324

To clarify exactly what we have estimated, we could write our model

$$y_t = .007 + .792 \left(y_{t-1} - .007\right) - .342\,\epsilon_{t-1} + .245\,\epsilon_{t-4} + \epsilon_t$$
$$\sigma_t^2 = .204\,\epsilon_{t-1}^2 + .695\,\sigma_{t-1}^2$$

where $y_t = \ln\left(\text{wpi}_t\right) - \ln\left(\text{wpi}_{t-1}\right)$.

The ARCH(1) coefficient, .204, is not significantly different from zero, but the ARCH(1) and GARCH(1) coefficients are significant collectively. If you doubt this, you can check this with test.

```
. test [ARCH]L1.arch [ARCH]L1.garch
 ( 1)  [ARCH]L.arch = 0
 ( 2)  [ARCH]L.garch = 0
           chi2(  2) =     84.92
         Prob > chi2 =    0.0000
```

(Note that for comparison, we fitted the model over the same sample used in the example in [TS] **arima**; Enders fits this GARCH model but over a slightly different sample.)

◁

❑ Technical Note

The rather ugly iteration log on the previous result is typical, as difficulty in converging is common in ARCH models. This is actually a fairly well-behaved likelihood for an ARCH model. The "switching optimization to . . . " messages are standard messages from the default optimization method for `arch`. The "backed up" messages are typical of BFGS stepping as the BFGS Hessian is often over optimistic, particularly during early iterations. These messages are nothing to be concerned about.

Nevertheless, watch out for the messages "BFGS stepping has contracted, resetting BFGS Hessian" and "backed up", which can flag problems that may result in an iteration log that goes on and on. Stata will never report convergence and will never report final results. The question is, when do you give up and press *Break*, and if you do, what do you do then?

If the "BFGS stepping has contracted" message occurs repeatedly (more than, say, five times), it often indicates that convergence will never be achieved. Literally, it means that the BFGS algorithm was stuck and reset its Hessian and take a steepest-descent step.

The "backed up" message, if it occurs repeatedly, also indicates problems, but only if the likelihood value is simultaneously not changing. If the message occurs repeatedly but the likelihood value is changing, as it did above, all is going well; it is just going slowly.

If you have convergence problems, you can specify options to assist the current maximization method or try a different method. Or, your model specification and data may simply lead to a likelihood that is not concave in the allowable region and thus cannot be maximized.

If you see the "backed up" message with no change in the likelihood, you can reset the gradient tolerance to a larger value. Specifying option `gtolerance(999)` disables gradient checking, allowing convergence to be declared more easily. This does not guarantee that convergence will be declared, and even if it is, the global maximum likelihood may not have been found.

You can also try to specify initial values.

Finally, you can try a different maximization method; see options discussed under the **Max options** tab above.

ARCH models are notorious for having convergence difficulties. Unlike in most estimators in Stata, it is not uncommon for convergence to require many steps or even to fail. This is particularly true of the explicitly nonlinear terms such as `aarch()`, `narch()`, `aparch()`, or `archm` (ARCH-in-mean), and of any model with several lags in the ARCH terms. There is not always a solution. You can try alternate maximization methods or different starting values, but if your data do not support your assumed ARCH structure, convergence simply may not be possible.

ARCH models can be susceptible to irrelevant regressors or unnecessary lags, whether in the specification of the conditional mean or in the conditional variance. In these situations, `arch` will often continue to iterate, making little to no improvement in the likelihood. We view this conservative approach as better than declaring convergence prematurely when the likelihood has not been fully maximized. `arch` is estimating the conditional form of second sample moments, often with flexible functions, and that is asking much of the data.

❑

❏ Technical Note

if *exp* and in *range* are interpreted differently with commands accepting time-series operators. The time-series operators are resolved *before* the conditions are tested, which may lead to some confusion. Note the results of the following `list` commands:

```
. list t y l.y in 5/10
```

	t	y	L.y
5.	1961q1	30.8	30.7
6.	1961q2	30.5	30.8
7.	1961q3	30.5	30.5
8.	1961q4	30.6	30.5
9.	1962q1	30.7	30.6
10.	1962q2	30.6	30.7

```
. keep in 5/10
(124 observations deleted)
. list t y l.y
```

	t	y	L.y
1.	1961q1	30.8	.
2.	1961q2	30.5	30.8
3.	1961q3	30.5	30.5
4.	1961q4	30.6	30.5
5.	1962q1	30.7	30.6
6.	1962q2	30.6	30.7

We have one additional lagged observation for y in the first case. `l.y` was resolved before the in restriction was applied. In the second case, the dataset no longer contains the value of y to compute the first lag. This means that

```
. arch y l.x, arch(1), if twithin(1930, 1990)
```

is not the same as

```
. keep if twithin(1930, 1990)
. arch y l.x, arch(1)
```

                      ❏

▷ Example 3: Asymmetric effects—EGARCH model

Continuing with the WPI data, we might be concerned that the economy as a whole responds differently to unanticipated increases in wholesale prices than it does to unanticipated decreases. Perhaps unanticipated increases lead to cash-flow issues that affect inventories and lead to more volatility. We can see if the data support this supposition by specifying an ARCH model that allows an asymmetric effect of "news"—innovations or unanticipated changes. One of the most popular such models is EGARCH (Nelson 1991). The full first-order EGARCH model for the WPI can be specified as

```
. arch D.ln_wpi, ar(1) ma(1 4) earch(1) egarch(1)

(setting optimization to BHHH)
Iteration 0:   log likelihood =   227.5251
Iteration 1:   log likelihood =  381.69177
  (output omitted )
Iteration 21:  log likelihood =  405.31453

ARCH family regression -- ARMA disturbances
```

Sample: 1960q2 to 1990q4

Number of obs   =   123
Wald chi2(3)   =   156.04
Prob > chi2   =   0.0000

Log likelihood = 405.3145

D.ln_wpi	Coef.	OPG Std. Err.	z	P>|z|	[95% Conf. Interval]	
ln_wpi						
_cons	.0087343	.0034006	2.57	0.010	.0020692	.0153994
ARMA						
ar						
L1.	.7692304	.0968317	7.94	0.000	.5794438	.9590169
ma						
L1.	-.3554775	.1265674	-2.81	0.005	-.603545	-.10741
L4.	.241473	.086382	2.80	0.005	.0721674	.4107787
ARCH						
earch						
L1.	.4064035	.1163528	3.49	0.000	.1783561	.6344509
earch_a						
L1.	.2467611	.1233453	2.00	0.045	.0050088	.4885134
egarch						
L1.	.8417318	.0704089	11.95	0.000	.7037329	.9797307
_cons	-1.488376	.6604486	-2.25	0.024	-2.782831	-.1939203

Our result for the variance is

$$\ln(\sigma_t^2) = -1.49 + .406\, z_{t-1} + .247 \left| z_{t-1} - \sqrt{2/\pi} \right| + .842 \ln(\sigma_{t-1}^2)$$

where $z_t = \epsilon_t / \sigma_t$, which is distributed as $N(0, 1)$.

This is a strong indication for a leverage effect. The positive L1.earch coefficient implies that positive innovations (unanticipated price increases) are more destabilizing than negative innovations. The effect appears quite strong (.406) and is substantially larger than the symmetric effect (.247). In fact, the relative scales of the two coefficients imply that the positive leverage completely dominates the symmetric effect.

This can readily be seen if we plot what is often referred to as the news-response or news-impact function. This curve shows the resulting conditional variance as a function of unanticipated news, in the form of innovations, that is, the conditional variance σ_t^2 as a function of ϵ_t. Thus we must evaluate σ_t^2 for various values of ϵ_t—say, -4 to 4—and then graph the result.

◁

▷ Example 4: Asymmetric power ARCH model

As an example of a frequently sampled, long-run series, consider the daily closing indices of the Dow Jones Industrial Average, variable dowclose. Only data after 1jan1953 are used to avoid the first half of the century, when the New York Stock Exchange was open for Saturday trading. The compound return of the series is used as the dependent variable and is graphed below.

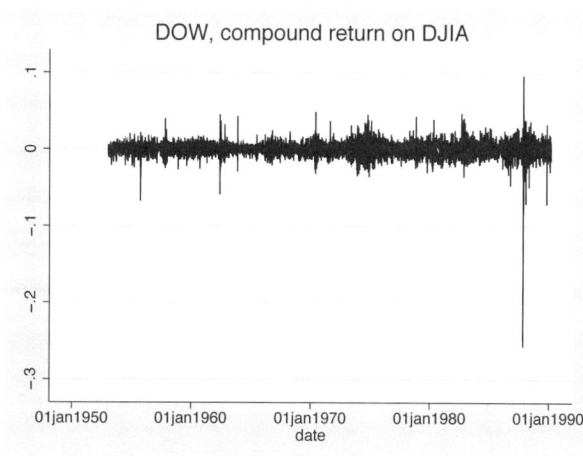

DOW, compound return on DJIA

We formed this difference by referring to D.ln_dow, but only after playing a trick. The series is daily, and each observation represents the Dow closing index for the day. Our data included a time variable recorded as a daily date. We wanted, however, to model the log differences in the series, and we wanted the span from Friday to Monday to appear as a single-period difference. That is, the day before Monday is Friday. Because our dataset was tsset with date, the span from Friday to Monday was 3 days. The solution was to create a second variable that sequentially numbered the observations. By tsseting the data with this new variable, we obtained the desired differences.

```
. generate t = _n
. tsset t
```

Now our data look like this:

```
. use http://www.stata-press.com/data/r9/dow1, clear
. generate dayofwk = dow(date)
. list date dayofwk t ln_dow D.ln_dow in 1/8
```

	date	dayofwk	t	ln_dow	D.ln_dow
1.	02jan1953	5	1	5.677096	.
2.	05jan1953	1	2	5.682899	.0058026
3.	06jan1953	2	3	5.677439	-.0054603
4.	07jan1953	3	4	5.672636	-.0048032
5.	08jan1953	4	5	5.671259	-.0013762
6.	09jan1953	5	6	5.661223	-.0100365
7.	12jan1953	1	7	5.653191	-.0080323
8.	13jan1953	2	8	5.659134	.0059433

```
. list date dayofwk t ln_dow D.ln_dow in -8/l
```

	date	dayofwk	t	ln_dow	D.ln_dow
9334.	08feb1990	4	9334	7.880188	.0016198
9335.	09feb1990	5	9335	7.881635	.0014472
9336.	12feb1990	1	9336	7.870601	-.011034
9337.	13feb1990	2	9337	7.872665	.0020638
9338.	14feb1990	3	9338	7.872577	-.0000877
9339.	15feb1990	4	9339	7.88213	.009553
9340.	16feb1990	5	9340	7.876863	-.0052676
9341.	20feb1990	2	9341	7.862054	-.0148082

The difference operator D spans weekends because the specified time variable t is not a true date and has a difference of 1 for all observations. We must leave this contrived time variable in place during estimation, or arch will be convinced that our dataset has gaps. If we were using calendar dates, we would indeed have gaps.

Ding, Granger, and Engle (1993) fitted an A-PARCH model of daily returns of the Standard and Poor's 500 (S&P 500) for 3jan1928 through 30aug1991. We will fit the same model for the Dow data shown above. The model includes an AR(1) term as well as the A-PARCH specification of conditional variance.

```
. arch D.ln_dow, ar(1) aparch(1) pgarch(1)

(setting optimization to BHHH)
Iteration 0:   log likelihood = 31138.796
Iteration 1:   log likelihood = 31350.762
Iteration 2:   log likelihood = 31351.092  (backed up)
 (output omitted )
ARCH family regression -- AR disturbances
```

```
Sample:  2 to 9341                        Number of obs    =      9340
                                          Wald chi2(1)     =    175.46
Log likelihood =  32273.56                Prob > chi2      =    0.0000
```

D.ln_dow	Coef.	OPG Std. Err.	z	P>|z|	[95% Conf. Interval]	
ln_dow						
_cons	.0001786	.0000875	2.04	0.041	7.16e-06	.00035
ARMA						
ar						
L1.	.1410944	.0106518	13.25	0.000	.1202172	.1619717
ARCH						
aparch						
L1.	.0626322	.0034307	18.26	0.000	.0559082	.0693563
aparch_e						
L1.	-.3645112	.0378488	-9.63	0.000	-.4386936	-.2903289
pgarch						
L1.	.9299024	.0030998	299.99	0.000	.923827	.9359779
_cons	7.19e-06	2.53e-06	2.84	0.004	2.23e-06	.0000121
POWER						
power	1.585143	.0629171	25.19	0.000	1.461828	1.708458

In the iteration log, we note that the final iteration reports the message "backed up". For most estimators, ending on a "backed up" message would be a cause for great concern, but not with `arch` or, for that matter, `arima`, as long as you do not specify option `gtolerance()`. `arch` and `arima`, by default, monitor the gradient and declare convergence only if, in addition to everything else, the gradient is sufficiently small.

The fitted model demonstrates substantial asymmetry, with the large negative `L1.aparch_e` coefficient indicating that the market responds with much more volatility to unexpected drops in returns (bad news) than it does to increases in returns (good news).

◁

▷ Example 5: ARCH model with constraints

Engle's (1982) original model, which sparked the interest in ARCH, provides an example requiring constraints. Most current ARCH specifications use GARCH terms to provide flexible dynamic properties without estimating an excessive number of parameters. The original model was limited to ARCH terms, and to help cope with the collinearity of the terms, a declining lag structure was imposed in the parameters. The conditional variance equation was specified as

$$\sigma_t^2 = \alpha_0 + \alpha(.4\,\epsilon_{t-1} + .3\,\epsilon_{t-2} + .2\,\epsilon_{t-3} + .1\,\epsilon_{t-4})$$
$$= \alpha_0 + .4\,\alpha\epsilon_{t-1} + .3\,\alpha\epsilon_{t-2} + .2\,\alpha\epsilon_{t-3} + .1\,\alpha\epsilon_{t-4}$$

From the earlier `arch` output, we know how the coefficients will be named. In Stata, the formula is

$$\sigma_t^2 = \texttt{[ARCH]\_cons} + .4\,\texttt{[ARCH]L1.arch}\,\epsilon_{t-1} + .3\,\texttt{[ARCH]L2.arch}\,\epsilon_{t-2}$$
$$+ .2\,\texttt{[ARCH]L3.arch}\,\epsilon_{t-3} + .1\,\texttt{[ARCH]L4.arch}\,\epsilon_{t-3}$$

We could specify these linear constraints any number of ways, but the following seems fairly intuitive; see [R] **constraint** for syntax.

```
. use http://www.stata-press.com/data/r9/wpi1, clear
. constraint define 1 (3/4)*[ARCH]l1.arch = [ARCH]l2.arch
. constraint define 2 (2/4)*[ARCH]l1.arch = [ARCH]l3.arch
. constraint define 3 (1/4)*[ARCH]l1.arch = [ARCH]l4.arch
```

The original model was fitted on UK inflation; we will again use the WPI data and retain our earlier specification of the mean equation, which differs from Engle's UK inflation model. With our constraints, we type

(Continued on next page)

```
. arch D.ln_wpi, ar(1) ma(1 4) arch(1/4) constraint(1/3)
(setting optimization to BHHH)
Iteration 0:   log likelihood =  396.80192
Iteration 1:   log likelihood =  399.07808
 (output omitted )
Iteration 9:   log likelihood =  399.46243

ARCH family regression -- ARMA disturbances
```

Sample: 1960q2 to 1990q4					Number of obs	=	123
					Wald chi2(3)	=	123.32
Log likelihood = 399.4624					Prob > chi2	=	0.0000

```
Constraints:
 ( 1)   .75 [ARCH]L.arch - [ARCH]L2.arch = 0
 ( 2)   .5 [ARCH]L.arch - [ARCH]L3.arch = 0
 ( 3)   .25 [ARCH]L.arch - [ARCH]L4.arch = 0
```

	Coef.	OPG Std. Err.	z	P>|z|	[95% Conf. Interval]	
ln_wpi						
_cons	.0077204	.0034531	2.24	0.025	.0009525	.0144883
ARMA						
ar						
L1.	.7388168	.1126811	6.56	0.000	.517966	.9596676
ma						
L1.	-.2559691	.1442861	-1.77	0.076	-.5387646	.0268264
L4.	.2528922	.1140185	2.22	0.027	.02942	.4763644
ARCH						
arch						
L1.	.2180138	.0737787	2.95	0.003	.0734101	.3626174
L2.	.1635103	.055334	2.95	0.003	.0550576	.2719631
L3.	.1090069	.0368894	2.95	0.003	.0367051	.1813087
L4	.0545034	.0184447	2.95	0.003	.0183525	.0906544
_cons	.0000483	7.66e-06	6.30	0.000	.0000333	.0000633

`L1.arch`, `L2.arch`, `L3.arch`, and `L4.arch` coefficients have the constrained relative sizes.

◁

Saved Results

arch saves in e():

Scalars

e(N)	number of observations	e(chi2)	χ^2
e(N_gaps)	number of gaps	e(p)	significance
e(condobs)	# of conditioning observations	e(tmin)	minimum time
e(k)	number of parameters	e(tmax)	maximum time
e(k_dv)	number of dependent variables	e(power)	φ for power ARCH terms
e(k_eq)	number of equations	e(rank)	rank of e(V)
e(archi)	$\sigma_0^2 = \epsilon_0^2$, priming values	e(ic)	number of iterations
e(df_m)	model degrees of freedom	e(rc)	return code
e(ll)	log likelihood	e(converged)	1 if converged, 0 otherwise

Macros

e(cmd)	arch	e(ma)	lags for moving-average terms
e(depvar)	name of dependent variable	e(ar)	lags for autoregressive terms
e(wtype)	weight type	e(arch)	lags for ARCH terms
e(wexp)	weight expression	e(archm)	ARCH-in-mean lags
e(title)	title in estimation output	e(archmexp)	ARCH-in-mean exp
e(eqnames)	names of equations	e(earch)	lags for EARCH terms
e(tmins)	formatted minimum time	e(egarch)	lags for EGARCH terms
e(tmaxs)	formatted maximum time	e(aarch)	lags for AARCH terms
e(mhet)	1 if multiplicative heteroskedasticity	e(narch)	lags for NARCH terms
e(chi2type)	Wald; type of model χ^2 test	e(aparch)	lags for APARCH terms
e(vce)	*vcetype* specified in vce()	e(nparch)	lags for NPARCH terms
e(vcetype)	title used to label Std. Err.	e(saarch)	lags for SAARCH terms
e(opt)	type of optimization	e(parch)	lags for PAARCH terms
e(ml_method)	type of ml method	e(tparch)	lags for TPARCH terms
e(user)	name of likelihood-evaluator program	e(abarch)	lags for ABARCH terms
e(technique)	maximization technique	e(tarch)	lags for TARCH terms
e(crittype)	optimization criterion	e(sdgarch)	lags for SDGARCH terms
e(properties)	b V	e(pgarch)	lags for PGARCH terms
e(estat_cmd)	program used to implement estat	e(garch)	lags for GARCH terms
e(predict)	program used to implement predict	e(cond)	flag if called by arima

Matrices

e(b)	coefficient vector	e(V)	variance–covariance matrix of
e(ilog)	iteration log (up to 20 iterations)		the estimators
e(gradient)	gradient vector		

Functions

e(sample)	marks estimation sample

Methods and Formulas

arch is implemented as an ado-file using the ml commands; see [R] **ml**. The robust variance computation is performed by _robust; see [P] **_robust**.

The mean equation for the model fitted by arch and with ARMA terms can be written as

$$y_t = \mathbf{x_t}\beta + \sum_{i=1}^{p} \psi_i g(\sigma_{t-i}^2) + \sum_{j=1}^{p} \rho_j \left\{ y_{t-j} - x_{t-j}\beta - \sum_{i=1}^{p} \psi_i g(\sigma_{t-j-i}^2) \right\}$$

$$+ \sum_{k=1}^{q} \theta_k \epsilon_{t-k} + \epsilon_t \qquad \text{(conditional mean)}$$

where

β are the regression parameters

ψ are the ARCH-in-mean parameters

ρ are the autoregression parameters

θ are the moving-average parameters

$g()$ is a general function, see option `archmexp()`

Any of the parameters in this full specification of the conditional mean may be zero. For example, the model need not have moving-average parameters ($\theta = 0$) or ARCH-in-mean parameters ($\psi = 0$). The variance equation will be one of the following:

$$\sigma^2 = \gamma_0 + A(\boldsymbol{\sigma}, \boldsymbol{\epsilon}) + B(\boldsymbol{\sigma}, \boldsymbol{\epsilon})^2 \qquad (1)$$

$$\ln \sigma_t^2 = \gamma_0 + C(\ln \boldsymbol{\sigma}, \mathbf{z}) + A(\boldsymbol{\sigma}, \boldsymbol{\epsilon}) + B(\boldsymbol{\sigma}, \boldsymbol{\epsilon})^2 \qquad (2)$$

$$\sigma_t^{\varphi} = \gamma_0 + D(\boldsymbol{\sigma}, \boldsymbol{\epsilon}) + A(\boldsymbol{\sigma}, \boldsymbol{\epsilon}) + B(\boldsymbol{\sigma}, \boldsymbol{\epsilon})^2 \qquad (3)$$

where $A(\boldsymbol{\sigma}, \boldsymbol{\epsilon})$, $B(\boldsymbol{\sigma}, \boldsymbol{\epsilon})$, $C(\ln \boldsymbol{\sigma}, \mathbf{z})$, and $D(\boldsymbol{\sigma}, \boldsymbol{\epsilon})$ are linear sums of the appropriate ARCH terms; see *Details of syntax* for more information. Equation (1) is used if no EGARCH or power ARCH terms are included in the model, (2) if EGARCH terms are included, and (3) if any power ARCH terms are included; see *Details of syntax*.

Priming values

The above model is recursive with potentially long memory. It is necessary to assume pre-estimation sample values for ϵ_t, ϵ_t^2, and σ_t^2 to begin the recursions, and the remaining computations are therefore conditioned on these priming values, which can be controlled using the `arch0()` and `arma0()` options. See options discussed under the **Priming** tab above.

The `arch0(xb0wt)` and `arch0(xbwt)` options compute a weighted sum of estimated disturbances with more weight on the early observations. With either of these options,

$$\sigma_{t_0-i}^2 = \epsilon_{t_0-i}^2 = (1 - .7) \sum_{t=0}^{T-1} .7^{T-t-1} \epsilon_{T-t}^2 \qquad \forall i$$

where t_0 is the first observation for which the likelihood is computed; see options discussed under the **Priming** tab above. The ϵ_t^2 are all computed from the conditional mean equation. If `arch0(xb0wt)` is specified, β, ψ_i, ρ_j, and θ_k are taken from initial regression estimates and held constant during optimization. If `arch0(xbwt)` is specified, the current estimates of β, ψ_i, ρ_j, and θ_k are used to compute ϵ_t^2 on every iteration. If any ψ_i is in the mean equation (ARCH-in-mean is specified), the estimates of ϵ_t^2 from the initial regression estimates are not consistent.

Likelihood from prediction error decomposition

The likelihood function for ARCH has a particularly simple form. Given priming (or conditioning) values of ϵ_t, ϵ_t^2, and σ_t^2, the mean equation above can be solved recursively for every ϵ_t (prediction error decomposition). Likewise, the conditional variance can be computed recursively for each observation by using the variance equation. Using these predicted errors, their associated variances, and the assumption that $\epsilon_t \sim N(0, \sigma_t^2)$, the log likelihood for each observation t is

$$\ln L_t = -\frac{1}{2} w_t \left\{ \ln(2\pi\sigma_t^2) + \sum_{t=t_0}^{T} \frac{\epsilon_t^2}{\sigma_t^2} \right\}$$

where $w_t = 1$ if weights are not specified.

Missing data

ARCH allows missing data or missing observations but does not attempt to condition on the surrounding data. If a dynamic component cannot be computed—ϵ_t, ϵ_t^2, and/or σ_t^2—its priming value is substituted. If a covariate, the dependent variable, or entire observation is missing, the observation does not enter the likelihood, and its dynamic components are set to their priming values for that observation. This is only acceptable asymptotically and should not be used with large amounts of missing data.

Robert Fry Engle (1942–) was born in Syracuse, New York. He earned degrees in physics and economics at Williams College and Cornell and then worked at MIT and the University of California, San Diego, before moving to New York University Stern School of Business in 2000. He was awarded the 2003 Nobel Prize in Economics for research on autoregressive conditional heteroskedasticity and is a leading expert in time series analysis, especially the analysis of financial markets.

References

Baum, C. F. 2000. sts15: Tests for stationarity of a time series. *Stata Technical Bulletin* 57: 36–39. Reprinted in *Stata Technical Bulletin Reprints*, vol. 10, pp. 356–360.

Baum, C. F. and R. Sperling. 2000. sts15.1: Tests for stationarity of a time series: Update. *Stata Technical Bulletin* 58: 35–36. Reprinted in *Stata Technical Bulletin Reprints*, vol. 10, pp. 360–362.

Baum, C. F. and V. L. Wiggins. 2000. sts16: Tests for long memory in a time series. *Stata Technical Bulletin* 57: 39–44. Reprinted in *Stata Technical Bulletin Reprints*, vol. 10, pp. 362–368.

Berndt, E. K., B. H. Hall, R. E. Hall, and J. A. Hausman. 1974. Estimation and inference in nonlinear structural models. *Annals of Economic and Social Measurement* 3/4: 653–665.

Black, F. 1976. Studies of stock price volatility changes. *Proceedings from the American Statistical Association, Business and Economics Statistics*, 653–665.

Bollerslev, T. 1986. Generalized autoregressive conditional heteroskedasticity. *Journal of Econometrics* 31: 307–327.

Bollerslev, T., R. Y. Chou, and K. F. Kroner. 1992. ARCH modeling in finance. *Journal of Econometrics* 52: 5–59.

Bollerslev, T., R. F. Engle, and D. B. Nelson. 1994. ARCH models. In *Handbook of Econometrics, Volume IV*, ed. R. F. Engle and D. L. McFadden. New York: Elsevier.

Bollerslev, T. and J. M. Wooldridge. 1992. Quasi maximum likelihood estimation and inference in dynamic models with time varying covariances. *Econometric Reviews* 11: 143–172.

Davidson, R. and J. G. MacKinnon. 1993. *Estimation and Inference in Econometrics*. Oxford: Oxford University Press.

——. 2004. *Econometric Theory and Methods*. Oxford: Oxford University Press.

Diebold, F. X. 2003. The *ET* Interview: Professor Robert F. Engle. *Econometric Theory* 19: 1159–1193.

Ding, Z., C. W. J. Granger, and R. F. Engle. 1993. A long memory property of stock market returns and a new model. *Journal of Empirical Finance* 1: 83–106.

Enders, W. 1995. *Applied Econometric Time Series*. New York: Wiley.

Engle, R. F. 1982. Autoregressive conditional heteroskedasticity with estimates of the variance of UK inflation. *Econometrica* 50: 987–1008.

——. 1990. Discussion: Stock market volatility and the crash of 87. *Review of financial studies* 3: 103–106.

Engle, R. F., D. M. Lilien, and R. P. Robins. 1987. Estimating time varying risk premia in the term structure: The ARCH-M model. *Econometrica* 55: 391–407.

Glosten, L. R., R. Jagannathan, and D. Runkle. 1993. On the relation between the expected value and the volatility of the nominal excess return on stocks. *Journal of Finance* 48: 1779–1801.

Greene, W. H. 2003. *Econometric Analysis*. 5th ed. Upper Saddle River, NJ: Prentice Hall.

Hamilton, J. D. 1994. *Time Series Analysis*. Princeton: Princeton University Press.

Harvey, A. C. 1989. *Forecasting, Structural Time Series Models and the Kalman Filter*. Cambridge: Cambridge University Press.

——. 1990. *The Econometric Analysis of Time Series*. 2nd ed. Cambridge, MA: MIT Press.

Higgins, M. L. and A. K. Bera. 1992. A class of nonlinear ARCH models. *International Economic Review* 33: 137–158.

Johnston, J. and J. DiNardo. 1997. *Econometric Methods*. 3rd ed. New York: McGraw–Hill.

Judge, G. G., W. E. Griffiths, R. C. Hill, H. Lütkepohl, and T.-C. Lee. 1985. *The Theory and Practice of Econometrics*. 2nd ed. New York: Wiley.

Kmenta, J. 1997. *Elements of Econometrics*. 2nd ed. Ann Arbor: University of Michigan Press.

Nelson, D. B. 1991. Conditional heteroskedasticity in asset returns: A new approach. *Econometrica* 59: 347–370.

Press, W. H., S. A. Teukolsky, W. T. Vetterling, and B. P. Flannery. 1992. *Numerical Recipes in C: The Art of Scientific Computing*. 2nd ed. Cambridge: Cambridge University Press.

Wooldridge, J. M. 2002. *Introductory Econometrics: A Modern Approach*. 2nd ed. Cincinnati, OH: South-Western.

Zakoian, J. M. 1990. Threshold heteroskedastic models. Unpublished manuscript. *CREST, INSEE*.

Also See

Complementary:	[TS] **arch postestimation**, [TS] **tsset**
Related:	[TS] **arima**, [TS] **var**, [TS] **var svar**, [TS] **vec**, [R] **regress**
Background:	[U] **11.1.10 Prefix commands**,
	[U] **11.4.3 Time-series varlists**,
	[U] **20 Estimation and postestimation commands**,
	[U] **24.3 Time-series dates**,
	[TS] **estimation options**,
	[R] **maximize**, [R] *vce_option*

Title

> **arch postestimation** — Postestimation tools for arch

Description

The following postestimation commands are available for `arch`:

command	description
estat	AIC, BIC, VCE, and estimation sample summary
estimates	cataloging estimation results
lincom	point estimates, standard errors, testing, and inference for linear combinations of coefficients
mfx	marginal effects or elasticities
nlcom	point estimates, standard errors, testing, and inference for nonlinear combinations of coefficients
predict	predictions, residuals, influence statistics, and other diagnostic measures
predictnl	point estimates, standard errors, testing, and inference for generalized predictions
test	Wald tests for simple and composite linear hypotheses
testnl	Wald tests of nonlinear hypotheses

See the corresponding entries in the *Stata Base Reference Manual* for details.

Syntax for predict

predict [*type*] *newvar* [*if*] [*in*] [, *statistic options*]

statistic	description
Main	
xb	predicted values for mean equation—the differenced series; the default
y	predicted values for the mean equation in y—the undifferenced series
variance	predicted values for the conditional variance
het	predicted values of the variance, considering only the multiplicative heteroskedasticity
residuals	residuals or predicted innovations
yresiduals	residuals or predicted innovations in y—the undifferenced series

These statistics are available both in and out of sample; type `predict ... if e(sample) ...` if wanted only for the estimation sample.

options	description
Options	
dynamic(*time_constant*)	how to handle the lags of y_t
at(*varname*$_\epsilon$ | #$_\epsilon$ *varname*$_{\sigma^2}$ | #$_{\sigma^2}$)	make static predictions
t0(*time_constant*)	set starting point for the recursions to *time_constant*
structural	calculate considering the structural component only

time_constant is a # or a time literal, such as d(1jan1995) or q(1995q1), etc.; see [U] **24.3 Time-series dates**.

Options for predict

Six statistics can be computed by using `predict` after `arch`: the predictions of the mean equation (option `xb`, the default), the undifferenced predictions of the mean equation (option `y`), the predictions of the conditional variance (option `variance`), the predictions of the multiplicative heteroskedasticity component of variance (option `het`), the predictions of residuals or innovations (option `residuals`), and the predictions of residuals or innovations in terms of y (option `yresiduals`). Given the dynamic nature of ARCH models and because the dependent variable might be differenced, there are alternative ways of computing each statistic. We can use all the data on the dependent variable available right up to the time of each prediction (the default, which is often called a one-step prediction), or we can use the data up to a particular time, after which the predicted value of the dependent variable is used recursively to make subsequent predictions (option `dynamic()`). Either way, we can consider or ignore the ARMA disturbance component, which is considered by default and is ignored if you specify option `structural`. We might also be interested in predictions at certain fixed points where we specify the prior values of ϵ_t and σ_t^2 (option `at()`).

    .. Main ..

`xb`, the default, calculates the predictions from the mean equation. If D.*depvar* is the dependent variable, these predictions are of D.*depvar* and not of *depvar* itself.

`y` specifies that predictions of *depvar* are to be made even if the model was specified in terms of, say, D.*depvar*.

`variance` calculates predictions of the conditional variance $\hat{\sigma}_t^2$.

`het` calculates predictions of the multiplicative heteroskedasticity component of variance.

`residuals` calculates the residuals. If no other options are specified, these are the predicted innovations ϵ_t; i.e., they include any ARMA component. If option `structural` is specified, these are the residuals from the mean equation, ignoring any ARMA terms; see `structural` below. The residuals are always from the estimated equation, which may have a differenced dependent variable; if *depvar* is differenced, they are not the residuals of the undifferenced *depvar*.

`yresiduals` calculates the residuals in terms of *depvar*, even if the model was specified in terms of, say, D.*depvar*. As with `residuals`, the `yresiduals` are computed from the model, including any ARMA component. If option `structural` is specified, any ARMA component is ignored and `yresiduals` are the residuals from the structural equation; see `structural` below.

    .. Options ..

`dynamic(`*time_constant*`)` specifies how lags of y_t in the model are to be handled. If `dynamic()` is not specified, actual values are used everywhere lagged values of y_t appear in the model to produce one-step-ahead forecasts.

`dynamic(`*time_constant*`)` produces dynamic (also known as recursive) forecasts. *time_constant* specifies when the forecast is to switch from one-step ahead to dynamic. In dynamic forecasts, references to y_t evaluate to the prediction of y_t for all periods at or after *time_constant*; they evaluate to the actual value of y_t for all prior periods.

`dynamic(10)`, for example, would calculate predictions where any reference to y_t with $t < 10$ evaluates to the actual value of y_t and any reference to y_t with $t \geq 10$ evaluates to the prediction of y_t. This means that one-step-ahead predictions would be calculated for $t < 10$ and dynamic predictions would be calculated thereafter. Depending on the lag structure of the model, the dynamic predictions might still reference some actual values of y_t.

In addition, you may specify dynamic(.) to have predict automatically switch from one-step to dynamic predictions at $p + q$, where p is the maximum AR lag and q is the maximum MA lag.

at($varname_\epsilon$ | #$_\epsilon$ $varname_{\sigma^2}$ | #$_{\sigma^2}$) makes static predictions. at() and dynamic() may not be specified together.

Specifying at() allows static evaluation of results for a given set of disturbances. This is useful, for instance, in generating the news response function. at() specifies two sets of values to be used for ϵ_t and σ_t^2, the dynamic components in the model. These specified values are treated as given. In addition any lagged values of *depvar* in the model are obtained from the real values of the dependent variable. All computations are based on actual data and the given values.

at() requires that you specify two arguments, which can be either a variable name or a number. The first argument supplies the values to be used for ϵ_t; the second supplies the values to be used for σ_t^2. If σ_t^2 plays no role in your model, the second argument may be specified as '.' to indicate missing.

t0(*time_constant*) specifies the starting point for the recursions to compute the predicted statistics; disturbances are assumed to be 0 for $t <$ t0(). The default is to set t0() to the minimum t observed in the estimation sample, meaning that observations before that are assumed to have disturbances of 0.

t0() is irrelevant if structural is specified because, in that case, all observations are assumed to have disturbances of 0.

t0(5), for example, would begin recursions at $t = 5$. If your data were quarterly, you might instead type t0(q(1961q2)) to obtain the same result.

Note that any ARMA component in the mean equation or GARCH term in the conditional-variance equation makes arch recursive and dependent on the starting point of the predictions. This includes one-step-ahead predictions.

structural makes the calculation considering the structural component only, ignoring any ARMA terms, and producing the steady-state equilibrium predictions.

Remarks

▷ Example 1

Continuing with our EGARCH model example (example 3) in [TS] **arch**, we can see that predict, at() calculates σ_t^2 given a set of specified innovations $(\epsilon_t, \epsilon_{t-1}, \ldots)$ and prior conditional variances $(\sigma_{t-1}^2, \sigma_{t-2}^2, \ldots)$. The syntax is

```
. predict newvar, variance at(epsilon sigma2)
```

epsilon and *sigma2* are either variables or numbers. Using *sigma2* is a little tricky because you specify values of σ_t^2, which predict is supposed to predict. *predict* does not simply copy variable *sigma2* into *newvar* but uses the lagged values contained in *sigma2* to produce the predicted value of σ_t^2. It does this for all t, and those results are stored in *newvar*. (If you are interested in dynamic predictions of σ_t^2, see *Options for predict*.)

We will generate predictions for σ_t^2, assuming that the lagged values of σ_t^2 are 1, and we will vary ϵ_t from -4 to 4. First, we will create variable et containing ϵ_t, and then we will create and graph the predictions:

```
. generate et = (_n-64)/15
. predict sigma2, variance at(et 1)
```

```
. line sigma2 et in 2/1, m(i) c(l) title(News response function)
```

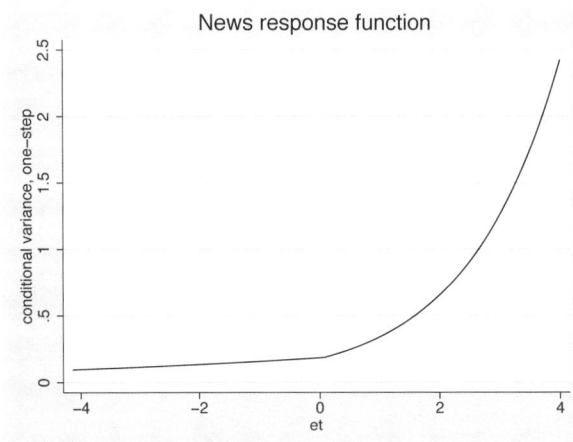

The positive asymmetry does indeed dominate the shape of the news response function. In fact, the response is a monotonically increasing function of news. The form of the response function shows that, for our simple model, only positive, unanticipated price increases have the destabilizing effect that we observe as larger conditional variances.

◁

▷ Example 2

Continuing with our ARCH model with constraints example (example 5) in [TS] **arch**, using lincom we can recover the α parameter from the original specification.

```
. lincom [ARCH]l1.arch/.4
 ( 1)   2.5 [ARCH]L.arch = 0
```

D.ln_wpi	Coef.	Std. Err.	z	P>|z|	[95% Conf. Interval]
(1)	.5450344	.1844468	2.95	0.003	.1835253 .9065436

Any of the arch parameters could be used to produce an identical estimate.

◁

Methods and Formulas

All postestimation commands listed above are implemented as ado-files.

Also See

Complementary: [TS] **arch**,

    [R] **estimates**, [R] **lincom**, [R] **mfx**, [R] **nlcom**, [R] **predictnl**,

    [R] **test**, [R] **testnl**

Background: [U] **13.5 Accessing coefficients and standard errors**,

    [U] **20 Estimation and postestimation commands**,

    [R] **estat**, [R] **predict**

Title

> **arima** — ARIMA, ARMAX, and other dynamic regression models

Syntax

Basic syntax for a regression model with ARMA *disturbances*

    arima *depvar* [*indepvars*] , ar(*numlist*) ma(*numlist*)

Basic syntax for an ARIMA(p, d, q) *model*

    arima *depvar* , arima($\#_p$,$\#_d$,$\#_q$)

Basic syntax for a multiplicative seasonal ARIMA$(p, d, q) \times (P, D, Q)_s$ *model*

    arima *depvar* , arima($\#_p$,$\#_d$,$\#_q$) sarima($\#_P$,$\#_D$,$\#_Q$,$\#_s$)

Full syntax

    arima *depvar* [*indepvars*] [*if*] [*in*] [*weight*] [, *options*]

options	description
Model	
<u>no</u>constant	suppress constant term
arima($\#_p$,$\#_d$,$\#_q$)	specify ARIMA(p, d, q) model for dependent variable
ar(*numlist*)	autoregressive terms of the structural model disturbance
ma(*numlist*)	moving-average terms of the structural model disturbance
<u>c</u>onstraints(*constraints*)	apply specified linear constraints
Model 2	
sarima($\#_P$,$\#_D$,$\#_Q$,$\#_s$)	specify period-$\#_s$ multiplicative seasonal ARIMA term
mar(*numlist*, $\#_s$)	multiplicative seasonal autoregressive term; may be repeated
mma(*numlist*, $\#_s$)	multiplicative seasonal moving-average term; may be repeated
Model 3	
<u>condition</u>	use conditional MLE instead of full MLE
<u>saves</u>pace	conserve memory during estimation
<u>diff</u>use	use diffuse prior for starting Kalman filter recursions
state0(*#* | *matname*)	use alternate state vector for starting Kalman filter recursions
p0(*#* | *matname*)	use alternate prior for starting Kalman recursions; seldom used
SE/Robust	
vce(*vcetype*)	*vcetype* may be <u>r</u>obust, oim, or opg
<u>r</u>obust	synonym for vce(robust)

Reporting

<u>level</u>(#)	set confidence level; default is level(95)
<u>det</u>ail	report list of gaps in time series

Max options

maximize_options	control the maximization process; seldom used

You must tsset your data before using arima; see [TS] **tsset**.

depvar and *indepvars* may contain time-series operators; see [U] **11.4.3 Time-series varlists**.

by, rolling, statsby, and xi may be used with arima; see [U] **11.1.10 Prefix commands**.

iweights are allowed; see [U] **11.1.6 weight**.

See [U] **20 Estimation and postestimation commands** for additional capabilities of estimation commands.

Description

arima fits univariate models with time-dependent disturbances. arima fits a model of *depvar* on *indepvars* where the disturbances are allowed to follow a linear autoregressive moving-average (ARMA) specification. The dependent and independent variables may be differenced or seasonally differenced to any degree. When independent variables are included in the specification, such models are frequently called ARMAX models; and when independent variables are not specified, they reduce to Box–Jenkins autoregressive integrated moving-average (ARIMA) models in the dependent variable. Multiplicative seasonal ARMAX and ARIMA models can also be fitted. Missing data are allowed and are handled using the Kalman filter and methods suggested by Harvey (1989 and 1993); see *Methods and Formulas*.

In the full syntax, *depvar* is the variable being modeled, and the structural or regression part of the model is specified in *indepvars*. ar() and ma() specify the lags of autoregressive and moving-average terms, respectively; and mar() and mma() specify the multiplicative seasonal autoregressive and moving-average terms, respectively.

arima allows time-series operators in the dependent variable and independent variable lists, and it is often convenient to make extensive use of these operators; see [U] **24.3 Time-series dates** for an extended discussion of time-series operators.

arima typed without arguments redisplays the previous estimates.

Options

     ⌐ Model ⌐───

noconstant; see [TS] **estimation options**.

arima($\#_p$,$\#_d$,$\#_q$) is an alternative, shorthand notation for specifying models with ARMA disturbances. The dependent variable and any independent variables are differenced $\#_d$ times, and 1 through $\#_p$ lags of autocorrelations and 1 through $\#_q$ lags of moving averages are included in the model. For example, the specification

    . arima D.y, ar(1/2) ma(1/3)

is equivalent to

    . arima y, arima(2,1,3)

The latter is easier to write for simple ARMAX and ARIMA models, but if gaps in the AR or MA lags are to be modeled, or if different operators are to be applied to independent variables, the first syntax is required.

ar(*numlist*) specifies the autoregressive terms of the structural model disturbance to be included in the model. For example, ar(1/3) specifies that lags of 1, 2, and 3 of the structural disturbance be included in the model; and ar(1 4) specifies that lags 1 and 4 be included, perhaps to account for additive quarterly effects.

If the model does not contain regressors, these terms can also be considered autoregressive terms for the dependent variable.

ma(*numlist*) specifies the moving-average terms to be included in the model. These are the terms for the lagged innovations (white-noise disturbances).

constraints(*constraints*); see [TS] **estimation options**.

If constraints are placed between structural model parameters and ARMA terms, the first few iterations may attempt steps into nonstationary areas. This can be ignored if the final solution is well within the bounds of stationary solutions.

Model 2

sarima(#$_P$,#$_D$,#$_Q$,#$_s$) is an alternative, shorthand notation for specifying the multiplicative seasonal components of models with ARMA disturbances. The dependent variable and any independent variables are lag-#$_s$ seasonally differenced #$_D$ times, and 1 through #$_P$ seasonal lags of autoregressive terms and 1 through #$_Q$ seasonal lags of moving-average terms are included in the model. For example, the specification

```
. arima DS12.y, ar(1/2) ma(1/3) mar(1/2,12) mma(1/2,12)
```

is equivalent to

```
. arima y, arima(2,1,3) sarima(2,1,2,12)
```

mar(*numlist*,#$_s$) specifies the lag-#$_s$ multiplicative seasonal autoregressive terms. For example, mar(1/2,12) requests that the first two lag-12 multiplicative seasonal autoregressive terms be included in the model.

mma(*numlist*,#$_s$) specified the lag-#$_s$ multiplicative seasonal moving-average terms. For example, mma(1 3,12) requests that the first and third (but not the second) lag-12 multiplicative seasonal moving-average terms be included in the model.

Model 3

condition specifies that conditional, rather than full, maximum likelihood estimates be produced. The pre-sample values for ϵ_t and μ_t are taken to be their expected value of zero, and the estimate of the variance of ϵ_t is taken to be constant over the entire sample; see Hamilton (1994, 132). This estimation method is not appropriate for nonstationary series but may be preferable for long series or for models that have one or more long AR or MA lags. diffuse, p0(), and state0() have no meaning for models fitted from the conditional likelihood and may not be specified with condition.

If the series is long and stationary and the underlying data-generating process does not have a long memory, estimates will be similar, whether estimated by unconditional maximum likelihood (the default), conditional maximum likelihood (condition), or maximum likelihood from a diffuse prior (diffuse).

In small samples, however, results of conditional and unconditional maximum likelihood may differ substantially; see Ansley and Newbold (1980). Whereas the default unconditional maximum likelihood estimates make the most use of sample information when all the assumptions of the model are met, Harvey (1989) and Ansley and Kohn (1985) argue for diffuse priors in many cases, particularly in ARIMA models corresponding to an underlying structural model.

The `condition` or `diffuse` options may also be preferred when the model contains one or more long AR or MA lags; this avoids inverting potentially large matrices (see `diffuse` below).

When `condition` is specified, estimation is performed by the `arch` command (see [TS] **arch**), and more control of the estimation process can be obtained by using `arch` directly.

`condition` cannot be specified if the model contains any multiplicative seasonal terms.

`savespace` specifies that memory use be conserved by retaining only those variables required for estimation. The original dataset is restored after estimation. This option is rarely used and should be used only if there is insufficient space to fit a model without the option. Note, however, that `arima` requires considerably more temporary storage during estimation than most estimation commands in Stata.

`diffuse` specifies that a diffuse prior (see Harvey 1989 or 1993) be used as a starting point for the Kalman filter recursions. Using `diffuse`, nonstationary models may be fitted with `arima` (see option `p0()` below; `diffuse` is equivalent to specifying `p0(1e9)`).

By default, `arima` uses the unconditional expected value of the state vector ξ_t (see *Methods and Formulas*) and the mean squared error (MSE) of the state vector to initialize the filter. When the process is stationary, this corresponds to the expected value and expected variance of a random draw from the state vector and produces unconditional maximum likelihood estimates of the parameters. When the process is not stationary, however, this default is not appropriate, and the unconditional MSE cannot be computed. For a nonstationary process, an alternative starting point must be used for the recursions.

In the absence of nonsample or pre-sample information, `diffuse` may be specified to start the recursions from a state vector of zero and a state MSE matrix corresponding to an effectively infinite variance on this initial state. This amounts to an uninformative and improper prior that is updated to a proper MSE as data from the sample become available; see Harvey (1989).

Note that nonstationary models may also correspond to models with infinite variance given a particular specification. This and other problems with nonstationary series make convergence difficult and sometimes impossible.

`diffuse` can also be useful if a model contains one or more long AR or MA lags. Computation of the unconditional MSE of the state vector (see *Methods and Formulas*) requires construction and inversion of a square matrix that is of dimension $\{\max(p, q + 1)\}^2$, where p and q are the maximum AR and MA lags, respectively. If $q = 27$, for example, we would require a 784-by-784 matrix. Estimation with `diffuse` does not require this matrix.

For large samples, there is little difference between using the default starting point and the `diffuse` starting point. Unless the series has a very long memory, the initial conditions affect the likelihood of only the first few observations.

`state0(#|matname)` is a rarely used option that specifies an alternate initial state vector, $\xi_{1|0}$ (see *Methods and Formulas*), for starting the Kalman filter recursions. If # is specified, all elements of the vector are taken to be #. The default initial state vector is `state0(0)`.

`p0(#|matname)` is a rarely specified option that can be used for nonstationary series or when an alternate prior for starting the Kalman recursions is desired (see `diffuse` above for a discussion of the default starting point and *Methods and Formulas* for background).

matname specifies a matrix to be used as the MSE of the state vector for starting the Kalman filter recursions—$\mathbf{P}_{1|0}$. Alternatively, a single number, #, may be supplied, and the MSE of the initial state vector $\mathbf{P}_{1|0}$ will have this number on its diagonal and all off-diagonal values set to zero.

This option may be used with nonstationary series to specify a larger or smaller diagonal for $\mathbf{P}_{1|0}$ than that supplied by diffuse. It may also be used in conjunction with state0() when you believe you have a better prior for the initial state vector and its MSE.

⌐ SE/Robust ⌐

vce(*vcetype*); see [R] ***vce_option***.

robust; see [TS] **estimation options**.

For state-space models in general and ARMAX and ARIMA models in particular, the robust or quasi-maximum likelihood estimates (QMLE) of variance are robust to symmetric non-normality in the disturbances, including, as a special case, heteroskedasticity. The robust variance estimates are not generally robust to functional misspecification of the structural or ARMA components of the model; see Hamilton (1994, 389) for a brief discussion.

⌐ Reporting ⌐

level(*#*); see [TS] **estimation options**.

detail specifies that a detailed list of any gaps in the series be reported, including gaps due to missing observations or missing data for the dependent variable or independent variables.

⌐ Max options ⌐

maximize_options: difficult, technique(*algorithm_spec*), iterate(*#*), [no]log, trace, gradient, showstep, hessian, shownrtolerance, tolerance(*#*), ltolerance(*#*), gtolerance(*#*), nrtolerance(*#*), nonrtolerance(*#*), from(*init_specs*); see [R] **maximize**.

These options are sometimes more important for ARIMA models than most maximum likelihood models because of potential convergence problems with ARIMA models, particularly if the specified model and the sample data imply a nonstationary model.

Several alternate optimization methods, such as Berndt–Hall–Hall–Hausman (BHHH) and Broyden–Fletcher–Goldfarb–Shanno (BFGS), are provided for arima models. Although arima models are not as difficult to optimize as ARCH models, their likelihoods are nevertheless generally not quadratic and often pose optimization difficulties; this is particularly true if a model is nonstationary or nearly nonstationary. Since each method approaches optimization differently, some problems can be successfully optimized by an alternate method when one method fails.

The following options are all related to maximization and are either particularly important in fitting ARIMA models or not available for most other estimators.

  technique(*algorithm_spec*) specifies the optimization technique to use to maximize the likelihood function.

    technique(bhhh) specifies the Berndt–Hall–Hall–Hausman (BHHH) algorithm.

    technique(dfp) specifies the Davidon–Fletcher–Powell (DFP) algorithm.

    technique(bfgs) specifies the Broyden–Fletcher–Goldfarb–Shanno (BFGS) algorithm.

    technique(nr) specifies Stata's modified Newton–Raphson (NR) algorithm.

  You can specify multiple optimization methods. For example,

    technique(bhhh 10 nr 20)

  requests that the optimizer perform 10 BHHH iterations, switch to Newton–Raphson for 20 iterations, switch back the BHHH for 10 more iterations, and so on.

  The default for arima is technique(bhhh 5 bfgs 10).

gtolerance(#) is a rarely used maximization option that specifies a threshold for the relative size of the gradient; see [R] **maximize**. The default gradient tolerance for arima is .05.

gtolerance(999) effectively disables the gradient criterion when convergence is difficult to achieve. If the optimizer becomes stuck with repeated "(backed up)" messages, the gradient probably still contains substantial values, but an uphill direction cannot be found for the likelihood. Using gtolerance(999) will often obtain results but it may be unclear whether the global maximum likelihood has been found. It is usually better to set the maximum number of iterations (see [R] **maximize**) to the point where the optimizer appears to be stuck and then inspect the estimation results.

from(*init_specs*) allows you to set the starting values of the model coefficients; see [R] **maximize** for a general discussion and syntax options.

The standard syntax for from() accepts a matrix, a list of values, or coefficient name value pairs; see [R] **maximize**. In addition, arima accepts from(armab0), which sets the starting value for all ARMA parameters in the model to zero prior to optimization.

ARIMA models may be sensitive to initial conditions and may have coefficient values that correspond to local maxima. The default starting values for arima are generally very good, particularly in large samples for stationary series.

Remarks

Remarks are presented under the headings

Introduction
ARIMA models
Multiplicative seasonal ARIMA models
ARMAX models
Dynamic forecasting

Introduction

arima fits both standard ARIMA models that are autoregressive in the dependent variable and structural models with ARMA disturbances. Good introductions to the former models can be found in Box, Jenkins, and Reinsel (1994), Hamilton (1994), Harvey (1993), Newton (1988), Diggle (1990), and many others. The latter models are developed fully in Hamilton (1994) and Harvey (1989), both of which provide extensive treatment of the Kalman filter (Kalman 1960) and the state-space form used by arima to fit the models.

Consider a first-order autoregressive moving-average process. Then arima estimates all the parameters in the model

$$y_t = \mathbf{x}_t \boldsymbol{\beta} + \mu_t \qquad \qquad \textit{structural equation}$$
$$\mu_t = \rho \mu_{t-1} + \theta \epsilon_{t-1} + \epsilon_t \qquad \qquad \textit{disturbance, } \mathrm{ARMA}(1,1)$$

where

ρ	is the first-order autocorrelation parameter
θ	is the first-order moving-average parameter
ϵ_t	$\sim$ *i.i.d.* $N(0, \sigma^2)$, meaning ϵ_t is a white-noise disturbance

You can combine the two equations and write a general ARMA(p, q) in the disturbances process as

$$y_t = \mathbf{x_t}\boldsymbol{\beta} + \rho_1(y_{t-1} - \mathbf{x}_{t-1}\boldsymbol{\beta}) + \rho_2(y_{t-2} - \mathbf{x}_{t-2}\boldsymbol{\beta}) + \cdots + \rho_p(y_{t-p} - \mathbf{x}_{t-p}\boldsymbol{\beta})$$
$$+ \theta_1\epsilon_{t-1} + \theta_2\epsilon_{t-2} + \cdots + \theta_q\epsilon_{t-q} + \epsilon_t$$

It is also common to write the general form of the ARMA model more succinctly using lag operator notation as

$$\rho(L^p)(y_t - \mathbf{x}_t\boldsymbol{\beta}) = \boldsymbol{\theta}(L^q)\epsilon_t \qquad \text{ARMA}(p,q)$$

where

$$\rho(L^p) = 1 - \rho_1 L - \rho_2 L^2 - \cdots - \rho_p L^p$$
$$\boldsymbol{\theta}(L^q) = 1 + \theta_1 L + \theta_2 L^2 + \cdots + \theta_q L^q$$

and $L^j y_t = y_{t-j}$.

For stationary series, full or unconditional maximum likelihood estimates are obtained via the Kalman filter. For nonstationary series, if some prior information is available, you can specify initial values for the filter using `state0()` and `p0()` as suggested by Hamilton (1994) or assume an uninformative prior using the option `diffuse` as suggested by Harvey (1989).

ARIMA models

Pure ARIMA models without a structural component do not have regressors and are often written as autoregressions in the dependent variable, rather than autoregressions in the disturbances from a structural equation. For example, an ARMA(1, 1) model can be written as

$$y_t = \alpha + \rho y_{t-1} + \theta \epsilon_{t-1} + \epsilon_t \qquad (1a)$$

Other than a scale factor for the constant term α, these models are equivalent to the ARMA in the disturbances formulation estimated by `arima`, though the latter are more flexible and allow a wider class of models.

To see this, replace $\mathbf{x}_t\boldsymbol{\beta}$ in the structural equation above with a constant term β_0 so that

$$\begin{aligned} y_t &= \beta_0 + \mu_t \\ &= \beta_0 + \rho\mu_{t-1} + \theta\epsilon_{t-1} + \epsilon_t \\ &= \beta_0 + \rho(y_{t-1} - \beta_0) + \theta\epsilon_{t-1} + \epsilon_t \\ &= (1-\rho)\beta_0 + \rho y_{t-1} + \theta\epsilon_{t-1} + \epsilon_t \end{aligned} \qquad (1b)$$

Equations (1a) and (1b) are equivalent, with $\alpha = (1-\rho)\beta_0$, so whether we consider an ARIMA model as autoregressive in the dependent variable or disturbances is immaterial. Our illustration can easily be extended from the ARMA(1, 1) case to the general ARIMA(p, d, q) case.

▷ Example 1: ARIMA model

Enders (1995, 106–110) considers an ARIMA model of the US Wholesale Price Index (WPI) using quarterly data over the period 1960q1 through 1990q4. The simplest ARIMA model that includes differencing and both autoregressive and moving-average components is the ARIMA(1,1,1) specification. We can fit this model using `arima` by typing

```
. use http://www.stata-press.com/data/r9/wpi1

. arima wpi, arima(1,1,1)

(setting optimization to BHHH)
Iteration 0:    log likelihood = -139.80133
Iteration 1:    log likelihood =  -135.6278
Iteration 2:    log likelihood = -135.41838
Iteration 3:    log likelihood = -135.36691
Iteration 4:    log likelihood = -135.35892
(switching optimization to BFGS)
Iteration 5:    log likelihood = -135.35471
Iteration 6:    log likelihood = -135.35135
Iteration 7:    log likelihood = -135.35132
Iteration 8:    log likelihood = -135.35131

ARIMA regression

Sample:  1960q2 to 1990q4              Number of obs     =        123
                                      Wald chi2(2)      =     310.64
Log likelihood = -135.3513            Prob > chi2       =     0.0000
```

D.wpi	Coef.	OPG Std. Err.	z	P>|z|	[95% Conf. Interval]
wpi						
_cons	.7498197	.3340968	2.24	0.025	.0950019 1.404637
ARMA						
ar						
L1.	.8742288	.0545435	16.03	0.000	.7673256 .981132
ma						
L1.	-.4120458	.1000284	-4.12	0.000	-.6080979 -.2159938
/sigma	.7250436	.0368065	19.70	0.000	.6529042 .7971829

Examining the estimation results, we see that the AR(1) coefficient is 0.874, the MA(1) coefficient is -0.412, and both are highly significant. The estimated standard deviation of the white-noise disturbance ϵ is 0.725.

This model also could have been fitted by typing

```
. arima D.wpi, ar(1) ma(1)
```

The D. placed in front of the dependent variable wpi is the Stata time-series operator for differencing. Thus we would be modeling the first difference in WPI from the second quarter of 1960 through the fourth quarter of 1990 since the first observation is lost due to differencing. This second syntax allows a richer choice of models.

◁

▷ Example 2: ARIMA model with additive seasonal effects

After examining first-differences of WPI, Enders chose a model of differences in the natural logarithms to stabilize the variance in the differenced series. The raw data and first-difference of the logarithms are graphed below.

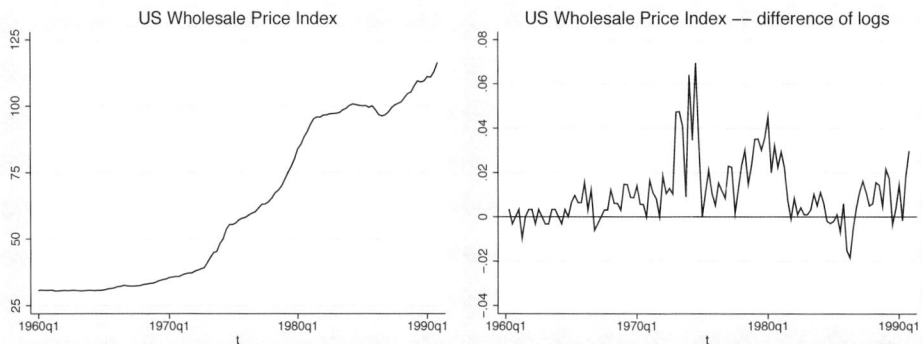

On the basis of the autocorrelations, partial autocorrelations (see graphs below), and the results of preliminary estimations, Enders identified an ARMA model in the log-differenced series.

```
. ac D.ln_wpi, ylabels(-.4(.2).6)
. pac D.ln_wpi, ylabels(-.4(.2).6)
```

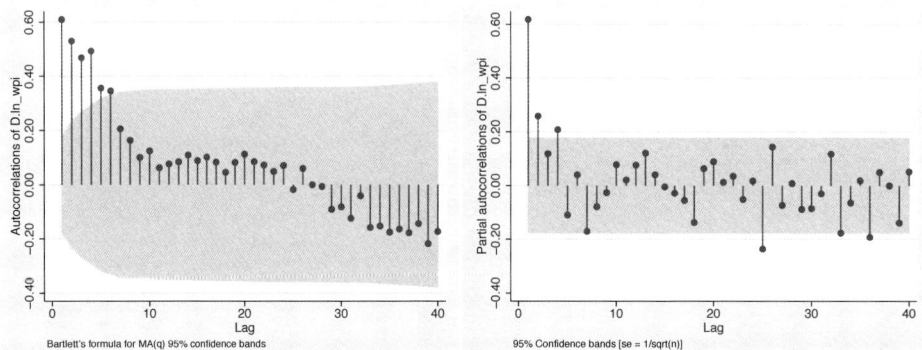

In addition to an autoregressive term and an MA(1) term, an MA(4) term is included to account for a remaining quarterly effect. Thus the model to be fitted is

$$\Delta \ln(wpi_t) = \beta_0 + \rho_1\{\Delta \ln(wpi_{t-1}) - \beta_0\} + \theta_1 \epsilon_{t-1} + \theta_4 \epsilon_{t-4} + \epsilon_t$$

We can fit this model using **arima** and Stata's standard difference operator:

```
. arima D.ln_wpi, ar(1) ma(1 4)
(setting optimization to BHHH)
Iteration 0:   log likelihood =  382.67447
Iteration 1:   log likelihood =  384.80754
Iteration 2:   log likelihood =  384.84749
Iteration 3:   log likelihood =  385.39213
Iteration 4:   log likelihood =  385.40983
(switching optimization to BFGS)
Iteration 5:   log likelihood =   385.9021
Iteration 6:   log likelihood =  385.95646
Iteration 7:   log likelihood =  386.02979
Iteration 8:   log likelihood =  386.03326
Iteration 9:   log likelihood =  386.03354
Iteration 10:  log likelihood =  386.03357
```

```
ARIMA regression

Sample:  1960q2 to 1990q4                  Number of obs    =       123
                                           Wald chi2(3)     =    333.60
Log likelihood =  386.0336                 Prob > chi2      =    0.0000
```

D.ln_wpi	Coef.	OPG Std. Err.	z	P>|z|	[95% Conf. Interval]	
ln_wpi						
_cons	.0110493	.0048349	2.29	0.022	.0015731	.0205255
ARMA						
ar						
L1.	.7806991	.0944946	8.26	0.000	.5954931	.965905
ma						
L1.	-.3990039	.1258753	-3.17	0.002	-.6457149	-.1522928
L4.	.3090813	.1200945	2.57	0.010	.0737003	.5444622
/sigma	.0104394	.0004702	22.20	0.000	.0095178	.0113609

In this final specification, the log-differenced series is still highly autocorrelated at a level of 0.781, though innovations have a negative impact in the ensuing quarter (-0.399) and a positive seasonal impact of 0.309 in the following year.

◁

❏ Technical Note

In one way, the results differ from most of Stata's estimation commands: the standard error of the coefficients is reported as OPG Std. Err. As noted in *Options*, the default standard errors and covariance matrix for arima estimates are derived from the outer product of gradients (OPG). This is one of three asymptotically equivalent methods of estimating the covariance matrix of the coefficients (only two of which are usually tractable to derive). Discussions and derivations of all three estimates can be found in Davidson and MacKinnon (1993), Greene (2003), and Hamilton (1994). Bollerslev, Engle, and Nelson (1994) suggest that the OPG estimates may be more numerically stable in time-series regressions when the likelihood and its derivatives depend on recursive computations, which is certainly the case for the Kalman filter. To date, we have not found any numerical instabilities in either estimate of the covariance matrix—subject to the stability and convergence of the overall model.

Most of Stata's estimation commands provide covariance estimates derived from the Hessian of the likelihood function. These alternate estimates can also be obtained from arima by specifying the oim option.

❏

Multiplicative seasonal ARIMA models

Many time series exhibit a periodic seasonal component, and seasonal ARIMA models, often abbreviated SARIMA, can be used in these cases. For example, monthly sales data for air conditioners have a strong seasonal component, with sales high in the summer months and low in the winter months.

In the previous example, we accounted for quarterly effects by fitting the model

$$(1 - \rho_1 L)\{\Delta \ln(wpi_t) - \beta_0\} = (1 + \theta_1 L + \theta_4 L^4)\epsilon_t$$

This is an *additive* seasonal ARIMA model, in the sense that the first- and fourth-order MA terms work in an additive way – notice the term $(1 + \theta_1 L + \theta_4 L^4)$.

An alternative way to handle the quarterly effect would be to fit a multiplicative seasonal ARIMA model. A multiplicative SARIMA model of order $(1, 1, 1) \times (0, 0, 1)_4$ for the $\ln(wpi_t)$ series is

$$(1 - \rho_1 L)\{\Delta \ln(wpi_t) - \beta_0\} = (1 + \theta_1 L)(1 + \theta_{4,1} L^4)\epsilon_t$$

or, upon expanding terms,

$$\Delta \ln(wpi_t) = \beta_0 + \rho_1\{\Delta \ln(wpi_t) - \beta_0\} + \theta_1\epsilon_{t-1} + \theta_{4,1}\epsilon_{t-4} + \theta_1\theta_{4,1}\epsilon_{t-5} + \epsilon_t \qquad (2)$$

In the notation $(1, 1, 1) \times (0, 0, 1)_4$, the $(1, 1, 1)$ means that there is one nonseasonal autoregressive term $(1 - \rho_1 L)$ and one nonseasonal moving-average term $(1 + \theta_1 L)$ and that the time series is first-differenced one time. The $(0, 0, 1)_4$ indicates that there is no lag-4 seasonal autoregressive term, that there is one lag-4 seasonal moving-average term $(1 + \theta_{4,1} L^4)$, and that the series is seasonally differenced zero times. This is a known as a *multiplicative* SARIMA model because the nonseasonal and seasonal factors work in a multiplicative fashion—notice the term $(1 + \theta_1 L)(1 + \theta_{4,1} L^4)$. Multiplying the terms imposes nonlinear constraints on the parameters of the 5th-order lagged values; `arima` imposes these constraints automatically.

To further clarify the notation, consider a $(2, 1, 1) \times (1, 1, 2)_4$ multiplicative SARIMA model:

$$(1 - \rho_1 L - \rho_2 L^2)(1 - \rho_{4,1} L^4)\Delta\Delta_4 z_t = (1 + \theta_1 L)(1 + \theta_{4,1} L^4 + \theta_{4,2} L^8)\epsilon_t \qquad (3)$$

where Δ denotes the difference operator $\Delta y_t = y_t - y_{t-1}$ and Δ_s denotes the lag-s seasonal difference operator $\Delta_s y_t = y_t - y_{t-s}$. Expanding equation (3) we have

$$\widetilde{z}_t = \rho_1\widetilde{z}_{t-1} + \rho_2\widetilde{z}_{t-2} + \rho_{4,1}\widetilde{z}_{t-4} - \rho_1\rho_{4,1}\widetilde{z}_{t-5} - \rho_2\rho_{4,1}\widetilde{z}_{t-6}$$
$$+ \theta_1\epsilon_{t-1} + \theta_{4,1}\epsilon_{t-4} + \theta_1\theta_{4,1}\epsilon_{t-5} + \theta_{4,2}\epsilon_{t-8} + \theta_1\theta_{4,2}\epsilon_{t-9}$$

where

$$\widetilde{z}_t = \Delta\Delta_4 z_t = \Delta(z_t - z_{t-4}) = z_t - z_{t-1} - (z_{t-4} - z_{t-5})$$

and $z_t = y_t - \mathbf{x}_t\boldsymbol{\beta}$ if regressors are included in the model, $z_t = y_t - \beta_0$ if just a constant term is included, and $z_t = y_t$ otherwise.

More generally, a $(p, d, q) \times (P, D, Q)_s$ multiplicative SARIMA model is

$$\boldsymbol{\rho}(L^p)\boldsymbol{\rho}_s(L^P)\Delta^d\Delta_s^D z_t = \boldsymbol{\theta}(L^q)\boldsymbol{\theta}_s(L^Q)$$

where

$$\boldsymbol{\rho}_s(L^P) = (1 - \rho_{s,1}L^s - \rho_{s,2}L^{2s} - \cdots - \rho_{s,P}L^{Ps})$$
$$\boldsymbol{\theta}_s(L^Q) = (1 + \theta_{s,1}L^s + \theta_{s,2}L^{2s} + \cdots + \theta_{s,Q}L^{Qs})$$

$\boldsymbol{\rho}(L^p)$ and $\boldsymbol{\theta}(L^q)$ were defined previously, Δ^d means apply the Δ operator d times, and similarly for Δ_s^D. Typically, d and D will be zero or one; and p, q, P, and Q will seldom be more than two or three. s will typically be four for quarterly data and twelve for monthly data. In fact, the model can be extended to include both monthly and quarterly seasonal factors, as we explain below.

If a plot of the data suggests that the seasonal effect is proportional to the mean of the series, then the seasonal effect is probably multiplicative; and a multiplicative SARIMA model may be appropriate. Box, Jenkins, and Reinsel (1994, §9.3.1) suggest starting with a multiplicative SARIMA model with any data that exhibits seasonal patterns then exploring nonmultiplicative SARIMA models if the multiplicative models do not fit the data well. On the other hand, Chatfield (2004, p. 14) suggests that taking the logarithm of the series will make the seasonal effect additive, in which case an additive SARIMA model as fit in the previous example would be appropriate. In short, the analyst should probably try both additive and multiplicative SARIMA models to see which provide better fits and forecasts.

As mentioned previously, unless the `diffuse` option is used `arima` must create square matrices of dimension $\{\max(p, q+1)\}^2$, where p and q are the maximum AR and MA lags, respectively; and the inclusion of long seasonal terms can make this dimension rather large. For example, with monthly data, you might fit a $(0, 1, 1) \times (0, 1, 2)_{12}$ SARIMA model. The maximum MA lag is $2 \times 12 + 1 = 25$, requiring a matrix with $26^2 = 676$ rows and columns.

▷ Example 3: Multiplicative SARIMA model

One of the most common multiplicative SARIMA specifications is the $(0, 1, 1) \times (0, 1, 1)_{12}$ "airline" model of Box, Jenkens, and Reinsel (1994, §9.2). The dataset `airline.dta` contains monthly international airline passenger data from January, 1949 through December, 1960. After first- and seasonally differencing the data, we do not suspect the presence of a trend component, so we use the `noconstant` option with `arima`:

```
. use http://www.stata-press.com/data/r9/air2, clear
(TIMESLAB: Airline passengers)

. generate lnair = ln(air)

. arima lnair, arima(0,1,1) sarima(0,1,1,12) noconstant

(setting optimization to BHHH)
Iteration 0:   log likelihood =   223.8437
Iteration 1:   log likelihood =  239.80405
Iteration 2:   log likelihood =  244.10265
Iteration 3:   log likelihood =  244.65895
Iteration 4:   log likelihood =  244.68945
(switching optimization to BFGS)
Iteration 5:   log likelihood =  244.69431
Iteration 6:   log likelihood =  244.69647
Iteration 7:   log likelihood =  244.69651
Iteration 8:   log likelihood =  244.69651

ARIMA regression

Sample:  14 to 144                         Number of obs      =       131
                                           Wald chi2(2)       =     84.53
Log likelihood =  244.6965                 Prob > chi2        =    0.0000
```

DS12.lnair	Coef.	OPG Std. Err.	z	P>|z|	[95% Conf. Interval]	
ARMA						
ma						
L1.	-.4018324	.0730307	-5.50	0.000	-.5449698	-.2586949
ARMA12						
ma						
L1.	-.5569342	.0963129	-5.78	0.000	-.745704	-.3681644
/sigma	.0367167	.0020132	18.24	0.000	.0327708	.0406625

Thus our model of the monthly number of international airline passengers is

$$\Delta\Delta_{12}\texttt{lnair}_t = -0.402\epsilon_{t-1} - 0.557\epsilon_{t-12} + 0.224\epsilon_{t-13} + \epsilon_t$$
$$\hat{\sigma} = 0.037$$

In equation (2), for example, the coefficient on ϵ_{t-13} is the product of the coefficients on the ϵ_{t-1} and ϵ_{t-12} terms $(0.224 \approx -0.402 \times -0.557)$. `arima` labeled the dependent variable `DS12.lnair` to indicate that it has applied the difference operator Δ and the lag-12 seasonal difference operator Δ_{12} to `lnair`; see [U] **11.4.3 Time-series varlists** for more information.

We could have fitted this model by typing

```
. arima DS12.lnair, ma(1) mma(1, 12) noconstant
```

For simple multiplicative models, using the `sarima()` option is easier, though this second syntax allows us to incorporate more complicated seasonal terms.

◁

The `mar()` and `mma()` options can be repeated, allowing us to control for multiple seasonal patterns. For example, we may have monthly sales data that exhibit a quarterly pattern as businesses purchase our product at the beginning of calendar quarters when new funds are budgeted, and our product is purchased more frequently in a few months of the year than in most others, even after we control for quarterly fluctuations. Thus we might choose to fit the model

$$(1-\rho L)(1-\rho_{4,1}L^4)(1-\rho_{12,1}L^{12})(\Delta\Delta_4\Delta_{12}\texttt{sales}_t - \beta_0) = (1+\theta L)(1+\theta_{4,1}L^4)(1+\theta_{12,1}L^{12})\epsilon_t$$

While this model looks rather complicated, estimating it using `arima` is straightforward:

```
. arima DS4S12.sales, ar(1) mar(1, 4) mar(1, 12) ma(1) mma(1, 4) mma(1, 12)
```

If we instead wanted to include two lags in the lag-4 seasonal AR term and the first and third (but not the second) term in the lag-12 seasonal MA term, we would type

```
. arima DS4S12.sales, ar(1) mar(1 2, 4) mar(1, 12) ma(1) mma(1, 4) mma(1 3, 12)
```

However note that models with multiple seasonal terms can be difficult to fit. Usually, a single seasonal factor with just one or two AR or MA terms is adequate.

ARMAX models

Thus far all of our examples have been pure ARIMA models in which the dependent variable was modeled solely as a function of its past values and disturbances. Additionally, `arima` can fit ARMAX models, which model the dependent variable in terms of a linear combination of independent variables, as well as an ARMA disturbance process. The `prais` command, for example, only allows you to control for AR(1) disturbances, while `arima` allows you to control for a much richer dynamic error structure. `arima` allows for both nonseasonal and seasonal ARMA components in the disturbances.

▷ Example 4: ARMAX model

As a simple example of a model including covariates, we can estimate an update of Friedman and Meiselman's (1963) equation representing the quantity theory of money. They postulate a straight-forward relationship between personal-consumption expenditures (`consump`) and the money supply as measured by M2 (`m2`).

$$\texttt{consump}_t = \beta_0 + \beta_1\texttt{m2}_t + \mu_t$$

Friedman and Meiselman fit the model over a period ending in 1956; we will refit the model over the period 1959q1 through 1981q4. We restrict our attention to the period prior to 1982 because the Federal Reserve manipulated the money supply extensively in the latter 1980s to control inflation, and the relationship between consumption and the money supply becomes much more complex during the latter part of the decade.

To demonstrate `arima`, we will include both an autoregressive term and a moving-average term for the disturbances in the model; the original estimates included neither. Thus we model the disturbance of the structural equation as

$$\mu_t = \rho\mu_{t-1} + \theta\epsilon_{t-1} + \epsilon_t$$

Following the original authors, the relationship is estimated on seasonally adjusted data, so there is no need to include seasonal effects explicitly. It might be preferable to obtain seasonally unadjusted data and simultaneously model the structural and seasonal effects.

We will restrict the estimation to the desired sample by using the `tin()` function in an `if` expression; see [D] **functions** and [U] **24.3 Time-series dates**. By leaving the first argument of `tin()` blank, we are including all available data up to and including the second date (1981q4). We fit the model by typing

```
. use http://www.stata-press.com/data/r9/friedman2, clear

. arima consump m2, ar(1) ma(1), if tin(, 1981q4)

  (output omitted)
Iteration 10:  log likelihood = -340.50774

ARIMA regression

Sample:  1959q1 to 1981q4                     Number of obs   =        92
                                              Wald chi2(3)    =   4394.80
Log likelihood = -340.5077                    Prob > chi2     =    0.0000
```

consump	Coef.	OPG Std. Err.	z	P>|z|	[95% Conf. Interval]
consump					
m2	1.122029	.0363563	30.86	0.000	1.050772 1.193286
_cons	-36.09872	56.56703	-0.64	0.523	-146.9681 74.77062
ARMA					
ar					
L1.	.9348486	.0411323	22.73	0.000	.8542308 1.015467
ma					
L1.	.3090592	.0885883	3.49	0.000	.1354293 .4826891
/sigma	9.655308	.5635157	17.13	0.000	8.550837 10.75978

We find a relatively small money velocity with respect to consumption (1.122) over this period, although consumption is only one facet of the income velocity. We also note a very large first-order autocorrelation in the disturbances, as well as a statistically significant first-order moving average.

We might be concerned that our specification has led to disturbances that are heteroskedastic or non-Gaussian. We refit the model using the `robust` option.

```
. arima consump m2, ar(1) ma(1) robust, if tin(, 1981q4)
 (output omitted)
Iteration 10:  log pseudolikelihood = -340.50774

ARIMA regression

Sample:  1959q1 to 1981q4                    Number of obs    =        92
                                             Wald chi2(3)     =   1176.26
Log pseudolikelihood = -340.5077             Prob > chi2      =    0.0000
```

consump	Coef.	Semi-robust Std. Err.	z	P>|z|	[95% Conf. Interval]	
consump						
m2	1.122029	.0433302	25.89	0.000	1.037103	1.206954
_cons	-36.09872	28.10478	-1.28	0.199	-91.18309	18.98565
ARMA						
ar						
L1.	.9348486	.0493428	18.95	0.000	.8381385	1.031559
ma						
L1.	.3090592	.1605359	1.93	0.054	-.0055854	.6237038
/sigma	9.655308	1.082639	8.92	0.000	7.533375	11.77724

We note a substantial increase in the estimated standard errors, and our once clearly significant moving-average term is now only marginally significant.

◁

Dynamic forecasting

An additional feature of the arima command is the ability to use predict afterward to make dynamic forecasts. For example, suppose that we wish to fit the regression model

$$y_t = \beta_0 + \beta_1 x_t + \rho y_{t-1} + \epsilon_t$$

using a sample of data from $t = 1 \ldots T$ and make forecasts beginning at time f.

If we use regress or prais to fit the model, then we can use predict to make one-step-ahead forecasts. That is, predict will compute

$$\widehat{y}_f = \widehat{\beta_0} + \widehat{\beta_1} x_f + \widehat{\rho} y_{f-1}$$

Most importantly, notice that here predict will use the actual value of y at period $f-1$ in computing the forecast for time f. Thus if we use regress or prais, we cannot make forecasts for any periods beyond $f = T + 1$ unless we have observed values for y for those periods.

If we instead fit our model using arima, then predict can produce dynamic forecasts using the Kalman filter. If we use the dynamic(f) option, then for period f predict will compute

$$\widehat{y}_f = \widehat{\beta_0} + \widehat{\beta_1} x_f + \widehat{\rho} y_{f-1}$$

using the observed value of y_{f-1} just as predict after regress or prais. However, for period $f + 1$ predict $newvar$, dynamic(f) will compute

$$\widehat{y}_{f+1} = \widehat{\beta_0} + \widehat{\beta_1} x_{f+1} + \widehat{\rho}\widehat{y}_f$$

using the *predicted* value of y_f instead of the observed value. Similarly, the period $f + 2$ forecast will be

$$\widehat{y}_{f+2} = \widehat{\beta_0} + \widehat{\beta_1} x_{f+2} + \widehat{\rho} \widetilde{y}_{f+1}$$

Of course, since our model includes the regressor x_t, we can only make forecasts through periods for which we have observations on x_t. However, for pure ARIMA models, we can compute dynamic forecasts as far beyond the final period of our dataset as desired.

For more information on `predict` after `arima`, see [TS] **arima postestimation**.

Saved Results

`arima` saves in `e()`:

Scalars

e(N)	number of observations	e(p)	significance
e(N_gaps)	number of gaps	e(tmin)	minimum time
e(k)	number of parameters	e(tmax)	maximum time
e(k_dv)	number of dependent variables	e(rank)	rank of e(V)
e(k_eq)	number of equations	e(ic)	number of iterations
e(k_1)	number of variables in first equation	e(rc)	return code
		e(converged)	1 if converged, 0 otherwise
e(df_m)	model degrees of freedom	e(ar_max)	maximum AR lag
e(ll)	log likelihood	e(ma_max)	maximum MA lag
e(chi2)	χ^2		

Macros

e(cmd)	arch	e(seasons)	seasonal lags in model
e(depvar)	name of dependent variable	e(unsta)	unstationary or blank
e(wtype)	weight type	e(chi2type)	Wald; type of model χ^2 test
e(wexp)	weight expression	e(vce)	*vcetype* specified in vce()
e(title)	title in estimation output	e(vcetype)	title used to label Std. Err.
e(eqnames)	names of equations	e(opt)	type of optimization
e(tmins)	formatted minimum time	e(ml_method)	type of ml method
e(tmaxs)	formatted maximum time	e(user)	name of likelihood-evaluator program
e(ma)	lags for moving-average terms	e(technique)	maximization technique
e(ar)	lags for autoregressive terms	e(crittype)	optimization criterion
e(mar*i*)	multiplicative AR terms and lag $i=1...$ (# seasonal AR terms)	e(properties)	b V
		e(estat_cmd)	program used to implement estat
e(mma*i*)	multiplicative MA terms and lag $i=1...$ (# seasonal MA terms)	e(predict)	program used to implement predict

Matrices

e(b)	coefficient vector	e(V)	variance–covariance matrix of the estimators
e(ilog)	iteration log (up to 20 iterations)		
e(gradient)	gradient vector		

Functions

e(sample)	marks estimation sample

Methods and Formulas

`arima` is implemented as an ado-file.

Estimation is by maximum likelihood using the Kalman filter via the prediction error decomposition; see Hamilton (1994), Gourieroux and Monfort (1997), or, in particular, Harvey (1989). Any of these sources will serve as excellent background for the fitting of these models using the state-space form; each also provides considerable detail on the method outlined below.

ARIMA model

The model to be fitted is

$$y_t = \mathbf{x}_t\boldsymbol{\beta} + \mu_t$$

$$\mu_t = \sum_{i=1}^{p} \rho_i\mu_{t-i} + \sum_{j=1}^{q} \theta_j\epsilon_{t-j} + \epsilon_t$$

which can be written as the single equation

$$y_t = \mathbf{x}_t\boldsymbol{\beta} + \sum_{i=1}^{p} \rho_i(y_{t-i} - x_{t-i}\boldsymbol{\beta}) + \sum_{j=1}^{q} \theta_j\epsilon_{t-j} + \epsilon_t$$

Some of the ρs and θs may be constrained to zero or, in the case of multiplicative seasonal models, the products of other parameters.

Kalman filter equations

We will roughly follow Hamilton's (1994) notation and write the Kalman filter

$$\boldsymbol{\xi}_t = \mathbf{F}\boldsymbol{\xi}_{t-1} + \mathbf{v}_t \qquad \qquad (state\ equation)$$

$$\mathbf{y}_t = \mathbf{A}'\mathbf{x}_t + \mathbf{H}'\boldsymbol{\xi}_t + \mathbf{w}_t \qquad \qquad (observation\ equation)$$

and

$$\begin{pmatrix} \mathbf{v}_t \\ \mathbf{w}_t \end{pmatrix} \sim N\left\{ \mathbf{0}, \begin{pmatrix} \mathbf{Q} & \mathbf{0} \\ \mathbf{0} & \mathbf{R} \end{pmatrix} \right\}$$

We maintain the standard Kalman filter matrix and vector notation, although for univariate models $\mathbf{y}_t$, $\mathbf{w}_t$, and $\mathbf{R}$ are scalars.

Kalman filter or state-space representation of the ARIMA model

A univariate ARIMA model can be cast in state-space form by defining the Kalman filter matrices as follows (see Hamilton 1994, or Gourieroux and Monfort 1997, for details):

$$F = \begin{bmatrix} \rho_1 & \rho_2 & \cdots & \rho_{p-1} & \rho_p \\ 1 & 0 & \cdots & 0 & 0 \\ 0 & 1 & \cdots & 0 & 0 \\ 0 & 0 & \cdots & 1 & 0 \end{bmatrix}$$

$$\mathbf{v}_t = \begin{bmatrix} \epsilon_{t-1} \\ 0 \\ \cdots \\ \cdots \\ \cdots \\ 0 \end{bmatrix}$$

$$\mathbf{A}' = \beta$$

$$\mathbf{H}' = \begin{bmatrix} 1 & \theta_1 & \theta_2 & \cdots & \theta_q \end{bmatrix}$$

$$\mathbf{w}_t = 0$$

Note that the Kalman filter representation does not require the moving-average terms to be invertible.

Kalman filter recursions

To demonstrate how missing data are handled, the updating recursions for the Kalman filter will be left in two steps. It is common to write the updating equations as a single step using the gain matrix $\mathbf{K}$. We will provide the updating equations with little justification; see the sources listed above for details.

As a linear combination of a vector of random variables, the state ξ_t can be updated to its expected value based on the prior state as

$$\xi_{t|t-1} = \mathbf{F}\xi_{t-1} + \mathbf{v}_{t-1} \tag{1}$$

This state is a quadratic form that has the covariance matrix

$$\mathbf{P}_{t|t-1} = \mathbf{F}\mathbf{P}_{t-1}\mathbf{F}' + \mathbf{Q} \tag{2}$$

The estimator of $\mathbf{y}_t$ is

$$\widehat{\mathbf{y}}_{t|t-1} = \mathbf{x}_t\beta + \mathbf{H}'\xi_{t|t-1}$$

which implies an innovation or prediction error

$$\widehat{\iota}_t = \mathbf{y}_t - \widehat{\mathbf{y}}_{t|t-1}$$

This value or vector has mean squared error (MSE)

$$\mathbf{M}_t = \mathbf{H}'\mathbf{P}_{t|t-1}\mathbf{H} + \mathbf{R}$$

Now the expected value of ξ_t conditional on a realization of $\mathbf{y}_t$ is

$$\xi_t = \xi_{t|t-1} + \mathbf{P}_{t|t-1}\mathbf{H}\mathbf{M}_t^{-1}\widehat{\iota}_t \tag{3}$$

with MSE

$$\mathbf{P}_t = \mathbf{P}_{t|t-1} - \mathbf{P}_{t|t-1}\mathbf{H}\mathbf{M}_t^{-1}\mathbf{H}'\mathbf{P}_{t|t-1} \tag{4}$$

This gives the full set of Kalman filter recursions.

Kalman filter initial conditions

When the series is stationary, conditional on $\mathbf{x}_t\beta$, the initial conditions for the filter can be considered a random draw from the stationary distribution of the state equation. The initial values of the state and the state MSE are the expected values from this stationary distribution. For an ARIMA model, these can be written as

$$\xi_{1|0} = \mathbf{0}$$

and

$$\text{vec}(\mathbf{P}_{1|0}) = (\mathbf{I}_{r^2} - \mathbf{F} \otimes \mathbf{F})^{-1}\text{vec}(\mathbf{Q})$$

where vec() is an operator representing the column matrix resulting from stacking each successive column of the target matrix.

If the series is not stationary, the initial state conditions do not constitute a random draw from a stationary distribution, and some other values must be chosen. Hamilton (1994) suggests that they be chosen based on prior expectations, while Harvey suggests a diffuse and improper prior having a state vector of $\mathbf{0}$ and an infinite variance. This corresponds to $\mathbf{P}_{1|0}$ with diagonal elements of ∞. Stata allows either approach to be taken for nonstationary series—initial priors may be specified with state0() and p0(), and a diffuse prior may be specified with diffuse.

Likelihood from prediction error decomposition

Given the outputs from the Kalman filter recursions and assuming that the state and observation vectors are Gaussian, the likelihood for the state space model follows directly from the resulting multivariate normal in the predicted innovations. The log likelihood for observation t is

$$\ln L_t = -\frac{1}{2}\left\{ \ln(2\pi) + \ln(|\mathbf{M}_t|) - \widehat{\iota}_t'\mathbf{M}_t^{-1}\widehat{\iota}_t \right\}$$

Missing data

Missing data, whether a missing dependent variable y_t, one or more missing covariates $\mathbf{x}_t$, or completely missing observations, are handled by continuing the state-updating equations without any contribution from the data; see Harvey (1989 and 1993). That is, (1) and (2) are iterated for every missing observation, while (3) and (4) are ignored. Thus for observations with missing data, $\xi_t = \xi_{t|t-1}$ and $\mathbf{P}_t = \mathbf{P}_{t|t-1}$. In the absence of any information from the sample, this effectively assumes that the prediction error for the missing observations is 0. Alternative methods of handling missing data based on the EM algorithm have been suggested, e.g., Shumway (1984, 1988).

George Edward Pelham Box (1919–) was born in Kent, England, and earned degrees in statistics at the University of London. Ater work in the chemical industry, he taught and researched at Princeton and the University of Wisconsin. His many major contributions to statistics include papers and books in Bayesian inference, robustness (a term he introduced to statistics), modeling strategy, experimental design and response surfaces, time series analysis, distribution theory, transformations and nonlinear estimation.

Gwilym Meirion Jenkins (1933–1982) was a British mathematician and statistician who spent his career in industry and academia, working for extended periods at Imperial College London and the University of Lancaster before running his own company. His interests were centered on time series and he collaborated with G. E. P. Box on what are often called Box–Jenkins models. The last years of Jenkins' life were marked by a slowly losing battle against Hodgkin's disease.

References

Ansley, C. F. and R. Kohn. 1985. Estimation, filtering and smoothing in state space models with incompletely specified initial conditions. *Annals of Statistics* 13: 1286–1316.

Ansley, C. F. and P. Newbold. 1980. Finite sample properties of estimators for autoregressive moving-average processes. *Journal of Econometrics* 13: 159–184.

Baum, C. F. 2000. sts15: Tests for stationarity of a time series. *Stata Technical Bulletin* 57: 36–39. Reprinted in *Stata Technical Bulletin Reprints*, vol. 10, pp. 356–360.

———. 2001. sts18: A test for long-range dependence in a time series. *Stata Technical Bulletin* 60: 37–39. Reprinted in *Stata Technical Bulletin Reprints*, vol. 10, pp. 370–373.

Baum, C. F. and R. Sperling. 2001. sts15.1: Tests for stationarity of a time series: Update. *Stata Technical Bulletin* 58: 35–36. Reprinted in *Stata Technical Bulletin Reprints*, vol. 10, pp. 360–362.

Baum, C. F. and V. L. Wiggins. 2000. sts16: Tests for long memory in a time series. *Stata Technical Bulletin* 57: 39–44. Reprinted in *Stata Technical Bulletin Reprints*, vol. 10, pp. 362–368.

Berndt, E. K., B. H. Hall, R. E. Hall, and J. A. Hausman. 1974. Estimation and inference in nonlinear structural models. *Annals of Economic and Social Measurement* 3/4: 653–665.

Bollerslev, T., R. F. Engle, and D. B. Nelson. 1994. ARCH Models. In *Handbook of Econometrics, Volume IV*, ed. R. F. Engle and D. L. McFadden. New York: Elsevier.

Box, G. E. P. 1983. G. M. Jenkins, 1933–1982. *Journal of the Royal Statistical Society, Series A* 146: 205–206.

Box, G. E. P., G. M. Jenkins, and G. C. Reinsel. 1994. *Time Series Analysis: Forecasting and Control.* 3rd ed. Englewood Cliffs, NJ: Prentice Hall.

Chatfield, C. 2004. *The Analysis of Time Series: An Introduction.* 6th ed. Boca Raton, FL: Chapman & Hall/CRC.

David, J. S. 1999. sts14: Bivariate Granger causality test. *Stata Technical Bulletin* 51: 40–41. Reprinted in *Stata Technical Bulletin Reprints*, vol. 9, pp. 350–351.

Davidson, R. and J. G. MacKinnon. 1993. *Estimation and Inference in Econometrics.* Oxford: Oxford University Press.

DeGroot, M. H. 1987. A conversation with George Box. *Statistical Science* 2: 239–258.

Diggle, P. J. 1990. *Time Series: A Biostatistical Introduction.* Oxford: Oxford University Press.

Enders, W. 1995. *Applied Econometric Time Series.* New York: Wiley.

Friedman, M. and D. Meiselman. 1963. The relative stability of monetary velocity and the investment multiplier in the United States, 1897–1958. In *Stabilization Policies*, Commission on Money and Credit. Englewood Cliffs, NJ: Prentice Hall.

Gourieroux, C. and A. Monfort. 1997. *Time Series and Dynamic Models.* Cambridge: Cambridge University Press.

Greene, W. H. 2003. *Econometric Analysis.* 5th ed. Upper Saddle River, NJ: Prentice Hall.

Hamilton, J. D. 1994. *Time Series Analysis.* Princeton: Princeton University Press.

Harvey, A. C. 1989. *Forecasting, Structural Time Series Models and the Kalman Filter.* Cambridge: Cambridge University Press.

———. 1993. *Time Series Models.* Cambridge, MA: MIT Press.

Hipel, K. W. and A. I. McLeod. 1994. *Time Series Modelling of Water Resources and Environmental Systems.* Amsterdam: Elsevier.

Kalman, R. E. 1960. A new approach to linear filtering and prediction problems. *Journal of Basic Engineering, Transactions of the ASME, Series D* 82: 35–45.

McDowell, A. W. 2002. From the help desk: Transfer functions. *Stata Journal* 2: 71–85.

———. 2004. From the help desk: Polynomial distributed lag models. *Stata Journal* 4: 180–189.

Newton, H. J. 1988. *TIMESLAB: A Time Series Analysis Laboratory.* Belmont, CA: Wadsworth & Brooks/Cole.

Press, W. H., S. A. Teukolsky, W. T. Vetterling, and B. P. Flannery. 1992. *Numerical Recipes in C: The Art of Scientific Computing.* 2nd ed. Cambridge: Cambridge University Press.

Shumway, R. H. 1984. Some applications of the EM algorithm to analyzing incomplete time series data. In *Time Series Analysis of Irregularly Observed Data*, ed. E. Parzen, 290–324. New York: Springer.

——. 1988. *Applied Statistical Time Series Analysis*. Upper Saddle River, NJ: Prentice Hall.

Also See

Complementary:	[TS] **arima postestimation**, [TS] **tsset**
Related:	[TS] **arch**, [TS] **prais**, [TS] **var**, [TS] **var svar**, [TS] **vec**, [R] **regress**
Background:	[U] **11.1.10 Prefix commands**, [U] **11.4.3 Time-series varlists**, [U] **20 Estimation and postestimation commands**, [U] **24.3 Time-series dates**, [TS] **estimation options**, [R] **maximize**, [R] *vce_option*

Title

> **arima postestimation** — Postestimation tools for arima

Description

The following postestimation commands are available for `arima`:

command	description
estat	AIC, BIC, VCE, and estimation sample summary
estimates	cataloging estimation results
lincom	point estimates, standard errors, testing, and inference for linear combinations of coefficients
nlcom	point estimates, standard errors, testing, and inference for nonlinear combinations of coefficients
predict	predictions, residuals, influence statistics, and other diagnostic measures
predictnl	point estimates, standard errors, testing, and inference for generalized predictions
test	Wald tests for simple and composite linear hypotheses
testnl	Wald tests of nonlinear hypotheses

See the corresponding entries in the *Stata Base Reference Manual* for details.

Syntax for predict

predict [*type*] *newvar* [*if*] [*in*] [, *statistic options*]

statistic	description
Main	
xb	predicted values for mean equation—the differenced series; the default
y	predicted values for the mean equation in y—the undifferenced series
mse	mean squared error of the predicted values
<u>r</u>esiduals	residuals or predicted innovations
<u>y</u>residuals	residuals or predicted innovations in y, reversing any time-series operators

These statistics are available both in and out of sample; type `predict ... if e(sample) ...` if wanted only for the estimation sample.

options	description
Options	
<u>d</u>ynamic(*time_constant*)	how to handle the lags of y_t
t0(*time_constant*)	set starting point for the recursions to *time_constant*
<u>str</u>uctural	calculate considering the structural component only

time_constant is a # or a time literal, such as d(1jan1995) or q(1995q1); see [U] **24.3 Time-series dates**.

Options for predict

Five statistics can be computed by using `predict` after `arima`: the predictions from the model (the default also given by `xb`), the predictions after reversing any time-series operators applied to the dependent variable (`y`), the MSE of `xb` (`mse`), the predictions of residuals or innovations (`residual`), and the predicted residuals or innovations in terms of y (`yresiduals`). Given the dynamic nature of the ARMA component and because the dependent variable might be differenced, there are alternative ways of computing each. We can use all the data on the dependent variable that is available right up to the time of each prediction (the default, which is often called a one-step prediction), or we can use the data up to a particular time, after which the predicted value of the dependent variable is used recursively to make subsequent predictions (`dynamic()`). Either way, we can consider or ignore the ARMA disturbance component (the component is considered by default and is ignored if you specify `structural`).

All calculations can be made in or out of sample.

<hr>

◄ Main └───

`xb`, the default, calculates the predictions from the model. If D.*depvar* is the dependent variable, these predictions are of D.*depvar* and not of *depvar* itself.

`y` specifies that predictions of *depvar* be made, even if the model was specified in terms of, say, D.*depvar*.

`mse` calculates the MSE of `xb`.

`residuals` calculates the residuals. If no other options are specified, these are the predicted innovations ϵ_t; i.e., they include the ARMA component. If `structural` is specified, these are the residuals μ_t from the structural equation; see `structural` below.

`yresiduals` calculates the residuals in terms of *depvar*, even if the model was specified in terms of, say, D.*depvar*. As with `residuals`, the `yresiduals` are computed from the model, including any ARMA component. If `structural` is specified, any ARMA component is ignored, and `yresiduals` are the residuals from the structural equation; see `structural` below.

<hr>

◄ Options └───

`dynamic(`*time_constant*`)` specifies how lags of y_t in the model are to be handled. If `dynamic()` is not specified, actual values are used everywhere lagged values of y_t appear in the model to produce one-step-ahead forecasts.

`dynamic(`*time_constant*`)` produces dynamic (also known as recursive) forecasts. *time_constant* specifies when the forecast is to switch from one-step ahead to dynamic. In dynamic forecasts, references to y_t evaluate to the prediction of y_t for all periods at or after *time_constant*; they evaluate to the actual value of y_t for all prior periods.

For example, `dynamic(10)` would calculate predictions in which any reference to y_t with $t < 10$ evaluates to the actual value of y_t and any reference to y_t with $t \geq 10$ evaluates to the prediction of y_t. This means that one-step-ahead predictions are calculated for $t < 10$ and dynamic predictions thereafter. Depending on the lag structure of the model, the dynamic predictions might still refer some actual values of y_t.

In addition, you may specify `dynamic(.)` to have `predict` automatically switch from one-step to dynamic predictions at $p + q$, where p is the maximum AR lag and q is the maximum MA lag.

`t0(`*time_constant*`)` specifies the starting point for the recursions to compute the predicted statistics; disturbances are assumed to be 0 for $t <$ `t0()`. The default is to set `t0()` to the minimum t

observed in the estimation sample, meaning that observations before that are assumed to have disturbances of 0.

t0() is irrelevant if structural is specified because, in that case, all observations are assumed to have disturbances of 0.

t0(5) would begin recursions at $t = 5$. If the data were quarterly, you might instead type t0(q(1961q2)) to obtain the same result.

Note that the ARMA component of arima models is recursive and depends on the starting point of the predictions. This includes one-step-ahead predictions.

structural specifies that the calculation be made considering the structural component only, ignoring the ARMA terms, producing the steady-state equilibrium predictions.

Remarks

We assume that you have already read [TS] **arima**. In this entry we illustrate some of the features of predict after fitting ARIMA, ARMAX, and other dynamic models using arima. In example 2 of [TS] **arima**, we fitted the model

$$\Delta \ln(wpi_t) = \beta_0 + \rho_1 \{\Delta \ln(wpi_{t-1}) - \beta_0\} + \theta_1 \epsilon_{t-1} + \theta_4 \epsilon_{t-4} + \epsilon_t$$

by typing

    . use http://www.stata-press.com/data/r9/wpi1
    . arima D.ln_wpi, ar(1) ma(1 4)
    (output omitted)

If we use the command

    . predict xb, xb

then Stata computes xb_t as

$$xb_t = \widehat{\beta}_0 + \widehat{\rho}_1 \{\Delta \ln(wpi_{t-1}) - \widehat{\beta}_0\} + \widehat{\theta}_1 \widehat{\epsilon}_{t-1} + \widehat{\theta}_4 \widehat{\epsilon}_{t-4}$$

where

$$\widehat{\epsilon}_{t-j} = \begin{cases} \Delta \ln(wpi_{t-j}) - xb_{t-j} & t - j > 0 \\ 0 & \text{otherwise} \end{cases}$$

meaning that predict *newvar*, xb calculates predictions using the metric of the dependent variable. In this example, the dependent variable represented *changes* in $\ln(wpi_t)$, and so the predictions are likewise for *changes* in that variable.

If we instead use

    . predict y, y

Stata computes y_t as $y_t = xb_t + \ln(wpi_{t-1})$ so that y_t represents the predicted *levels* of $\ln(wpi_t)$. In general, predict *newvar*, y will reverse any time-series operators applied to the dependent variable during estimation.

If we want to ignore the ARMA error components when making predictions, we use the structural option,

    . predict xbs, xb structural

which generates $\text{xbs}_t = \widehat{\beta}_0$ since there are no regressors in this model, and

```
. predict ys, y structural
```

generates $\text{ys}_t = \widehat{\beta}_0 + \ln(wpi_{t-1})$

▷ Example 1: Dynamic forecasts

An attractive feature of the `arima` command is the ability to make dynamic forecasts. In example 4 of [TS] **arima**, we fitted the model

$$\text{consump}_t = \beta_0 + \beta_1 \text{m2}_t + \mu_t$$

$$\mu_t = \rho\mu_{t-1} + \theta\epsilon_{t-1} + \epsilon_t$$

First we refit the model using data up through the first quarter of 1978, then we will evaluate the one-step-ahead and dynamic forecasts.

```
. use http://www.stata-press.com/data/r9/friedman2
. keep if time <= q(1981q4)
. arima consump m2 if tin(, 1978q1), ar(1) ma(1)
```
(output omitted)

To make one-step-ahead forecasts, we type

```
. predict chat, y
(52 missing values generated)
```

(Since our dependent variable did not contain any time-series operators, we could have instead used `predict chat, xb` and accomplish the same thing.) We will also make dynamic forecasts, switching from observed values of consump to forecasted values at the first quarter of 1978.

```
. predict chatdy, dynamic(q(1978q1)) y
(52 missing values generated)
```

The following graph compares the forecasted values to the observed values for the first few years following the estimation sample:

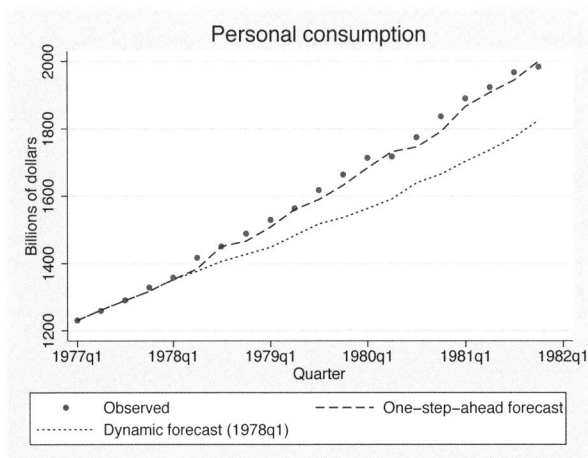

The one-step-ahead forecasts never deviate far from the observed values, though over time the dynamic forecasts have larger errors. To understand why that is the case, rewrite the model as

$$\text{consump}_t = \beta_0 + \beta_1 \text{m2}_t + \rho \mu_{t-1} + \theta \epsilon_{t-1} + \epsilon_t$$

$$= \beta_0 + \beta_1 \text{m2}_t + \rho \left(\text{consump}_{t-1} - \beta_0 - \beta_1 \text{m2}_{t-1} \right) + \theta \epsilon_{t-1} + \epsilon_t$$

This shows that the forecasted value of consumption at time t depends on the value of consumption at time $t-1$. When making the one-step-ahead forecast for period t, we know the actual value of consumption at time $t-1$. On the other hand, with the dynamic(q(1978q1)) option, the forecasted value of consumption for period 1978q1 is based on the observed value of consumption in period 1977q4, but the forecast for 1978q2 is based on the forecast value for 1978q1, the forecast for 1978q3 is based on the forecast value for 1978q2, and so on. Thus, with dynamic forecasts, prior forecast errors accumulate over time. The following graph illustrates this.

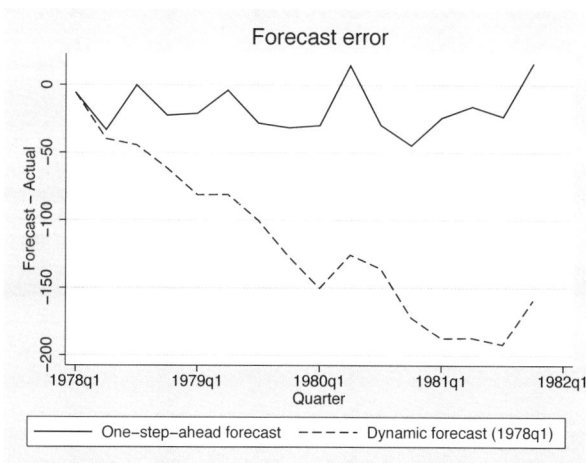

◁

Methods and Formulas

All postestimation commands listed above are implemented as ado-files.

Also See

Complementary:	[TS] **arima**,
	[R] **estimates**, [R] **lincom**, [R] **nlcom**,
	[R] **predictnl**, [R] **test**, [R] **testnl**
Background:	[U] **13.5 Accessing coefficients and standard errors**,
	[U] **20 Estimation and postestimation commands**,
	[R] **estat**, [R] **predict**

Title

corrgram — Tabulate and graph autocorrelations

Syntax

Autocorrelations, partial autocorrelations, and Portmanteau (Q) statistics

corrgram *varname* [*if*] [*in*] [, *corrgram_options*]

Graph of autocorrelations

ac *varname* [*if*] [*in*] [, *ac_options*]

Graph of partial autocorrelations

pac *varname* [*if*] [*in*] [, *pac_options*]

corrgram_options	description
Main	
<u>l</u>ags(#)	calculate # autocorrelations
noplot	suppress character-based plots
yw	calculate partial autocorrelations using Yule–Walker equations

ac_options	description
Main	
<u>l</u>ags(#)	calculate # autocorrelations
generate(*newvar*)	generate a variable to hold the autocorrelations; implies nograph
<u>l</u>evel(#)	set confidence level; default is level(95)
fft	calculate autocorrelation using Fourier transforms
Plot	
marker_options	change look of markers (color, size, etc.)
marker_label_options	add marker labels; change look or position
line_options	change look of dropped lines
CI plot	
ciopts(*area_options*)	affect rendition of the confidence bands
Add plot	
addplot(*plot*)	add other plots to the generated graph
Y-Axis, X-Axis, Title, Caption, Legend, Overall	
twoway_options	any options other than by() documented in [G] *twoway_options*

70

pac_options	description
Main	
<u>lags</u>(#)	calculate # partial autocorrelations
<u>g</u>enerate(newvar)	generate a variable to hold the partial autocorrelations; implies nograph
srv	include standardized residual variances in graph
yw	calculate partial autocorrelations using Yule–Walker equations
<u>l</u>evel(#)	set confidence level; default is level(95)
Plot	
marker_options	change look of markers (color, size, etc.)
marker_label_options	add marker labels; change look or position
line_options	change look of dropped lines
CI plot	
<u>ciopts</u>(area_options)	affect rendition of the confidence bands
SRV plot	
<u>srvopts</u>(marker_options)	affect rendition of the plotted standardized residual variances (SRVs)
Add plot	
<u>addplot</u>(plot)	add other plots to the generated graph
Y-Axis, X-Axis, Title, Caption, Legend, Overall	
twoway_options	any options other than by() documented in [G] **twoway_options**

You must tsset your data before using corrgram, ac, or pac; see [TS] **tsset**. In addition, the time series must be dense (nonmissing and no gaps in the time variable) in the sample if you specify the fft option. *varname* may contain time-series operators; see [U] **11.4.3 Time-series varlists**.

Description

corrgram produces a table of the autocorrelations, partial autocorrelations, and Portmanteau (Q) statistics. It also displays a character-based plot of the autocorrelations and partial autocorrelations. See [TS] **wntestq** for more information on the Q statistic.

ac produces a correlogram (a graph of autocorrelations) with pointwise confidence intervals based on Bartlett's formula for MA(q) processes.

pac produces a partial correlogram (a graph of partial autocorrelations) with confidence intervals calculated using a standard error of $1/\sqrt{n}$. The residual variances for each lag may optionally be included on the graph.

Options

┌─ Main ┐

lags(#) specifies the number of autocorrelations to calculate. The default is to use $\min([n/2]-2, 40)$ where $[n/2]$ is the greatest integer less than or equal to $n/2$.

noplot prevents the character-based plots from being in the listed table of autocorrelations and partial autocorrelations.

generate(*newvar*) specifies a new variable to contain the autocorrelation (ac command) or partial autocorrelation (pac command) values. This option is required if the nograph option is used.

   nograph (implied when using generate() in the dialog box) prevents ac and pac from constructing a graph. This option requires the generate() option.

srv (pac only) specifies that the standardized residual variances be plotted with the partial autocorrelations. srv cannot be used if yw is used.

yw specifies that the partial autocorrelations be calculated using the Yule–Walker equations instead of using the default regression-based technique. yw cannot be used if srv is used.

level(#) specifies the confidence level, as a percentage, for the confidence bands in the ac or pac graph. The default is level(95) or as set by set level; see [R] **level**.

fft (ac only) specifies that the autocorrelations be calculated using two Fourier transforms. This technique can be faster than simply iterating over the requested number of lags.

___
⌐ Plot ⌐
___

marker_options, *marker_label_options*, and *line_options* affect the rendition of the plotted autocorrelations (with ac) or partial autocorrelations (with pac).

   marker_options specify the look of markers plotted at the data correlation points. This look includes the marker symbol, the marker size, its color and outline; see [G] ***marker_options***.

   marker_label_options specify if and how the markers are to be labeled; see
      [G] ***marker_label_options***.

   line_options specify the look of the dropped lines, including pattern, width, and color; see
      [G] ***line_options***.

___
⌐ CI plot ⌐
___

ciopts(*area_options*) affect the rendition of the confidence bands; see [G] ***area_options***.

___
⌐ SRV plot ⌐
___

srvopts(*marker_options*) affect the rendition of the plotted standardized residual variances; see
   [G] ***marker_options***. This option implies the srv option.

   SRVs are available only with pac.

___
⌐ Add plot ⌐
___

addplot(*plot*) adds specified plots to the generated graph; see [G] ***addplot_option***.

___
⌐ Y-Axis, X-Axis, Title, Caption, Legend, Overall ⌐
___

twoway_options are any of the options documented in [G] ***twoway_options***, excluding by(). These include options for titling the graph (see [G] ***title_options***) and saving the graph to disk (see
   [G] ***saving_option***).

Remarks

corrgram tabulates autocorrelations, partial autocorrelations, and Portmanteau (Q) statistics and plots the autocorrelations and partial autocorrelations. The Q statistics are the same as those produced by [TS] **wntestq**. ac produces graphs of the autocorrelations, and pac produces graphs of the partial autocorrelations.

▷ Example 1

Here we use the international airline passengers dataset (Box, Jenkins, and Reinsel 1994, Series G). This dataset has 144 observations on the monthly number of international airline passengers from 1949 through 1960. We can list the autocorrelations and partial autocorrelations using corrgram.

```
. use http://www.stata-press.com/data/r9/air2
(TIMESLAB: Airline passengers)

. corrgram air, lags(20)
```

| | | | | | -1   0   1 | -1   0   1 |
LAG	AC	PAC	Q	Prob>Q	[Autocorrelation]	[Partial Autocor]
1	0.9480	0.9589	132.14	0.0000		
2	0.8756	-0.3298	245.65	0.0000		
3	0.8067	0.2018	342.67	0.0000		
4	0.7526	0.1450	427.74	0.0000		
5	0.7138	0.2585	504.8	0.0000		
6	0.6817	-0.0269	575.6	0.0000		
7	0.6629	0.2043	643.04	0.0000		
8	0.6556	0.1561	709.48	0.0000		
9	0.6709	0.5686	779.59	0.0000		
10	0.7027	0.2926	857.07	0.0000		
11	0.7432	0.8402	944.39	0.0000		
12	0.7604	0.6127	1036.5	0.0000		
13	0.7127	-0.6660	1118	0.0000		
14	0.6463	-0.3846	1185.6	0.0000		
15	0.5859	0.0787	1241.5	0.0000		
16	0.5380	-0.0266	1289	0.0000		
17	0.4997	-0.0581	1330.4	0.0000		
18	0.4687	-0.0435	1367	0.0000		
19	0.4499	0.2773	1401.1	0.0000		
20	0.4416	-0.0405	1434.1	0.0000		

(Continued on next page)

We can use `ac` to produce a graph of the autocorrelations.

```
. ac air, lags(20)
```

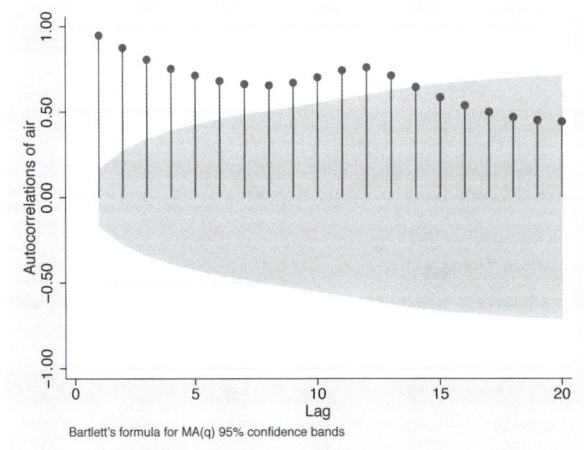

Bartlett's formula for MA(q) 95% confidence bands

The data probably has a trend component as well as a seasonal component. First-differencing will mitigate the effects of the trend, and seasonal differencing will help control for seasonality. To accomplish this, we can use Stata's time-series operators. Here we graph the partial autocorrelations after controlling for trends and seasonality. We also use `srv` to include the standardized residual variances.

```
. pac DS12.air, lags(20) srv
```

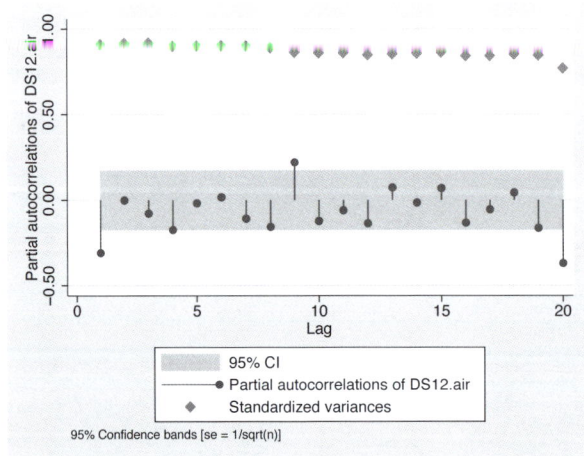

95% Confidence bands [se = 1/sqrt(n)]

See [U] **11.4.3 Time-series varlists** for more information about time-series operators.

◁

Saved Results

corrgram saves in r():

Scalars
    r(lags)   number of lags

Matrices
    r(AC)   vector of autocorrelations
    r(PAC)   vector of partial autocorrelations
    r(Q)   vector of Q statistics
    r(ac#) AC for lab #
    r(pac#) PAC for lab #
    r(q#) Q for lag #

Methods and Formulas

corrgram, ac, and pac are implemented as ado-files.

Box, Jenkins, and Reinsel (1994); Newton (1988); Chatfield (1996); and Hamilton (1994) provide excellent descriptions of correlograms. Newton (1988) also discusses the calculation of the various quantities.

The autocovariance function for a time series $x_1, x_2, \ldots, x_n$ is defined for $|v| < n$ as

$$\widehat{R}(v) = \frac{1}{n} \sum_{i=1}^{n-|v|} (x_i - \overline{x})(x_{i+v} - \overline{x})$$

where $\overline{x}$ is the sample mean, and the autocorrelation function is then defined as

$$\widehat{\rho}_v = \frac{\widehat{R}(v)}{\widehat{R}(0)}$$

The variance of $\widehat{\rho}_v$ is given by Bartlett's formula for MA(q) processes. From Brockwell and Davis (2002, 94), we have

$$\operatorname{Var}(\widehat{\rho}_v) = \begin{cases} 1/n & v = 1 \\ \frac{1}{n}\left\{1 + 2\sum_{i=1}^{v-1} \widehat{\rho}^2(i)\right\} & v > 1 \end{cases}$$

The partial autocorrelation at lag v measures the correlation between x_t and x_{t+v} after the effects of $x_{t+1}, \ldots, x_{t+v-1}$ have been removed. By default, corrgram and pac use a regression-based method to estimate it. We run an OLS regression of x_t on $x_{t-1}, \ldots, x_{t-v}$ and a constant term. The estimated coefficient on x_{t-v} is our estimate of the vth partial autocorrelation. The residual variance is the estimated variance of that regression, which we then standardize by dividing by $\widehat{R}(0)$.

If the yw option is specified, corrgram and pac use the Yule–Walker equations to estimate the partial autocorrelations. Following Enders (1995, 83), let ϕ_{vv} denote the vth partial autocorrelation coefficient. We then have

$$\widehat{\phi}_{11} = \widehat{\rho}_1$$

and for $v > 1$

$$\widehat{\phi}_{vv} = \frac{\widehat{\rho}_v - \sum\limits_{j=1}^{v-1} \widehat{\phi}_{v-1,j} \widehat{\rho}_{v-j}}{1 - \sum\limits_{j=1}^{v-1} \widehat{\phi}_{v-1,j} \widehat{\rho}_j}$$

and

$$\widehat{\phi}_{vj} = \widehat{\phi}_{v-1,j} - \widehat{\phi}_{vv} \widehat{\phi}_{v-1,v-j} \qquad j = 1, 2, \ldots, v-1$$

Unlike the regression-based method, the Yule–Walker equations-based method ensures that the first sample partial autocorrelation equal the first sample autocorrelation coefficient, as must be true in the population; see Greene (2003, 618).

McCullough (1998) discusses alternative methods of estimating ϕ_{vv}; he finds that relative to other methods, such as linear regression, the Yule–Walker equations-based method performs poorly, in part because it is susceptible to numerical error. Box, Jenkins, and Reinsel (1994, 68) also caution against the use of the Yule–Walker equations-based method, especially with data which is nearly nonstationary.

Acknowledgment

The ac and pac commands are based on the ac and pac commands written by Sean Becketti (1992), a past editor of the *Stata Technical Bulletin*.

References

Becketti, S. 1992. sts1: Autocorrelation and partial autocorrelation graphs. *Stata Technical Bulletin* 5: 27–28. Reprinted in *Stata Technical Bulletin Reprints*, vol. 1, pp. 221–223.

Box, G. E. P., G. M. Jenkins, and G. C. Reinsel. 1994. *Time Series Analysis: Forecasting and Control*. 3rd ed. Englewood Cliffs, NJ: Prentice Hall.

Brockwell, P. J., and R. A. Davis. 2002. *Introduction to Time Series and Forecasting*. 2nd ed. New York: Springer.

Chatfield, C. 1996. *The Analysis of Time Series: An Introduction*. 5th ed. London: Chapman & Hall.

Enders, W. 1995. *Applied Econometric Time Series*. New York: Wiley.

Greene, W. H. 2003. *Econometric Analysis*. 5th ed. Upper Saddle River, NJ: Prentice Hall.

Hamilton, J. D. 1994. *Time Series Analysis*. Princeton: Princeton University Press.

McCullough, B. D. 1998. Algorithm choice for (partial) autocorrelation functions. *Journal of Economic and Social Measurement* 24: 265–278.

Newton, H. J. 1988. *TIMESLAB: A Time Series Laboratory*. Pacific Grove, CA: Wadsworth & Brooks/Cole.

Also See

Complementary:	[TS] **tsset**, [TS] **wntestq**
Related:	[TS] **pergram**
Background:	*Stata Graphics Reference Manual*

Title

> **cumsp** — Cumulative spectral distribution

Syntax

cumsp *varname* [*if*] [*in*] [, *options*]

options	description
Main	
generate(*newvar*)	create *newvar* holding distribution values
Plot	
connect_options	affect rendition of the plotted points connected by lines
Add plot	
addplot(*plot*)	add other plots to the generated graph
Y-Axis, X-Axis, Title, Caption, Legend, Overall	
twoway_options	any options other than by() documented in [G] *twoway_options*

You must tsset your data before using cumsp; see [TS] **tsset**. In addition, the time series must be dense (nonmissing with no gaps in the time variable) in the sample specified.

varname may contain time-series operators; see [U] **11.4.3 Time-series varlists**.

Description

cumsp plots the cumulative sample spectral-distribution function evaluated at the natural frequencies for a (dense) time series.

Options

⌐ **Main** ⌐

generate(*newvar*) specifies a new variable to contain the estimated cumulative spectral-distribution values.

⌐ **Plot** ⌐

connect_options affect the rendition of the plotted points connected by lines; see [G] *connect_options*.

⌐ **Add plot** ⌐

addplot(*plot*) provides a way to add other plots to the generated graph; see [G] *addplot_option*.

⌐ **Y-Axis, X-Axis, Title, Caption, Legend, Overall** ⌐

twoway_options are any of the options documented in [G] *twoway_options* excluding by(). These include options for titling the graph (see [G] *title_options*) and options for saving the graph to disk (see [G] *saving_option*).

Remarks

▷ Example 1

Here we use the international airline passengers dataset (Box, Jenkins, and Reinsel 1994, Series G). This dataset has 144 observations on the monthly number of international airline passengers from 1949 through 1960. In the cumulative sample spectral distribution function for these data, we also request a vertical line at frequency 1/12. Since the data are monthly, there will be a pronounced jump in the cumulative sample spectral-distribution plot at the 1/12 value if there is an annual cycle in the data.

```
. use http://www.stata-press.com/data/r9/air2
(TIMESLAB: Airline passengers)
. cumsp air, xline(.083333333)
```

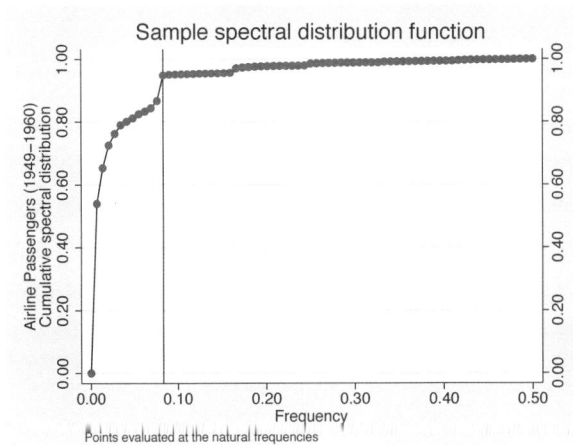

The cumulative sample spectral-distribution function clearly illustrates the annual cycle.

◁

Methods and Formulas

cumsp is implemented as an ado-file.

A time series of interest is decomposed into a unique set of sinusoids of various frequencies and amplitudes.

A plot of the sinusoidal amplitudes versus the frequencies for the sinusoidal decomposition of a time series gives us the spectral density of the time series. If we calculate the sinusoidal amplitudes for a discrete set of "natural" frequencies $(1/n, 2/n, \ldots, q/n)$, we obtain the periodogram.

Let $x(1), \ldots, x(n)$ be a time series, and let $\omega_k = (k-1)/n$ denote the natural frequencies for $k = 1, \ldots, [n/2] + 1$ where $[\,]$ indicates the greatest integer function. Define

$$C_k^2 = \frac{1}{n^2} \left| \sum_{t=1}^{n} x(t) e^{2\pi i (t-1)\omega_k} \right|^2$$

A plot of nC_k^2 versus ω_k is then called the periodogram.

The sample spectral density may then be defined as $\widehat{f}(\omega_k) = nC_k^2$.

If we let $\widehat{f}(\omega_1), \ldots, \widehat{f}(\omega_Q)$ be the sample spectral density function of the time series evaluated at the frequencies $\omega_j = (j-1)/Q$ for $j = 1, \ldots, Q$ and we let $q = [Q/2] + 1$, then

$$\widehat{F}(\omega_k) = \frac{\displaystyle\sum_{i=1}^{k} \widehat{f}(\omega_j)}{\displaystyle\sum_{i=1}^{q} \widehat{f}(\omega_j)}$$

is the sample spectral-distribution function of the time series.

References

Box, G. E. P., G. M. Jenkins, and G. C. Reinsel. 1994. *Time Series Analysis: Forecasting and Control.* 3rd ed. Englewood Cliffs, NJ: Prentice Hall.

Newton, H. J. 1988. *TIMESLAB: A Time Series Laboratory.* Pacific Grove, CA: Wadsworth & Brooks/Cole.

Also See

Complementary:	[TS] **tsset**
Related:	[TS] **corrgram**, [TS] **pergram**
Background:	*Stata Graphics Reference Manual*

Title

> **dfgls** — DF-GLS unit-root test

Syntax

dfgls *varname* [*if*] [*in*] [, *options*]

options	description
Main	
<u>m</u>axlag(#)	use # as the highest lag order for Dickey–Fuller GLS regressions
notrend	series is stationary around a mean instead of around a linear time trend
ers	present interpolated critical values from Elliott, Rothenberg, and Stock (1996)

You must tsset your data before using dfgls; see [TS] **tsset**.

varname may contain time-series operators; see [U] **11.4.3 Time-series varlists**.

Description

dfgls performs a modified Dickey–Fuller *t*-test for a unit root in which the series has been transformed by a generalized least-squares regression.

Options

> **Main**

maxlag(#) sets the value of k, the highest lag order for the first-differenced, detrended variable in the Dickey–Fuller regression. By default, dfgls sets k according to the method proposed by Schwert (1989); i.e., dfgls sets $k_{max} = \text{int}[12\{(T+1)/100\}^{0.25}]$.

notrend specifies that the alternative hypothesis be that the series is stationary around a mean instead of around a linear time trend. By default, a trend is included.

ers specifies that dfgls should present interpolated critical values from tables presented by Elliott, Rothenberg, and Stock (1996), which they obtained from simulations. See *Critical values* under *Methods and Formulas* for details.

Remarks

dfgls tests for a unit root in a time series. It performs the modified Dickey–Fuller *t*-test (known as the DF-GLS test) proposed by Elliott, Rothenberg, and Stock (1996). Essentially, the test is an augmented Dickey–Fuller test, similar to the test performed by Stata's **dfuller** command, except that the time series is transformed via a generalized least squares (GLS) regression before performing the test. Elliott, Rothenberg, and Stock and subsequent studies have shown that this test has significantly greater power than the previous versions of the augmented Dickey–Fuller test.

dfgls performs the DF-GLS test for the series of models that include 1 to k lags of the first-differenced, detrended variable, where k can be set by the user or by the method described in Schwert (1989). Stock and Watson (2003, 549–552) provide an excellent discussion of the methodology.

As discussed in [TS] **dfuller**, the augmented Dickey–Fuller test involves fitting a regression of the form

$$\Delta y_t = \alpha + \beta y_{t-1} + \delta t + \zeta_1 \Delta y_{t-1} + \zeta_2 \Delta y_{t-2} + \ldots + \zeta_k \Delta y_{t-k} + \epsilon_t$$

and then testing the null hypothesis $H_0 : \beta = 0$. The DG-GLS test is performed analogously but on GLS-detrended data. The null hypothesis of the test is that y_t is a random walk, possibly with drift. There are two possible alternative hypotheses: y_t is stationary about a linear time trend, or y_t is stationary with a possibly nonzero mean but with no linear time trend. The default is to use the former. To specify the latter alternative, use the `notrend` option.

▷ Example 1

Here we use the German macroeconomic dataset and test whether the natural log of income exhibits a unit root. We use the default options with `dfgls`.

```
. use http://www.stata-press.com/data/r9/lutkepohl
(Quarterly SA West German macro data, Bil DM, from Lutkepohl 1993 Table E.1)

. dfgls linvestment
DF-GLS for linvestment                                Number of obs =     80
Maxlag = 11 chosen by Schwert criterion
                 DF-GLS tau      1% Critical     5% Critical     10% Critical
    [lags]      Test Statistic     Value           Value           Value

      11           -2.925          -3.610          -2.763          -2.489
      10           -2.671          -3.610          -2.798          -2.523
       9           -2.766          -3.610          -2.832          -2.555
       8           -3.259          -3.610          -2.865          -2.587
       7           -3.536          -3.610          -2.898          -2.617
       6           -3.115          -3.610          -2.929          -2.646
       5           -3.054          -3.610          -2.958          -2.674
       4           -3.016          -3.610          -2.986          -2.699
       3           -2.071          -3.610          -3.012          -2.723
       2           -1.675          -3.610          -3.035          -2.744
       1           -1.752          -3.610          -3.055          -2.762
Opt Lag (Ng-Perron seq t) =  7 with RMSE   .0388771
Min SC    = -6.169137 at lag  4 with RMSE   .0398949
Min MAIC  = -6.136371 at lag  1 with RMSE   .0440319
```

The null of a unit root is rejected at the 5% level for lags 4–8 and 11 and at the 10% level for lags 9 and 10. For the sake of comparison, we also test for a unit root in log income using `dfuller` with two different lag specifications. We need to use the `trend` option with `dfuller` because it is not included by default.

```
. dfuller linvestment, lag(4) trend
Augmented Dickey-Fuller test for unit root            Number of obs  =     87

                            ------- Interpolated Dickey-Fuller -------
                Test         1% Critical     5% Critical     10% Critical
              Statistic        Value           Value           Value

 Z(t)          -3.133          -4.069          -3.463          -3.158

MacKinnon approximate p-value for Z(t) = 0.0987
```

```
. dfuller linvestment, lag(7) trend
Augmented Dickey-Fuller test for unit root          Number of obs   =          84
                                  ─────────── Interpolated Dickey-Fuller ───────────
                      Test          1% Critical         5% Critical        10% Critical
                   Statistic           Value               Value               Value
─────────────────────────────────────────────────────────────────────────────────────
   Z(t)            -3.994             -4.075              -3.466              -3.160
─────────────────────────────────────────────────────────────────────────────────────
MacKinnon approximate p-value for Z(t) = 0.0090
```

The results from `dfuller` also support rejecting the null hypothesis, though the p-value with four lags is quite close to 0.10. That the `dfuller` results are not quite as strong as those produced by `dfgls` is not surprising since the DF-GLS test with a trend has been shown to be more powerful than the standard augmented Dickey–Fuller test.

◁

Saved Results

`dfgls` saves in `r()`:

Scalars
  r(maxlag)   highest lag order k
  r(N)   number of observations
  r(sclag)   lag chosen by Schwarz criterion
  r(maiclag)   lag chosen by modified AIC method
  r(optlag)   lag chosen by sequential-t method

Matrices
  r(results) k, MAIC, SIC, RMSE, and DF-GLS statistics

Methods and Formulas

`dfgls` is implemented as an ado-file.

`dfgls` tests for a unit root. There are two possible alternative hypotheses: y_t is stationary around a linear trend, or y_t is stationary with no linear time trend. Under the first alternative hypothesis, the DF-GLS test is performed by first estimating the intercept and trend via GLS. The GLS estimation is performed by generating the new variables, $\widetilde{y}_t$, x_t, and z_t, where

$$\widetilde{y}_1 = y_1$$
$$\widetilde{y}_t = y_t - \alpha^* y_{t-1}, \qquad t = 2, \ldots, T$$
$$x_1 = 1$$
$$x_t = 1 - \alpha^*, \qquad t = 2, \ldots, T$$
$$z_1 = 1$$
$$z_t = t - \alpha^*(t - 1)$$

and $\alpha^* = 1 - (13.5/T)$. An OLS regression is then estimated for the equation

$$\widetilde{y}_t = \delta_0 x_t + \delta_1 z_t + \epsilon_t$$

The OLS estimators $\widehat{\delta}_0$ and $\widehat{\delta}_1$ are then used to remove the trend from y_t; i.e., we generate

$$y^* = y_t - (\widehat{\delta}_0 + \widehat{\delta}_1 t)$$

Finally, we perform an augmented Dickey–Fuller test on the transformed variable by fitting the OLS regression

$$\Delta y_t^* = \alpha + \beta y_{t-1}^* + \sum_{j=1}^{k} \zeta_j \Delta y_{t-j}^* + \epsilon_t$$

and then test the null hypothesis $H_0 : \beta = 0$ using tabulated critical values.

To perform the DF-GLS test under the second alternative hypothesis, we proceed as before but define $\alpha^* = 1 - (7/T)$, eliminate z from the GLS regression, compute $y^* = y_t - \delta_0$, fit the augmented Dickey–Fuller regression using the newly transformed variable, and perform a test of the null hypothesis that $\beta = 0$ using the tabulated critical values.

dfgls reports the DF-GLS statistic and its critical values obtained from the regression in (1) for $k \in \{1, 2, \ldots, k_{\max}\}$. By default, dfgls sets $k_{\max} = \text{int}[12\{(T + 1)/100\}^{0.25}]$ as proposed by Schwert (1989), although you can override this choice with another value. The sample size available with $k_{\max}$ lags is used in all the regressions. Since there are $k_{\max}$ lags of the first-differenced series, $k_{\max} + 1$ observations are lost, leaving $T - k_{\max}$ observations. dfgls requires that the sample of $T + 1$ observations on $y_t = (y_0, y_1, \ldots, y_T)$ have no gaps.

dfgls reports the results of three different methods for choosing which value of k to use. These methods are (1) the Ng–Perron sequential t, (2) the minimum Schwarz information criterion (SIC), and (3) the Ng–Perron modified Akaike information criterion (MAIC). While the SIC has a long history in time-series modeling, the Ng–Perron sequential t was developed by Ng and Perron (1995), and the MAIC was developed by Ng and Perron (2000).

The SIC can be calculated using either the log likelihood or the sum-of-squared errors from a regression; dfgls uses the latter definition. Specifically, for each k

$$\text{SIC} = \ln(\widehat{\text{rmse}}^2) + (k + 1) \frac{\ln(T - k_{\max})}{(T - k_{\max})}$$

where

$$\widehat{\text{rmse}} = \frac{1}{(T - k_{\max})} \sum_{t=k_{\max}+1}^{T} \widehat{e}_t^2$$

dfgls reports the value of the smallest SIC and the k that produced it.

Ng and Perron (1995) derived a sequential-t algorithm for choosing k:

   i. Set $n = 0$ and run the regression in (2) with all $k_{\max} - n$ lags. If the coefficient on $\beta_{k_{\max}}$ is significantly different from zero at level α, choose k to $k_{\max}$. Otherwise, continue on to ii.

   ii. If $n < k_{\max}$, set $n = n + 1$ and continue on to iii. Otherwise, set $k = 0$ and stop.

   iii. Run the regression in (2) with $k_{\max} - n$ lags. If the coefficient on $\beta_{k_{\max}-n}$ is significantly different from zero at level α, choose k to $k_{\max} - n$. Otherwise, return to ii.

Following Ng and Perron (1995), dfgls uses $\alpha = 10\%$. dfgls reports the k selected by this sequential-t algorithm and the $\widehat{\text{rmse}}$ from the regression.

Method (3) is based on choosing k to minimize the MAIC. The MAIC is calculated as

$$\text{MAIC}(k) = \ln(\widehat{\text{rmse}}^2) + \frac{2\{\tau(k) + k\}}{T - k_{\max}}$$

where

$$\tau(k) = \frac{1}{\widehat{\text{rmse}}^2} \widehat{\beta}_0^2 \sum_{t=k_{\max}+1}^{T} \widetilde{y}_t^2$$

and $\widetilde{y}$ was defined previously.

Critical values

By default, `dfgls` uses the 5% and 10% critical values computed from the response surface analysis of Cheung and Kim (1995). Since Cheung and Kim (1995) did not present results for the 1% case, the 1% critical values are always interpolated from the critical values presented by ERS.

ERS presented critical values, obtained from simulations, for the DF-GLS test with a linear trend and showed that the critical values for the mean-only DF-GLS test were the same as those for the ADF test. If `dfgls` is run with the `ers` option, `dfgls` will present interpolated critical values from these tables. The method of interpolation is standard. For the trend case, below 50 observations and above 200 there is no interpolation; the values for 50 and ∞ are reported from the tables. For a value N that lies between two values in the table, say, N_1 and N_2, with corresponding critical values CV_1 and CV_2, the critical value

$$\text{cv} = CV_1 + \frac{N - N_1}{N_1}(CV_2 - CV_1)$$

is presented. The same method is used for the mean-only case, except that interpolation is possible for values between 50 and 500.

Acknowledgments

We wish to thank Christopher Baum of Boston College and Richard Sperling of Boston College for a previous version of `dfgls`.

References

Cheung, Y. and K. S. Lai. 1995. Lag order and critical values of a modified Dickey–Fuller test. *Oxford Bulletin of Economics and Statistics* (57)3: 411–419.

Dickey, D. A. and W. A. Fuller. 1979. Distribution of the estimators for autoregressive time series with a unit root. *Journal of the American Statistical Association* 74: 427–431.

Elliott, G., T. Rothenberg, and J. H. Stock. 1996. Efficient tests for an autoregressive unit root. *Econometrica* 64: 813–836.

Ng, S. and P. Perron. 1995. Unit root tests in ARMA models with data-dependent methods for the selection of the truncation lag. *Journal of the American Statistical Association* 90: 268–281.

———. 2000. Lag length selection and the construction of unit root tests with good size and power. Working paper. Department of Economics, Boston College.

Schwert, G. W. 1989. Tests for unit roots: A Monte Carlo investigation. *Journal of Business and Economic Statistics* 2: 147–159.

Stock, J. H. and M. W. Watson. 2003. *Introduction to Econometrics.* Boston: Addison–Wesley.

Also See

Related: [TS] **dfuller**, [TS] **pperron**, [TS] **tsset**

Title

> **dfuller** — Augmented Dickey–Fuller unit-root test

Syntax

dfuller *varname* [*if*] [*in*] [, *options*]

options	description
Main	
<u>noc</u>onstant	suppress constant term in regression
<u>trend</u>	include trend term in regression
<u>drift</u>	include drift term in regression
<u>regress</u>	display regression table
<u>lags</u>(#)	include # lagged differences

You must tsset your data before using dfuller; see [TS] **tsset**.

varname may contain time-series operators; see [U] **11.4.3 Time-series varlists**.

Description

dfuller performs the augmented Dickey–Fuller test that a variable follows a unit-root process. The null hypothesis is that the variable contains a unit root, and the alternative is that the variable was generated by a stationary process. You may optionally exclude the constant, include a trend term, and include lagged values of the difference of the variable in the regression.

Options

⌐ Main ⌐

noconstant suppresses the constant term (intercept) in the model and indicates that the process under the null hypothesis is a random walk without drift. noconstant cannot be used with the trend or drift options.

trend specifies that a trend term be included in the associated regression and that the process under the null hypothesis is a random walk, perhaps with drift. This option may not be used with the noconstant or drift options.

drift indicates that the process under the null hypothesis is a random walk with nonzero drift. This option may not be used with the noconstant or trend options.

regress specifies that the associated regression table appear in the output. By default, the regression table is not produced.

lags(#) specifies the number of lagged difference terms to include in the covariate list.

86

Remarks

Dickey and Fuller (1979) developed a procedure for testing whether a variable has a unit root or, equivalently, that the variable follows a random walk. Hamilton (1994, 528–529) describes the four different cases to which the augmented Dickey–Fuller test can be applied. In all cases, the null hypothesis is that the variable has a unit root. They differ in whether the null hypothesis includes a drift term and whether the regression used to obtain the test statistic includes a constant term and time trend.

The true model is assumed to be

$$y_t = \alpha + y_{t-1} + u_t$$

where u_t is an independently and identically distributed zero-mean error term. In cases one and two, presumably $\alpha = 0$, which is a random walk without drift. In cases three and four, we allow for a drift term by letting α be unrestricted.

The Dickey–Fuller test involves fitting the model

$$y_t = \alpha + \rho y_{t-1} + \delta t + u_t$$

by ordinary least squares (OLS), perhaps setting $\alpha = 0$ or $\delta = 0$. However, such a regression is likely to be plagued by serial correlation. To control for that, the augmented Dickey–Fuller test instead fits a model of the form

$$\Delta y_t = \alpha + \beta y_{t-1} + \delta t + \zeta_1 \Delta y_{t-1} + \zeta_2 \Delta y_{t-2} + \cdots + \zeta_k \Delta y_{t-k} + \epsilon_t \tag{1}$$

where k is the number of lags specified in the `lags()` option. The `noconstant` option removes the constant term α from this regression, and the `trend` option includes the time trend δt, which by default is not included. Testing $\beta = 0$ is equivalent to testing $\rho = 1$, or, equivalently, that y_t follows a unit root process.

In the first case, the null hypothesis is that y_t follows a random walk without drift, and (1) is fitted without the constant term α and the time trend δt. The second case has the same null hypothesis as the first, except that we include α in the regression. In both cases, the population value of α is zero under the null hypothesis. In the third case, we hypothesize that y_t follows a unit root with drift, so that the population value of α is nonzero; we do not include the time trend in the regression. Finally, in the fourth case, the null hypothesis is that y_t follows a unit root with or without drift so that α is unrestricted, and we include a time trend in the regression.

The following table summarizes the four cases.

Case	Process under null hypothesis	Regression restrictions	dfuller option
1	Random walk without drift	$\alpha = 0, \delta = 0$	`noconstant`
2	Random walk without drift	$\delta = 0$	(default)
3	Random walk with drift	$\delta = 0$	`drift`
4	Random walk with or without drift	(none)	`trend`

Except in the third case, the t-statistic used to test $H_0 : \beta = 0$ does not have a standard distribution. Hamilton (1994, Chapter 17) derives the limiting distributions, which are different for each of the three other cases. The critical values reported by `dfuller` are interpolated based on the tables in Fuller (1976). MacKinnon (1994) shows how to approximate the p-values based on a regression surface, and `dfuller` also reports that p-value. In the third case, where the regression includes a constant term and under the null hypothesis the series has a nonzero drift parameter α, the t-statistic has the usual t distribution; `dfuller` reports the one-sided critical values and p-value for the test of H_0 against the alternative $H_A : \beta < 0$, which is equivalent to $\rho < 1$.

Deciding which case to use involves a combination of theory and visual inspection of the data. If economic theory favors a particular null hypothesis, the appropriate case can be chosen based on that. If a graph of the data shows an upward trend over time, then case four may be preferred. If the data do not show a trend but do have a nonzero mean, then case two would be a valid alternative.

▷ Example 1

In this example, we examine the international airline passengers dataset from Box, Jenkins, and Reinsel (1994, Series G). This dataset has 144 observations on the monthly number of international airline passengers from 1949 through 1960. Because the data show a clear upward trend, we use the `trend` option with `dfuller` to include a constant and time trend in the augmented Dickey–Fuller regression.

```
. use http://www.stata-press.com/data/r9/air2
(TIMESLAB: Airline passengers)
. dfuller air, lags(3) trend regress
```

Augmented Dickey-Fuller test for unit root   Number of obs   =   140

	Test Statistic	1% Critical Value	Interpolated Dickey-Fuller 5% Critical Value	10% Critical Value
Z(t)	-6.936	-4.027	-3.445	-3.145

MacKinnon approximate p-value for Z(t) = 0.0000

D.air	Coef.	Std. Err.	t	P>|t|	[95% Conf. Interval]
air						
L1.	-.5217089	.0752195	-6.94	0.000	-.67048	-.3729379
LD.	.5572871	.0799894	6.97	0.000	.399082	.7154923
L2D.	.095912	.0876692	1.09	0.276	-.0774825	.2693065
L3D.	.14511	.0879922	1.65	0.101	-.0289232	.3191433
_trend	1.407534	.2098378	6.71	0.000	.9925118	1.822557
_cons	44.49164	7.78335	5.72	0.000	29.09753	59.88575

Here we can overwhelmingly reject the null hypothesis of a unit root at all common significance levels. Looking at the regression output, the estimated β of -0.522 implies that $\rho = (1 - 0.522) = 0.478$. Experiments with fewer or more lags in the augmented regression yield the same conclusion.

◁

▷ Example 2

In this example, we use the German macroeconomic dataset to determine whether the log of consumption follows a unit root. We will again use the `trend` option, since consumption grows over time.

```
. use http://www.stata-press.com/data/r9/lutkepohl, clear
(Quarterly SA West German macro data, Bil DM, from Lutkepohl 1993 Table E.1)
. dfuller lconsumption, lags(4) trend
Augmented Dickey-Fuller test for unit root          Number of obs   =      87
```

	Test Statistic	1% Critical Value	Interpolated Dickey-Fuller 5% Critical Value	10% Critical Value
Z(t)	-1.318	-4.069	-3.463	-3.158

MacKinnon approximate p-value for Z(t) = 0.8834

As we might expect based on economic theory, here we cannot reject the null hypothesis that log consumption exhibits a unit root. Again using different numbers of lag terms yield the same conclusion.

◁

Saved Results

dfuller saves in r():

Scalars

r(N)	number of observations	r(Zt)	Dickey–Fuller test statistic
r(lags)	number of lagged differences	r(p)	MacKinnon approximate p-value (if there is a constant or trend in associated regression)

Methods and Formulas

dfuller is implemented as an ado-file.

In the OLS estimation of an AR(1) process with Gaussian errors,

$$y_t = \rho y_{t-1} + \epsilon_t$$

where ϵ_t are independently and identically distributed as $N(0, \sigma^2)$ and $y_0 = 0$, the OLS estimate (based on an n-observation time series) of the autocorrelation parameter ρ is given by

$$\widehat{\rho}_n = \frac{\sum_{t=1}^{n} y_{t-1} y_t}{\sum_{t=1}^{n} y_t^2}$$

If $|\rho| < 1$, then

$$\sqrt{n}(\widehat{\rho}_n - \rho) \rightarrow N(0, 1 - \rho^2)$$

If this result were valid when $\rho = 1$, the resulting distribution would have a variance of zero. When $\rho = 1$, the OLS estimate $\widehat{\rho}$ still converges in probability to one, though we need to find a suitable nondegenerate distribution so that we can perform hypothesis tests of $H_0 : \rho = 1$. Hamilton (1994, chapter 17) provides a superb exposition of the requisite theory.

To compute the test statistics, we fit the augmented Dickey–Fuller regression

$$\Delta y_t = \alpha + \beta y_{t-1} + \delta t + \sum_{j=1}^{k} \zeta_j \Delta y_{t-j} + e_t$$

via OLS where, depending on the options specified, the constant term α or time trend δt is omitted and k is the number of lags specified in the `lags()` option. The test statistic for $H_0 : \beta = 0$ is $Z_t = \widehat{\beta}/\widehat{\sigma}_\beta$ where $\widehat{\sigma}_\beta$ is the standard error of $\widehat{\beta}$.

The critical values included in the output are linearly interpolated from the table of values that appears in Fuller (1976), and the MacKinnon approximate p-values use the regression surface published in MacKinnon (1994).

David Alan Dickey (1945–) was born in Ohio and took degrees in mathematics at Miami University and a Ph.D. in statistics at Iowa State University in 1976 as a student of Wayne Fuller. He works at North Carolina State University and specializes in time-series analysis.

Wayne Arthur Fuller (1931–) was born in Iowa, took three degrees at Iowa State University and then served on the faculty between 1959 and 2001. He has made many distinguished contributions to time series, measurement-error models, survey sampling, and econometrics.

References

Box, G. E. P., G. M. Jenkins, and G. C. Reinsel. 1994. *Time Series Analysis: Forecasting and Control*. 3rd ed. Englewood Cliffs, NJ: Prentice Hall.

Dickey, D. A. and W. A. Fuller. 1979. Distribution of the estimators for autoregressive time series with a unit root. *Journal of the American Statistical Association* 74: 427–431.

Fuller, W. A. 1976. *Introduction to Statistical Time Series*. New York: Wiley.

Hamilton, J. D. 1994. *Time Series Analysis*. Princeton: Princeton University Press.

MacKinnon, J. G. 1994. Approximate asymptotic distribution functions for unit-root and cointegration tests. *Journal of Business and Economic Statistics* 12: 167–176.

Newton, H. J. 1988. *TIMESLAB: A Time Series Laboratory*. Pacific Grove, CA: Wadsworth & Brooks/Cole.

Also See

Complementary: [TS] **tsset**

Related: [TS] **dfgls**, [TS] **pperron**

Title

> **estimation options** — Estimation options

Description

This entry describes the options common to many estimation commands. Not all the options documented below work with all estimation commands; see the documentation for the particular estimation command. If an option is listed there, it is applicable.

Options

> ⌐ Model ⌐

noconstant suppresses the constant term (intercept) in the model.

constraints(*numlist* | *matname*) specifies the linear constraints to be applied during estimation. The default is to perform unconstrained estimation. constraints(*numlist*) specifies the constraints by number after they have been defined using the constraint command; see [R] **constraint**. See [R] **reg3** for the use of constraints in multiple-equation contexts.

constraints(*matname*) specifies a matrix containing the constraints; see [P] **makecns**.

constraints(*clist*) is used by some estimation commands, such as mlogit, where *clist* has the form $\#\left[-\#\right]\left[,\,\#\left[-\#\right]\,\ldots\right]$.

> ⌐ SE/Robust ⌐

robust specifies that the Huber/White/sandwich estimator of variance be used in place of the traditional calculation; see [U] **20.14 Obtaining robust variance estimates**. robust combined with cluster() allows observations that are not independent within cluster (although they must be independent between clusters).

If the command allows pweights and you specify them, robust is implied; see [U] **20.16 Weighted estimation**.

vce(robust) is a synonym for robust.

cluster(*varname*) specifies that the observations are independent across groups (clusters), but not necessarily within groups. *varname* specifies to which group each observation belongs, e.g., cluster(personid) in data with repeated observations on individuals. cluster() affects the estimated standard errors and variance–covariance matrix of the estimators (VCE), but not the estimated coefficients; see [U] **20.14 Obtaining robust variance estimates**.

cluster() implies robust; specifying robust cluster() is equivalent to typing cluster() by itself.

> ⌐ Reporting ⌐

level(*#*) specifies the confidence level, as a percentage, for confidence intervals. The default is level(95) or as set by set level; see [U] **20.6 Specifying the width of confidence intervals**.

Also See

Background: [U] **20 Estimation and postestimation commands**,

[TS] **time series**

Title

> **fcast compute** — Compute dynamic forecasts of dependent variables after var, svar, or vec

Syntax

After var *and* svar

> fcast <u>c</u>ompute *prefix* [, *options*₁]

After vec

> fcast <u>c</u>ompute *prefix* [, *options*₂]

prefix is the prefix appended to the names of the dependent variables to create the names of the variables holding the dynamic forecasts.

*options*₁	description
Main	
<u>step</u>(#)	set # periods to forecast; default is step(1)
<u>dynam</u>ic(*time_constant*)	begin dynamic forecasts at *time_constant*
<u>est</u>imates(*estname*)	use previously saved results *estname*; default is to use active results
replace	replace existing forecast variables that have the same prefix
Std. Errors	
nose	suppress asymptotic standard errors
bs	obtain standard errors from bootstrapped residuals
bsp	obtain standard errors from parametric bootstrap
bscentile	estimate bounds using centiles of bootstrapped dataset
<u>reps</u>(#)	perform # bootstrap replications; default is reps(200)
<u>nod</u>ots	suppress the usual dot after each bootstrap replication
<u>sav</u>ing(*filename*[, replace])	save bootstrap results as *filename*; use replace to overwrite existing *filename*
Reporting	
<u>level</u>(#)	set confidence level; default is level(95)

(Continued on next page)

options₂	description
Main	
step(#)	set # periods to forecast; default is step(1)
dynamic(*time_constant*)	begin dynamic forecasts at *time_constant*
estimates(*estname*)	use previously saved results *estname*; default is to use active results
replace	replace existing forecast variables that have the same prefix
differences	save dynamic predictions of the first-differenced variables
Std. Errors	
nose	suppress asymptotic standard errors
Reporting	
level(#)	set confidence level; default is level(95)

Default is to use asymptotic standard errors if no options are specified.

fcast compute can be used only after var, svar, and vec; see [TS] **var**, [TS] **var svar**, and [TS] **vec**.

You must tsset your data before using fcast compute; see [TS] **tsset**.

Description

fcast compute produces dynamic forecasts of the dependent variables in a model previously fit by var, svar, or vec. fcast compute creates new variables and, if necessary, extends the time frame of the dataset to contain the prediction horizon.

Options

┌─── Main ──

step(#) specifies the number of periods to be forecast. The default is 1.

dynamic(*time_constant*) specifies the period to begin the dynamic forecasts. The default is the period after the last observation in the estimation sample. The dynamic() option accepts either a Stata date function that returns an integer or an integer that corresponds to a date using the current tsset format. dynamic() must specify a date in or beyond the estimation sample.

estimates(*estname*) specifies that fcast compute use the estimation results stored as *estname*. By default, fcast compute uses the active estimation results. See [R] **estimates** for more information about saving and restoring previously obtained estimates.

replace causes fcast compute to replace the variables in memory with the specified predictions.

differences specifies that fcast compute also save dynamic predictions of the first-differenced variables. differences can only be specified with vec estimation results.

> ⌐ Std. Errors ⌐

nose specifies that the asymptotic standard errors of the forecasted levels and, thus, the asymptotic confidence intervals for the levels not be calculated. By default, the asymptotic standard errors and the asymptotic confidence intervals of the forecasted levels are calculated.

bs specifies that `fcast compute` use confidence bounds estimated by a simulation method based on bootstrapping the residuals.

bsp specifies that `fcast compute` use confidence bounds estimated via simulation in which the innovations are drawn from a multivariate normal distribution.

bscentile specifies that `fcast compute` use centiles of the bootstrapped dataset to estimate the bounds of the confidence intervals. By default, `fcast compute` uses the estimated standard errors and the quantiles of the standard normal distribution determined by `level()`.

reps(#) gives the number of repetitions used in the simulations. The default is 200.

nodots specifies that no dots be displayed while obtaining the simulation-based standard errors. By default, for each replication, a dot is displayed.

saving(*filename*[, replace]) specifies the name of the file to hold the dataset that contains the bootstrap replications. The `replace` option overwrites any file with this name.

    replace indicates that *filname* be overwritten, if it exists. This option is not shown in the dialog box.

> ⌐ Reporting ⌐

level(#) specifies the confidence level, as a percentage, for confidence intervals. The default is `level(95)` or as set by `set level`; see [U] **20.6 Specifying the width of confidence intervals**.

Remarks

Researchers frequently use VARs and VECMs to construct dynamic forecasts. `fcast compute` computes dynamic forecasts of the dependent variables in a VAR or VECM previously fit by `var`, `svar`, or `vec`. If you are interested in conditional, one-step-ahead predictions, use `predict` (see [TS] **var**, [TS] **var svar**, and [TS] **vec**).

To obtain and analyze dynamic forecasts, you (1) fit a model, (2) use `fcast compute` to compute the dynamic forecasts, and (3) use `fcast graph` to graph the results.

▷ Example 1

Typing

```
. use http://www.stata-press.com/data/r9/lutkepohl
. var dlincome dlconsumption dlinvestment if qtr < q(1979q1)
. fcast compute m2_, step(8)
. fcast graph m2_dlincome m2_dlinvestment m2_dlconsumption, observed
```

fits a VAR with two lags, computes eight-step dynamic predictions for each of the endogenous variables, and produces the graph

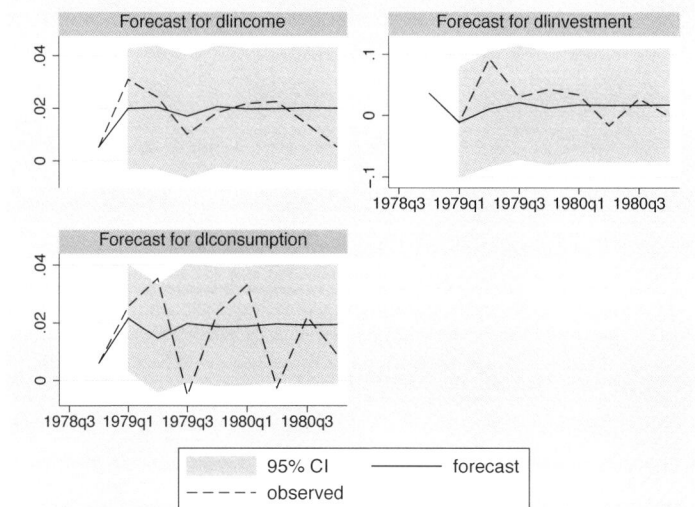

The graph shows that the model is better at predicting changes in income and investment than in consumption. The graph also shows how quickly the predictions from the two-lag model settle down to their mean values.

◁

fcast compute creates new variables in the dataset. If there are K dependent variables in the previously fit model, fcast compute generates $4K$ new variables:

K new variables that hold the forecasted levels, named by appending the specified prefix to the name of the original variable.

K estimated lower bounds for the forecast interval, named by appending the specified prefix and the suffix "_LB" to the name of the original variable.

K estimated upper bounds for the forecast interval, named by appending the specified prefix and the suffix "_UB" to the name of the original variable.

K estimated standard errors of the forecast, named by appending the specified prefix and the suffix "_SE" to the name of the original variable.

If you specify options so that fcast compute does not calculate standard errors, the $3K$ variables that hold them and the bounds of the confidence intervals are not generated.

If the model previously fit is a VECM, specifying differences generates an additional K variables that hold the forecasts of the first-differences of the dependent variables, named by appending the prefix "*prefix*D_" to the name of the original variable.

▷ Example 2

Plots of the forecasts from different models along with the observations from a hold-out sample can provide insights to their relative forecasting performance. Continuing the previous example,

```
. var dlincome dlconsumption dlinvestment if qtr < q(1979q1), lags(1/6)
. fcast compute m6_, step(8)
. graph twoway line m6_dlinvestment m2_dlinvestment dlinvestment qtr
> if m6_dlinvestment < ., legend(cols(1))
```

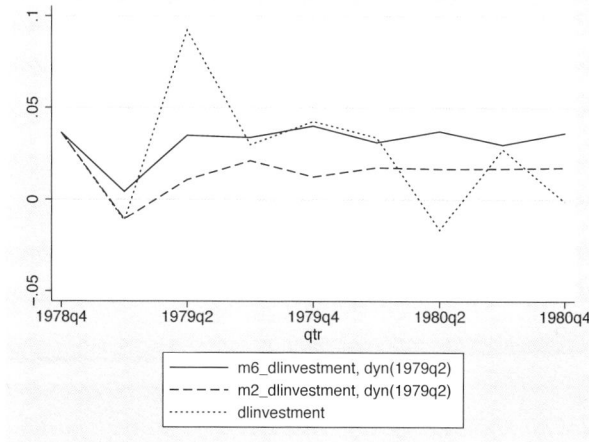

The model with six lags predicts changes in investment better than the two-lag model in some periods but markedly worse in other periods.

◁

Methods and Formulas

fcast compute is implemented as an ado-file.

Predictions after var and svar

Recall that a VAR with endogenous variables $\mathbf{y}_t$ and exogenous variables $\mathbf{x}_t$ can be written as

$$\mathbf{y}_t = \mathbf{v} + \mathbf{A}_1\mathbf{y}_{t-1} + \cdots + \mathbf{A}_p\mathbf{y}_{t-p} + \mathbf{B}\mathbf{x}_t + \mathbf{u}_t$$

where

$$t = 1, \ldots, T$$
$\mathbf{y}_t = (y_{1t}, \ldots, y_{Kt})'$ is a $K \times 1$ random vector,
the $\mathbf{A}_i$ are fixed $(K \times K)$ matrices of parameters,
$\mathbf{x}_t$ is an $(M \times 1)$ vector of exogenous variables,
$\mathbf{B}$ is a $(K \times M)$ matrix of coefficients,
$\mathbf{v}$ is a $(K \times 1)$ vector of fixed parameters, and
$\mathbf{u}_t$ is assumed to be white noise; that is,
$$E(\mathbf{u}_t) = \mathbf{0}_K$$
$$E(\mathbf{u}_t\mathbf{u}_t') = \mathbf{\Sigma}$$
$$E(\mathbf{u}_t\mathbf{u}_s') = \mathbf{0}_K \text{ for } t \neq s$$

fcast compute will dynamically predict the variables in the vector $\mathbf{y}_t$ conditional on p initial values of the endogenous variables and any exogenous $\mathbf{x}_t$. Adopting the notation from Lütkepohl (1993, 334) to fit the case at hand, the optimal h-step ahead forecast of $\mathbf{y}_{t+h}$ conditional on $\mathbf{x}_t$ is

$$\mathbf{y}_t(h) = \widehat{\mathbf{v}} + \widehat{\mathbf{A}}_1 \mathbf{y}_t(h-1) + \cdots + \widehat{\mathbf{A}}_p \mathbf{y}_t(h-p) + \widehat{\mathbf{B}} \mathbf{x}_t \tag{3}$$

If there are no exogenous variables, (3) becomes

$$\mathbf{y}_t(h) = \widehat{\mathbf{v}} + \widehat{\mathbf{A}}_1 \mathbf{y}_t(h-1) + \cdots + \widehat{\mathbf{A}}_p \mathbf{y}_t(h-p)$$

When there are no exogenous variables, `fcast compute` can compute the asymptotic confidence bounds.

As shown by Lütkepohl (1993, 177–178), the asymptotic estimator of the covariance matrix of the prediction error is given by

$$\widehat{\boldsymbol{\Sigma}}_{\widehat{y}}(h) = \widehat{\boldsymbol{\Sigma}}_y(h) + \frac{1}{T} \widehat{\boldsymbol{\Omega}}(h) \tag{4}$$

where

$$\widehat{\boldsymbol{\Sigma}}_y(h) = \sum_{i=0}^{h-1} \widehat{\boldsymbol{\Phi}}_i \widehat{\boldsymbol{\Sigma}} \widehat{\boldsymbol{\Phi}}_i'$$

$$\widehat{\boldsymbol{\Omega}}(h) = \frac{1}{T} \sum_{t=0}^{T} \left\{ \sum_{i=0}^{h-1} \mathbf{Z}_t' \left(\widehat{\mathbf{B}}' \right)^{h-1-i} \otimes \widehat{\boldsymbol{\Phi}}_i \right\} \widehat{\boldsymbol{\Sigma}}_\beta \left\{ \sum_{i=0}^{h-1} \mathbf{Z}_t' \left(\widehat{\mathbf{B}}' \right)^{h-1-i} \otimes \widehat{\boldsymbol{\Phi}}_i \right\}' \tag{5}$$

$$\widehat{\mathbf{B}} = \begin{bmatrix} 1 & \mathbf{0} & \mathbf{0} & \ldots & \mathbf{0} & \mathbf{0} \\ \widehat{\mathbf{v}} & \widehat{\mathbf{A}}_1 & \widehat{\mathbf{A}}_2 & \ldots & \widehat{\mathbf{A}}_{p-1} & \widehat{\mathbf{A}}_p \\ \mathbf{0} & \mathbf{I}_K & \mathbf{0} & \ldots & \mathbf{0} & \mathbf{0} \\ \mathbf{0} & \mathbf{0} & \mathbf{I}_K & & \mathbf{0} & \mathbf{0} \\ \vdots & \vdots & & \ddots & & \vdots \\ \mathbf{0} & \mathbf{0} & \mathbf{0} & \ldots & \mathbf{I}_K & \mathbf{0} \end{bmatrix}$$

$$\mathbf{Z}_t = (1, \mathbf{y}_t', \ldots, \mathbf{y}_{t-p-1}')'$$

$$\widehat{\boldsymbol{\Phi}}_0 = \mathbf{I}_K$$

$$\widehat{\boldsymbol{\Phi}}_i = \sum_{j=1}^{i} \widehat{\boldsymbol{\Phi}}_{i-j} \widehat{\mathbf{A}}_j \qquad i = 1, 2, \ldots$$

$$\widehat{\mathbf{A}}_j = \mathbf{0} \quad \text{for } j > p$$

$\widehat{\boldsymbol{\Sigma}}$ is the estimate of the covariance matrix of the innovations, and $\widehat{\boldsymbol{\Sigma}}_\beta$ is the estimated VCE of the coefficients in the VAR. The formula in (5) is general enough to handle the case in which constraints are placed on the coefficients in the VAR(p).

Equation (4) is made up of two terms. $\widehat{\boldsymbol{\Sigma}}_y(h)$ is the estimated mean squared error (MSE) of the forecast. $\widehat{\boldsymbol{\Sigma}}_y(h)$ estimates the error in the forecast arising from the unseen innovations. $T^{-1}\widehat{\boldsymbol{\Omega}}(h)$ estimates the error in the forecast that is due to using estimated coefficients instead of the true coefficients. As the sample size grows, uncertainty with respect to the coefficient estimates decreases, and $T^{-1}\widehat{\boldsymbol{\Omega}}(h)$ goes to zero.

Assuming that $\mathbf{y}_t$ is normally distributed, the bounds for the asymptotic $(1 - \alpha)100\%$ interval around the forecast for the kth component of $\mathbf{y}_t$, h periods ahead are

$$\widehat{\mathbf{y}}_{k,t}(h) \pm z_{(\frac{\alpha}{2})}\widehat{\sigma}_k(h) \tag{5}$$

where $\widehat{\sigma}_k(h)$ is the kth diagonal element of $\widehat{\mathbf{\Sigma}}_{\widehat{y}}(h)$.

Specifying the `bs` option causes the standard errors to be computed via simulation using bootstrapped residuals. Both `var` and `svar` contain estimators for the coefficients of a VAR that are conditional on the first p observations on the endogenous variables in the data. Similarly, these algorithms are conditional on the first p observations of the endogenous variables in the data. However, the simulation-based estimates of the standard errors are also conditional on the estimated coefficients. The asymptotic standard errors are not conditional on the coefficient estimates because the second term on the right-hand side of (4) accounts for the uncertainty arising from using estimated parameters.

For a simulation with R repetitions, this method uses the following algorithm:

1. Fit the model and save the estimated coefficients.

2. Use the estimated coefficients to calculate the residuals.

3. Repeat steps 3a to 3c R times.

   3a. Draw a simple random sample with replacement of size $T + h$ from the residuals. When the tth observation is drawn, all K residuals are selected, preserving any contemporaneous correlation among the residuals.

   3b. Use the sampled residuals, p initial values of the endogenous variables, any exogenous variables, and the estimated coefficients to construct a new sample dataset.

   3c. Save the simulated endogenous variables for the h forecast periods in the bootstrapped dataset.

4. For each endogenous variable and each of the forecast periods, the simulated standard error is the estimated standard error of the R simulated forecasts. By default, the upper and lower bounds of the $(1 - \alpha)100\%$ are estimated using the simulation-based estimates of the standard errors and the normality assumption, as in (5). If the `bscentile` option is specified, the sample centiles for the upper and lower bounds of the R simulated forecasts are used for the upper and lower bounds of the confidence intervals.

If the `bsp` option is specified, a parametric simulation algorithm is used. Specifically, everything is as above with the exception that 3a is replaced by 3a(bsp) as follows:

   3a(bsp). Draw $T + h$ observations from a multivariate normal distribution with covariance matrix $\widehat{\mathbf{\Sigma}}$.

The algorithm above assumes that h forecast periods come after the original sample of T observations. If the h forecast periods lie within the original sample, smaller simulated datasets are sufficient.

Dynamic forecasts after vec

Methods and Formulas of [TS] **vec** discusses how to obtain the one-step predicted differences and levels. `fcast compute` uses the previous dynamic predictions as inputs for subsequent dynamic predictions.

Following Lütkepohl (1993, section 11.3.1), `fcast compute` uses

$$\widehat{\Sigma}_{\hat{y}}(h) = \left(\frac{T}{T-d}\right) \sum_{i=0}^{h-1} \widehat{\Phi}_i \widehat{\Omega} \widehat{\Phi}_i$$

where the $\widehat{\Phi}_i$ are the estimated matrices of impulse–response functions, T is the number of observations in the sample, d is the number of degrees of freedom, and $\widehat{\Omega}$ is the estimated cross-equation variance matrix. The formulas for d and $\widehat{\Omega}$ are given in *Methods and Formulas* of [TS] **vec**.

The estimated standard errors at step h are the square roots of the diagonal elements of $\widehat{\Sigma}_{\hat{y}}(h)$.

Following Lütkepohl (1993), the estimated forecast-error variance does not consider parameter uncertainty. As the sample size gets infinitely large, the importance of parameter uncertainty diminishes to zero.

References

Hamilton, J. D. 1994. *Time Series Analysis.* Princeton: Princeton University Press.

Lütkepohl, H. 1993. *Introduction to Multiple Time Series Analysis.* 2nd ed. New York: Springer.

Also See

Complementary:	[TS] **fcast graph**, [TS] **var**, [TS] **var svar**, [TS] **varbasic**, [TS] **vec**
Related:	[TS] **irf**
Background:	[U] **11.4.3 Time-series varlists**,
	[TS] **var intro**, [TS] **vec intro**

Title

> **fcast graph** — Graph forecasts of variables computed by fcast compute

Syntax

> fcast graph *varlist* [*if*] [*in*] [, *options*]

where *varlist* contains one or more forecasted variables generated by fcast compute.

options	description
Main	
differences	graph forecasts of the first-differenced variables (vec only)
noci	suppress confidence bands
observed	include observed values of the predicted variables
Forecast plot	
cline_options	affect rendition of the forecast lines
CI plot	
ciopts(*area_options*)	affect rendition of the confidence bands
Observed plot	
obopts(*cline_options*)	affect rendition of the observed values
Y-Axis, T-Axis, Title, Caption, Legend, Overall	
twoway_options	any options other than by() documented in [G] *twoway_options*
By	
byopts(*by_option*)	affect appearance of the combined graph; see [G] *by_option*

Description

fcast graph graphs dynamic forecasts of the endogenous variables from a VAR(p) or VECM that has already been obtained from fcast compute; see [TS] **fcast compute**.

Options

> Main

differences specifies that the forecasts of the first-differenced variables be graphed. This option is only available with forecasts computed by fcast compute after vec. differences implies noci.

noci specifies that the confidence intervals be suppressed. By default, the confidence intervals are included.

observed specifies that observed values of the predicted variables be included in the graph. By default, observed values are not graphed.

Forecast plot

---

cline_options affect the rendition of the plotted lines corresponding to the forecast; see [G] ***cline_options***.

CI plot

---

ciopts(*area_options*) affect the rendition of the confidence bands for the forecasts; see [G] ***area_options***.

Observed plot

---

obopts(*cline_options*) affect the rendition of the observed values of the predicted variables; see [G] ***cline_options***. This option implies the observed option.

Y-Axis, T-Axis, Title, Caption, Legend, Overall

---

twoway_options are any of the options documented in [G] ***twoway_options***, excluding by().

By

---

byopts(*by_option*) are documented in [G] ***by_option***. These options affect the appearance of the combined graph.

Remarks

fcast graph graphs dynamic forecasts created by fcast compute.

▷ Example 1

In this example, we use a cointegrating VECM to model the state-level unemployment rates in Missouri, Indiana, Kentucky, and Illinois, and we graph the forecasts against a six-month holdout sample.

```
. use http://www.stata-press.com/data/r9/urates, clear
. vec missouri indiana kentucky illinois if t < m(2003m7), trend(rconstant)
> rank(2) lags(4)
  (output omitted )
. fcast compute m1_, step(6)
```

```
. fcast graph m1_missouri m1_indiana m1_kentucky m1_illinois, observed
```

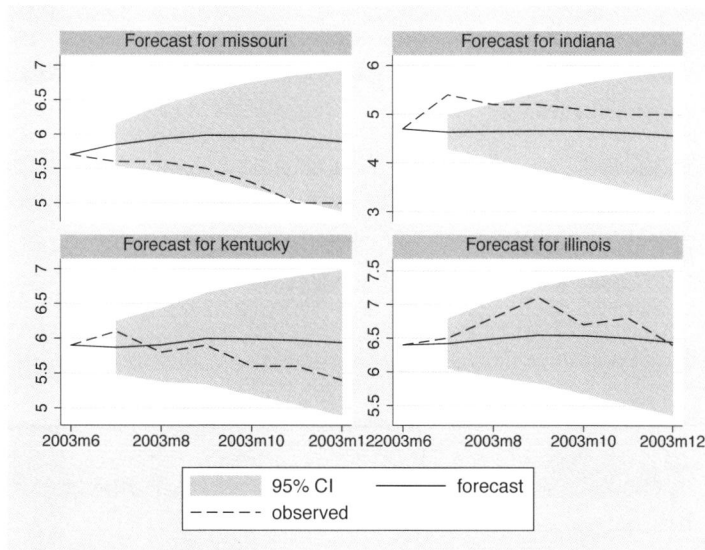

Because the 95% confidence bands for the predicted unemployment rates in Missouri and Indiana do not contain all of their observed values, the model does not reliably predict these unemployment rates.

◁

Methods and Formulas

fcast graph is implemented as an ado-file.

Also See

Complementary:	[TS] **fcast compute**, [TS] **var**, [TS] **var svar**, [TS] **varbasic**, [TS] **vec**
Related:	[TS] **irf**
Background:	[U] **11.4.3 Time-series varlists**,
	[TS] **var intro**, [TS] **vec intro**

Title

> **haver** — Load data from Haver Analytics database

Syntax

Describe contents of a Haver dataset

> haver <u>des</u>cribe *filename* $\left[\, , \, \underline{\text{det}}\text{ail} \,\right]$

Describe specified variables in a Haver dataset

> haver <u>des</u>cribe *varlist* using *filename* $\left[\, , \, \underline{\text{det}}\text{ail} \,\right]$

Load Haver dataset

> haver use *filename* $\left[\, , \, use\_options \,\right]$

Load specified variables from a Haver dataset

> haver use *varlist* using *filename* $\left[\, , \, use\_options \,\right]$

use_options	description
<u>tin</u>($\left[constant\right]$, $\left[constant\right]$)	load data within specified date range
<u>tw</u>ithin($\left[constant\right]$, $\left[constant\right]$)	same as tin(), except exclude the end points of range
<u>tv</u>ar(*varname*)	create time variable *varname*
<u>hm</u>issing(*misval*)	record missing values as *misval*
<u>f</u>ill	include observations with missing data in resulting dataset and record missing values for them
clear	clear data in memory before loading the Haver dataset

Description

Haver Analytics (*www.haver.com*) provides economic and financial databases in the form of `.dat` files to which you can purchase access. The `haver` command allows you to use those datasets with Stata. The `haver` command is provided only with Stata for Windows.

`haver describe` describes the contents of a Haver file.

`haver use` loads the specified variables from a Haver file into Stata's memory.

If *filename* is specified without a suffix, `.dat` is assumed.

Option for use with haver describe

`detail` specifies that a detailed report of all information available on the variables be presented. By default, the Haver concepts data type, number of observations, aggregate type, difference type, magnitude, and date modified are not reported.

Options for use with haver use

tin([*constant*], [*constant*]) specifies the date range of the data to be loaded. *constant* refers to a date constant specified in the usual way, e.g., tin(1jan1999, 31dec1999), which would mean from and including 1 January 1999 up to and including 31 December.

twithin([*constant*], [*constant*]) functions the same as tin(), except that the end points of the range will be excluded in the data loaded.

tvar(*varname*) specifies the name of the time variable Stata will create; the default is tvar(time). The tvar() variable is the name of the variable that you would use to tsset the data after loading, except that is unnecessary because haver use automatically tssets the data for you.

hmissing(*missval*) specifies which of Stata's 27 missing values (., .a, ..., .z) is to be recorded when there are missing values in the Haver dataset.

Two kinds of missing values can occur when loading a Haver dataset. One is created when a variable's data does not span the entire time series, and these missing values are always recorded as . by Stata. The other corresponds to an actual missing value recorded in the Haver format. hmissing() sets what is recorded when this second kind of missing value is encountered. You could specify hmissing(.h) to cause Stata's .h code to be recorded in this case.

The default is to store . for both kinds of missing values. See [U] **12.2.1 Missing values**.

fill specifies that observations with no data be left in the resulting dataset formed in memory and that missing values be recorded for them. The default is to exclude such observations, which can result in the loaded time-series dataset having gaps.

Specifying fill has the same effect as issuing the tsfill command after loading; see [TS] **tsfill**.

clear clears the data in memory before loading the Haver dataset.

Remarks

Remarks are presented under the headings

Installation
Determining the contents of a Haver dataset
Loading a Haver dataset
Combining variables
Using subsets

Installation

Haver Analytics (*www.haver.com*) provides over 100 economic and financial databases in the form of .dat files to which you can purchase access. The haver command allows you to use those datasets with Stata.

haver is provided only with Stata for Windows, and it uses the DLXAPI32.DLL file, which Haver provides. That file is probably already installed. If it is not, Stata will complain that "DLXAPI32.DLL could not be found".

Stata accesses this file via the Windows operating system if it is installed where Windows expects. If Windows cannot find the file, the location can be specified via Stata's set command:

```
. set haverdll "C:\Windows\system32\DLXAPI32.DLL"
```

Determining the contents of a Haver dataset

haver describe displays the contents of a Haver dataset. If no varlist is specified, all variables are described:

```
. haver describe dailytst
Dataset:        dailytst (use)
```

Variable	Description	Time span	Format	Group	Source
FDB6	6-Month Eurod..	02jan1996-01aug2003	Daily	F01	FRB
FFED	Federal Funds..	01jan1996-01aug2003	Daily	F01	FRB
FFED2	Federal Funds..	01jan1988-29dec1995	Daily	F70	FRB
FFED3	Federal Funds..	01jan1980-31dec1987	Daily	F70	FRB
FFED4	Federal Funds..	03jan1972-31dec1979	Daily	F70	FRB
FFED5	Federal Funds..	01jan1964-31dec1971	Daily	F70	FRB
FFED6	Federal Funds..	02jan1962-31dec1963	Daily	F70	FRB

Above, we describe the Haver dataset dailytst.dat, which we already have on our computer and in our current directory.

By default, each line of the output corresponds to one Haver variable. Specifying detail would display additional information about each of the variables, and specifying the optional varlist allows us to restrict the output to the variables that interest us:

```
. haver describe FDB6 using dailytst, detail
```

FDB6	6-Month Eurodollar Deposits (London Bid) (% p.a.)

```
        Frequency: Daily            Time span: 02jan1996 - 01aug2003
        Group: F01                  Number of Observations: 1979
        Source: FRB                 Date Modified: 04aug2003
        Agg Type: 1                 Decimal Precision: 2
        Difference Type: 1          Magnitude: 0
        Data Type: %
```

Loading a Haver dataset

haver use loads Haver datasets. If no varlist is specified, all variables are loaded:

```
. haver use dailytst
. list in 1/5
```

	FDB6	FFED	FFED2	FFED3	FFED4	FFED5	FFED6	time
1.	.	.	.	.	.	.	2.75	02jan1962
2.	.	.	.	.	.	.	2.5	03jan1962
3.	.	.	.	.	.	.	2.75	04jan1962
4.	.	.	.	.	.	.	2.5	05jan1962
5.	.	.	.	.	.	.	2	08jan1962

The entire dataset contains 10,849 observations. The variables FDB6 through FFED5 contain missing because there is no information about them for these first few dates; later in the dataset, nonmissing values for them will be found.

haver use requires that the variables loaded all be of the same time frequency (e.g., annual, quarterly, etc.) Haver's weekly data is treated by Stata as daily data with a seven-day interval, so weekly data is loaded as daily data with gaps.

Haver datasets include the following information about each variable:

1. Source — the Haver code associated with the source for the data

2. Group — the Haver group to which the variable belongs

3. DataType — whether the variable is a ratio, index, etc.

4. AggType — the Haver code for the level of aggregation

5. DifType — the Haver code for the type of difference

6. DecPrecision — the number of decimals to which the variable is recorded

7. Magnitude — the Haver magnitude code

8. DateTimeMod — the date the series was last modified by Haver

When a variable is loaded, this information is stored in variable characteristics (see [P] **char**). Those characteristics can be viewed using `char list`:

```
. char list FDB6[]
FDB6[Source]:              FRB
FDB6[Group]:               F01
FDB6[DataType]:            %
FDB6[AggType]:             1
FDB6[DifType]:             1
FDB6[DecPrecision]:        2
FDB6[Magnitude]:           0
FDB6[DateTimeMod]:         04aug2003
```

Combining variables

In fact, the variables FFED through FFED6 all contain information about the same thing (the Federal Funds rate) but over different times:

```
. haver describe dailytst
Dataset:      dailytst (use)
```

Variable	Description	Time span	Format	Group	Source
FDB6	6-Month Eurod..	02jan1996-01aug2003	Daily	F01	FRB
FFED	Federal Funds..	01jan1996-01aug2003	Daily	F01	FRB
FFED2	Federal Funds..	01jan1988-29dec1995	Daily	F70	FRB
FFED3	Federal Funds..	01jan1980-31dec1987	Daily	F70	FRB
FFED4	Federal Funds..	03jan1972-31dec1979	Daily	F70	FRB
FFED5	Federal Funds..	01jan1964-31dec1971	Daily	F70	FRB
FFED6	Federal Funds..	02jan1962-31dec1963	Daily	F70	FRB

In Haver datasets, variables may contain up to 2,100 data points, and if more are needed, Haver Analytics breaks the variables into multiple variables. It is your responsibility to put them back together.

To determine when variables need to be combined, look at the description and time span. To combine variables, use **egen**'s `rowfirst()` function and then drop the originals:

```
. egen fedfunds = rowfirst(FFED FFED2 FFED3 FFED4 FFED5 FFED6)
. drop FFED*
```

Alternatively, you can combine variables by hand:

```
. gen fedfunds = FFED
(8869 missing values generated)
. replace fedfunds = FFED2 if fedfunds >= .
(2086 real changes made)
. replace fedfunds = FFED3 if fedfunds >= .
(2088 real changes made)
. replace fedfunds = FFED4 if fedfunds >= .
(2086 real changes made)
. replace fedfunds = FFED5 if fedfunds >= .
(2088 real changes made)
. replace fedfunds = FFED6 if fedfunds >= .
(521 real changes made)
. drop FFED*
```

Using subsets

You can use a subset of the Haver dataset by specifying the variables to be loaded or the observations to be included:

```
. haver use FFED FFED2 using dailytst, tin(1jan1990,31dec1999)
(data from dailytst.dat loaded)
```

Above, we load only the variables FFED and FFED2 and the observations from 1jan1990 to 31dec1999.

Also See

Complementary:	[TS] **tsset**
Related:	[D] **infile**, [D] **insheet**, [D] **odbc**

Title

> **irf** — Create and analyze IRFs and FEVDs

Syntax

irf *subcommand* ... $\left[\, , \, ... \, \right]$

subcommand	description
create	create IRF file containing IRFs and FEVDs
set	set the active IRF file
graph	graph results from active file
cgraph	create combined graphs of multiple IRFs or FEVDs
ograph	create overlaid graphs of multiple IRFs or FEVDs
table	create table of IRF and/or FEVD from active file
ctable	create combined tables of multiple IRFs and/or FEVDs
describe	describe contents of active file
add	add results from an IRF file to the active IRF file
drop	drop IRF results from active file
rename	rename IRF results within a file

IRF stands for impulse–response function; FEVD stands for forecast-error variance decomposition.

irf can be used only after var, svar, or vec; see [TS] **var**, [TS] **var svar**, and [TS] **vec**.

See [TS] **irf create**, [TS] **irf set**, [TS] **irf graph**, [TS] **irf cgraph**, [TS] **irf ograph**, [TS] **irf table**, [TS] **irf ctable**, [TS] **irf describe**, [TS] **irf add**, [TS] **irf drop**, and [TS] **irf rename** for details about subcommands.

Description

irf creates and manipulates IRF files that contain estimates of the impulse–response functions (IRFs) and forecast-error variance decompositions (FEVDs) created after estimation by var, svar, or vec.

IRFs and FEVDs are described below, and the process of analyzing them is outlined. After reading this entry, please see [TS] **irf create**.

Remarks

An impulse–response function (IRF) measures the effect of a shock to an endogenous variable on itself or on another endogenous variable; see Lütkepohl (1993, 43–56) and Hamilton (1994, 318–323) for formal definitions. Of the many types of IRFs, irf create estimates the five most important: simple IRFs; orthogonalized IRFs; cumulative IRFs; cumulative, orthogonalized IRFs; and structural IRFs.

The forecast-error variance decomposition (FEVD) measures the fraction of the forecast-error variance of an endogenous variable that can be attributed to orthogonalized shocks to itself or to another endogenous variable; see Lütkepohl (1993, 56–58) and Hamilton (1994, 323–324) for formal definitions. Of the many types of FEVDs, irf create estimates the two most important: Cholesky and structural.

To analyze IRFs and FEVDs in Stata, you (1) fit a model, (2) use `irf create` to estimate the IRFs and FEVDs and store them in a file, and (3) use `irf graph` or any of the other `irf` analysis commands to examine results:

```
. use http://www.stata-press.com/data/r9/lutkepohl
(Quarterly SA West German macro data, Bil DM, from Lutkepohl 1993 Table E.1)
. var dlinvestment dlincome dlconsumption if qtr<=q(1978q4), lags(1/2) dfk
  (output omitted)
. irf create order1, step(10) set(myirf1)
(file myirf1.irf created)
(file myirf1.irf now active)
(file myirf1.irf updated)
. irf graph oirf, impulse(dlincome) response(dlconsumption)
```

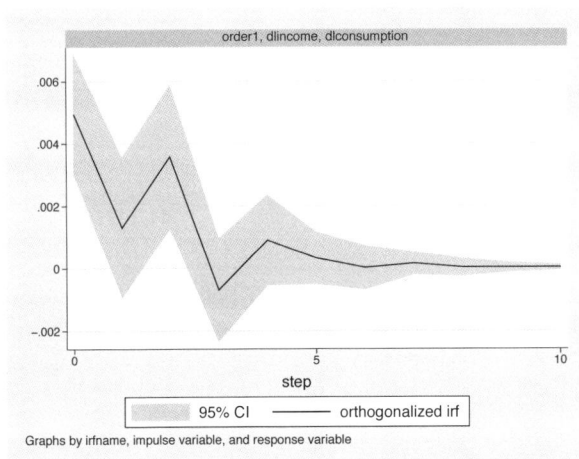

Multiple sets of IRFs and FEVDs can be placed in the same file, with each set of results in a file bearing a distinct name. The `irf create` command above created file `myirf1.irf` and put one set of results in it, named `order1`. The `order1` results include estimates of the simple IRFs; orthogonalized IRFs; cumulative IRFs; cumulative, orthogonalized IRFs; and Cholesky FEVDs.

Below we use the same estimated `var` but use a different Cholesky ordering to create a second set of IRF results, which we will store as `order2` in the same file, and then we will graph both results:

```
. irf create order2, step(10) order(dlincome dlinvestment dlconsumption)
(file myirf1.irf updated)
. irf graph oirf, irf(order1 order2) impulse(dlincome) response(dlconsumption)
```

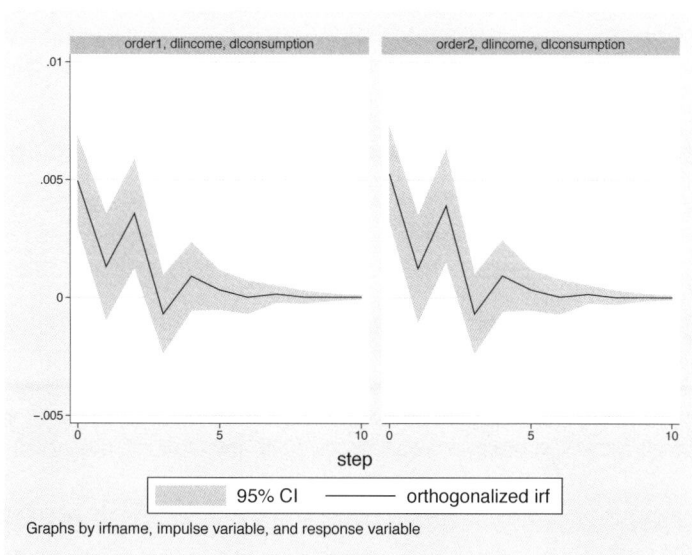

Graphs by irfname, impulse variable, and response variable

We have compared results for one model under two different identification schemes. We could just as well have compared results of two different models. We now use `irf table` to display the results tabularly:

```
. irf table oirf, irf(order1 order2) impulse(dlincome) response(dlconsumption)
```

Results from order1 order2

step	(1) oirf	(1) Lower	(1) Upper
0	.004934	.003016	.006852
1	.001309	-.000931	.003549
2	.003573	.001285	.005862
3	-.000692	-.002333	.00095
4	.000905	-.000541	.002351
5	.000328	-.0005	.001156
6	.000021	-.000675	.000717
7	.000154	-.000206	.000515
8	.000026	-.000248	.0003
9	.000026	-.000121	.000174
10	.000026	-.000061	.000113

(*Continued on next page*)

step	(2) oirf	(2) Lower	(2) Upper
0	.005244	.003252	.007237
1	.001235	-.001011	.003482
2	.00391	.001542	.006278
3	-.000677	-.002347	.000993
4	.00094	-.000576	.002456
5	.000341	-.000518	.001201
6	.000042	-.000693	.000777
7	.000161	-.000218	.00054
8	.000027	-.000261	.000315
9	.00003	-.000125	.000184
10	.000027	-.000065	.00012

```
95% lower and upper bounds reported
(1) irfname = order1, impulse = dlincome, and response = dlconsumption
(2) irfname = order2, impulse = dlincome, and response = dlconsumption
```

Both the table and the graph show that the two orthogonalized impulse–response functions are essentially the same. In both functions, an increase in the orthogonalized shock to `dlincome` causes a short series of increases in `dlconsumption` that dies out after four or five periods.

References

Hamilton, J. D. 1994. *Time Series Analysis*. Princeton: Princeton University Press.

Lütkepohl, H. 1993. *Introduction to Multiple Time Series Analysis*. 2nd ed. New York: Springer.

Also See

Complementary:	[TS] **var**, [TS] **var svar**, [TS] **varbasic**, [TS] **vec**;
	[TS] **irf add**, [TS] **irf cgraph**, [TS] **irf create**, [TS] **irf ctable**,
	[TS] **irf describe**, [TS] **irf drop**, [TS] **irf graph**, [TS] **irf ograph**,
	[TS] **irf rename**, [TS] **irf set**, [TS] **irf table**
Background:	[U] **11.4.3 Time-series varlists**,
	[TS] **var intro**, [TS] **vec intro**

Title

irf add — Add results from an IRF file to the active IRF file

Syntax

irf a̲dd { _all | [*newname=*] *oldname* ... } , using (*irf_filename*)

Description

irf add copies results from one IRF file to another—from the specified using() file to the active IRF file, set by irf set.

Option

using (*irf_filename*) specifies the file from which results are to be obtained and is required. If *irf_filename* is specified without an extension, .irf is assumed.

Remarks

If you have not read [TS] **irf**, please do so.

Results are copied from the specified file to the active file; use irf set to make a file active.

```
. irf add _all, using(luta)
no irf file active
r(111);
. irf set newfile
(file newfile.irf created)
(file newfile.irf now active)
. irf add _all, using(luta)
(file newfile.irf updated)
```

IRF results to be copied may be specified explicitly,

```
. irf add riley drukker1, using(desfile)
(file newfile.irf updated)
```

and IRF results may be renamed:

```
. irf add drukker2=drukker, using(desfile2)
(file newfile.irf updated)
```

This last command copied IRF results drukker from desfile2.irf to newfile.irf, where the results are now named drukker2.

Methods and Formulas

irf add is implemented as an ado-file.

Also See

Complementary:	[TS] **var**, [TS] **var svar**, [TS] **varbasic**, [TS] **vec**; [TS] **irf cgraph**, [TS] **irf create**, [TS] **irf ctable**, [TS] **irf describe**, [TS] **irf drop**, [TS] **irf graph**, [TS] **irf ograph**, [TS] **irf rename**, [TS] **irf set**, [TS] **irf table**
Background:	[U] **11.4.3 Time-series varlists**, [TS] **irf**, [TS] **var intro**, [TS] **vec intro**

Title

> **irf cgraph** — Combine graphs of impulse–response functions and FEVDs

Syntax

irf <u>c</u>graph $(spec_1)$ $\left[(spec_2) \dots \left[(spec_N) \right] \right]$ $\left[, options \right]$

where $(spec_k)$ is

$(irfname\ impulsevar\ responsevar\ stat\ \left[,\ spec\_options \right])$

irfname is the name of a set of IRF results in the active IRF file, *impulsevar* is an endogenous variable name, *responsevar* is an endogenous variable name, and *stat* is one or more statistics from the list below:

stat	description
Main	
irf	impulse–response function
oirf	orthogonalized impulse–response function
cirf	cumulative impulse–response function
coirf	cumulative, orthogonalized impulse–response function
fevd	Cholesky forecast-error variance decomposition
sirf	structural impulse–response function
sfevd	structural forecast-error variance decomposition

Notes: (1) No statistic may appear more than once.
       (2) If confidence intervals are included (the default), only two statistics may be included.
       (3) If confidence intervals are suppressed (option noci), up to four statistics may be included.

options	description
Main	
set(*filename*)	make *filename* active
Options	
combine_options	affect appearance of combined graph
Y-Axis, X-Axis, Title, Caption, Legend, Overall	
twoway_options	any options other than by() documented in [G] ***twoway_options***
* *spec_options*	level, steps, and rendition of plots and their CIs
† <u>individual</u>	graph each combination individually

* *spec_options* appear on multiple tabs in the dialog box.

† individual is not shown in the dialog box.

spec_options	description
Main	
noci	suppress confidence bands
Options	
level(*#*)	set confidence level; default is level(95)
lstep(*#*)	use *#* for first step
ustep(*#*)	use *#* for maximum step
Plot 1, Plot 2, Plot 3, Plot 4	
plot#opts(*line_options*)	affect rendition of the line plotting the *#* *stat*
CI 1, CI 2	
ci#opts(*area_options*)	affect rendition of the confidence interval for the *#* *stat*

spec_options may be specified within a graph specification, globally, or in both. When specified in a graph specification, the *spec_options* affect only the specification in which they are used. When supplied globally, the *spec_options* affect all graph specifications. When supplied in both places, options in the graph specification take precedence.

Description

irf cgraph makes a graph or a combined graph of IRF results. Each block within a pair of matching parentheses—each (*spec_k*)—specifies the information for a specific graph. irf cgraph combines these graphs into one image, unless option individual is also specified, in which case separate graphs for each block are created.

To become familiar with this command, we recommend that you type db irf cgraph.

Options

    └ Main ┘

noci suppresses graphing the confidence interval for each statistic. noci is assumed when the model was fitted by vec because no confidence intervals were estimated.

set(*filename*) specifies the file to be made active; see [TS] **irf set**. If set() is not specified, the active file is used.

    └ Options ┘

level(*#*) specifies the default confidence level, as a percentage, for confidence intervals, when they are reported. The default is level(95) or as set by set level; see [U] **20.6 Specifying the width of confidence intervals**. Note that the value set of an overall level() can be overridden by the level() inside a (*spec_k*).

lstep(*#*) specifies the first step, or period, to be included in the graph. lstep(0) is the default.

ustep(*#*), *# ≥ 1*, specifies the maximum step, or period, to be included in the graph.

combine_options affect the appearance of the combined graph; see [G] **graph combine**. *combine_options* cannot be specified with individual.

> Plot 1, Plot 2, Plot 3, Plot 4

plot1opts(*cline_options*), ..., plot4opts(*cline_options*) affect the rendition of the plotted statistics. plot1opts() affects the rendition of the first statistic; plot2opts(), the second; and so on. *cline_options* are as described in [G] *cline_options*.

> CI 1, CI 2

ci1opts1(*area_options*) and ci2opts2(*area_options*) affect the rendition of the confidence intervals for the first (ci1opts()) and second (ci2opts()) statistics. See [TS] **irf graph** for a complete description of this option and [G] *area_options* for the suboptions that change the look of the CI.

> Y-Axis, X-Axis, Title, Caption, Legend, Overall

twoway_options are any of the options documented in [G] *twoway_options* excluding by(). These include options for titling the graph (see [G] *title_options*) and options for saving the graph to disk (see [G] *saving_option*).

The following option is available with irf cgraph but is not shown in the dialog box:

individual specifies that each graph be displayed individually. By default, irf cgraph combines the subgraphs into a single image.

Remarks

If you have not read [TS] **irf**, please do so.

The relationship between irf cgraph and irf graph is syntactically and conceptually the same as that between irf ctable and irf table; see [TS] **irf ctable** for a complete description of the syntax.

irf cgraph is much the same as using irf graph to make individual graphs and then using graph combine to put them together. If you cannot use irf cgraph to do what you want, consider the other approach.

▷ Example 1

You have previously issued the commands

```
. use http://www.stata-press.com/data/r9/lutkepohl
. mat a = (., 0, 0\0,.,0\.,.,.)
. mat b = I(3)
. svar dlinvestment dlincome dlconsumption, aeq(a) beq(b)
. irf create modela, set(results3) step(8)
. svar dlincome dlinvestment dlconsumption, aeq(a) beq(b)
. irf create modelb, step(8)
```

(*Continued on next page*)

You now type

```
. irf cgraph (modela dlincome dlconsumption oirf sirf)
>            (modelb dlincome dlconsumption oirf sirf)
>            (modela dlincome dlconsumption fevd sfevd, lstep(1))
>            (modelb dlincome dlconsumption fevd sfevd, lstep(1)),
>            title("Results from modela and modelb")
```

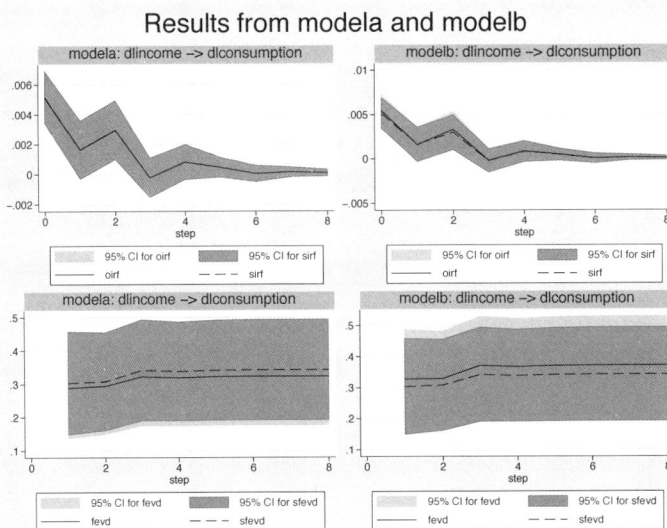

◁

Saved Results

irf cgraph saves in r():

Scalars
    r(k)   number of specific graph commands
Macros
    r(individual)   individual, if individual is specified
    r(save)   save filename and replace option for the combined graph, if specified
    r(name)   name and replace option for the combined graph, if specified
    r(title)   title of the combined graph, if individual is not specified
    r(save$j$)   save filename and replace option for the jth graph command, if specified
    r(name$j$)   name and replace option for the jth graph command, if specified
    r(title$j$)   title for the jth graph
    r(ci$j$)   level applied to the jth confidence interval or noci
    r(response$j$)   response in the jth command
    r(impulse$j$)   impulse in the jth command
    r(irfname$j$)   irfname in the jth command
    r(stats$j$)   stats specified in the jth command

Methods and Formulas

`irf cgraph` is implemented as an ado-file.

Also See

Complementary:	[TS] **var**, [TS] **var svar**, [TS] **vec**; [TS] **irf add**, [TS] **irf create**, [TS] **irf describe**, [TS] **irf drop**, [TS] **irf rename**, [TS] **irf set**
Related:	[TS] **irf ctable**, [TS] **irf graph**, [TS] **irf ograph**, [TS] **irf table**
Background:	[U] **11.4.3 Time-series varlists**, [TS] **irf**, [TS] **var intro**, [TS] **vec intro**

Title

irf create — Obtain impulse–response functions and FEVDs

Syntax

After var

    irf create *irfname* [, *var_options*]

After svar

    irf create *irfname* [, *svar_options*]

After vec

    irf create *irfname* [, *vec_options*]

irfname is any valid name that does not exceed 15 characters in length.

var_options	description
Main	
set(*filename*[, replace])	make *filename* active
replace	replace *irfname* if it already exists
step(*#*)	set forecast horizon to *#*; default is step(8)
order(*varlist*)	specify Cholesky ordering of endogenous variables
estimates(*estname*)	use previously saved results *estname*; default is to use active results
Std. Errors	
nose	do not calculate standard errors
bs	obtain standard errors from bootstrapped residuals
bsp	obtain standard errors from parametric bootstrap
reps(*#*)	use *#* bootstrap replications; default reps(200)
nodots	do not display "." for each bootstrap replication
bsaving(*filename*[, replace])	save bootstrap results in *filename*

svar_options	description
Main	
set(*filename*[, replace])	make *filename* active
replace	replace *irfname* if it already exists
step(*#*)	set forecast horizon to *#*; default is step(8)
estimates(*estname*)	use previously saved results *estname*; default is to use active results

Std. Errors

nose	do not calculate standard errors
bs	obtain standard errors from bootstrapped residual
bsp	obtain standard errors from parametric bootstrap
reps(#)	use # bootstrap replications; default reps(200)
nodots	do not display "." for each bootstrap replication
bsaving(filename[, replace])	save bootstrap results in filename

vec_options	description
Main	
set(filename[, replace])	make filename active
replace	replace irfname if it already exists
step(#)	set forecast horizon to #; default is step(8)
estimates(estname)	use previously saved results estname; default is to use active results

The default is to use asymptotic standard errors if no options are specified.

irf create is for use after fitting a model with the var, svar, or vec commands; see [TS] **var**, [TS] **var svar**, and [TS] **vec**.

You must tsset your data before using var, svar, or vec and, hence, before using irf create; see [TS] **tsset**.

Description

irf create estimates multiple sets of impulse–response functions (IRFs) and forecast-error variance decompositions (FEVDs) following estimation by var, svar, or vec. These estimates and their standard errors are known collectively as IRF results and are stored in an IRF file under the specified *irfname*.

The following types of impulse–response functions are stored:

simple IRFs	following svar, var, or vec
orthogonalized IRFs	following svar, var, or vec
cumulative IRFs	following svar, var, or vec
cumulative, orthogonalized IRFs	following svar, var, or vec
structural IRFs	following svar only

The following types of forecast-error variance decompositions are stored:

Cholesky FEVDs	following svar, var, or vec
structural FEVDs	following svar only

Once you have created a set of IRF results, use the other irf commands to analyze them.

Options

> Main

set(filename[, replace]) specifies the IRF file to be used. If set() is not specified, the active IRF file is used; see [TS] **irf set**.

If set() is specified, the specified file becomes the active file, just as if you had issued an irf set command.

replace specifies that the results stored under *irfname* may be replaced, if they already exist. IRF results are saved in files, and a single file may contain multiple IRF results.

step(#) specifies the step (forecast) horizon; the default is 8 periods.

order(*varlist*) is allowed only following estimation by var; it specifies the Cholesky ordering of the endogenous variables to be used when estimating the orthogonalized impulse–response functions. By default, the order in which the variables were originally specified on the var command is used.

estimates(*estname*) specifies that estimation results previously estimated by svar, var, or vec, and stored by estimates, be used. This option is rarely specified; see [R] **estimates**.

_____⌐ Std. Errors ⌐_____

nose, bs, and bsp are alternatives that specify how (whether) standard errors are to be calculated. If none of these options is specified, asymptotic standard errors are calculated, except in two cases: following estimation by vec and following estimation by svar in which long-run constraints were applied. In those two cases, the default is as if nose were specified, although in the second case, you could specify bs or bsp. Following estimation by vec, standard errors are simply not available.

nose specifies that no standard errors be calculated.

bs specifies that standard errors be calculated by bootstrapping the residuals. bs may not be specified if there are gaps in the data.

bsp specifies that standard errors be calculated via a multivariate-normal parametric bootstrap. bsp may not be specified if there are gaps in the data.

reps(#), nodots, and bsaving(*filename*[, replace]) are relevant only if bs or bsp is specified.

reps(*#*), *#* > 50, specifies the number of bootstrap replications to be performed. reps(200) is the default.

nodots specifies that dots not be displayed each time irf create performs a bootstrap replication.

bsaving(*filename*[, replace]) specifies that file *filename* be created and the bootstrap replications be saved in it. New file *filename* is just a .dta dataset than can be subsequently loaded using use. If *filename* is specified without an extension, .dta is assumed.

Remarks

If you have not read [TS] **irf**, please do so. An introductory example using IRFs is presented there.

Remarks are presented under the headings

       Introductory examples
       Technical aspects of IRF files
       IRFs and FEVDs
       IRF results for VARs
           An introduction to impulse–response functions for VARs
           An introduction to forecast-error variance decompositions for VARs
       IRF results for VECMs
           An introduction to impulse–response functions for VECMs
           An introduction to forecast-error variance decompositions for VECMs

Introductory examples

▷ Example 1: After var

Below we compare bootstrapped and asymptotic standard errors for a specific forecast-error variance decomposition (FEVD). We begin by fitting a VAR(2) model to the Lütkepohl data (we use the `var` command). We next use the `irf create` command twice, first to create results with asymptotic standard errors (saved under the name `asymp`) and then to recreate the same results again, this time with bootstrapped standard errors (saved under the name `bs`). Because bootstrapping is a random process, we set the random-number seed (`set seed 123456`) before using `irf create` the second time; this makes our results reproducible. Finally, we compare results using the IRF analysis command `irf ctable`.

```
. use http://www.stata-press.com/data/r9/lutkepohl
(Quarterly SA West German macro data, Bil DM, from Lutkepohl 1993 Table E.1)
. var dlinvestment dlincome dlconsumption if qtr>=q(1961q2) & qtr <= q(1978q4),
> lags(1/2)
  (output omitted )
. irf create asymp, step(8) set(results1)
(file results1.irf created)
(file results1.irf now active)
(file results1.irf updated)
. set seed 123456
. irf create bs, step(8) bs reps(250) nodots
(file results1.irf updated)
. irf ctable (asymp dlincome dlconsumption fevd)
> (bs dlincome dlconsumption fevd), noci stderror
```

step	(1) fevd	(1) S.E.	(2) fevd	(2) S.E.
0	0	0	0	0
1	.282135	.087373	.282135	.104073
2	.278777	.083782	.278777	.096954
3	.33855	.090006	.33855	.100452
4	.339942	.089207	.339942	.099085
5	.342813	.090494	.342813	.099326
6	.343119	.090517	.343119	.09934
7	.343079	.090499	.343079	.099325
8	.34315	.090569	.34315	.099368

```
(1) irfname = asymp, impulse = dlincome, and response = dlconsumption
(2) irfname = bs, impulse = dlincome, and response = dlconsumption
```

Point estimates are, of course, the same. The bootstrap estimates of the standard errors, however, are larger than the asymptotic estimates, which suggests that the sample size of 71 is not large enough for the distribution of the estimator of the forecast-error variance decomposition to be well approximated by the asymptotic distribution. In this case, we would expect the bootstrapped confidence interval to be more reliable than the confidence interval based on the asymptotic standard error.

◁

❏ Technical Note

The details of the bootstrap algorithms are given in *Methods and Formulas*. These algorithms are conditional on the first p observations, where p is the order of the fitted VAR. (In an SVAR model, p is the order of the VAR that underlies the SVAR.) The bootstrapped estimates are conditional on the first p observations, just as the estimators of the coefficients in VAR models are conditional on the first p observations. With bootstrapped standard errors (option bs), the p initial observations are used with resampling the residuals to produce the bootstrap samples used for estimation. With the more parametric bootstrap (option bsp), the p initial observations are used with draws from a multivariate-normal distribution with variance–covariance matrix $\widehat{\Sigma}$ to generate the bootstrap samples.

❏

❏ Technical Note

For var and svar e() results, irf uses $\widehat{\Sigma}$, the estimated variance matrix of the disturbances, in computing the asymptotic standard errors of all the functions. The point estimates of the orthogonalized impulse–response functions, the structural impulse–response functions, and all the variance decompositions also depend on $\widehat{\Sigma}$. As discussed in [TS] **var**, var and svar use the ML estimator of this matrix by default, but they have option dfk, which will instead use an estimator that includes a small-sample correction. Specifying dfk when the model is fitted—when the var or svar commands are given—changes the estimate of $\widehat{\Sigma}$ and will change subsequent IRF results that depend on it.

❏

▷ Example 2: After vec

While all IRFs and orthogonalized IRFs (OIRFs) from models with stationary variables will taper off to zero, some of the IRFs and OIRFs from models with first-difference stationary variables will not. This is the key difference between IRFs and OIRFs from systems of stationary variables fit by var or svar and those obtained from systems of first-difference stationary variables fit by vec. When the effect of the innovations dies out over time, the shocks are said to be transitory. In contrast, when the effect does not taper off, shocks are said to be permanent.

In this example, we look at the OIRF from one of the VECMs fit to the unemployment-rate data analyzed in [TS] **vec**. We see that an orthogonalized shock to Indiana has a permanent effect on the unemployment rate in Missouri:

```
. use http://www.stata-press.com/data/r9/urates
. vec missouri indiana kentucky illinois, trend(rconstant) rank(2) lags(4)
> noetable
  (output omitted )
. irf create vec1, set(vecirfs) step(50)
(file vecirfs.irf created)
(file vecirfs.irf now active)
(file vecirfs.irf updated)
```

Now we can use irf graph to graph the OIRF of interest:

. irf graph oirf, impulse(indiana) response(missouri)

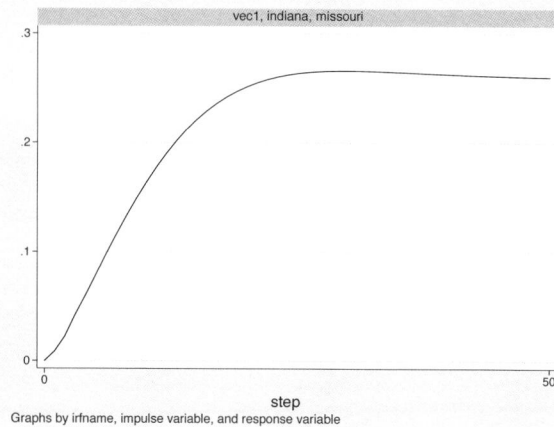

Graphs by irfname, impulse variable, and response variable

The graph shows that the estimated OIRF converges to a positive asymptote, which indicates that an orthogonalized innovation to the unemployment rate in Indiana has a permanent effect on the unemployment rate in Missouri.

◁

Technical aspects of IRF files

The following section is included for programmers wishing to extend the irf system.

irf create estimates a series of impulse–response functions and their standard errors. Although these estimates are stored in an IRF file, most users will never need to look at the contents of this file. The IRF commands fill in, analyze, present, and manage IRF results.

IRF files are just Stata datasets that have names ending in .irf instead of .dta. The dataset in the file has a nested panel structure.

Variable irfname contains the *irfname* specified by the user. Variable impulse records the name of the endogenous variable whose innovations are the impulse. Variable response records the name of the endogenous variable that is responding to the innovations. In a model with K endogenous variables, there are K^2 combinations of impulse and response. Variable step records the periods for which these estimates were computed.

Below is a catalog of the statistics that irf create estimates and the variable names under which they are stored in the IRF file.

(Continued on next page)

Statistic	Name
impulse–response functions	irf
orthogonalized impulse–response functions	oirf
cumulative impulse–response functions	cirf
cumulative, orthogonalized impulse–response functions	coirf
Cholesky forecast-error decomposition	fevd
structural impulse–response functions	sirf
structural forecast-error decomposition	sfevd
standard error of the impulse–response functions	stdirf
standard error of the orthogonalized impulse–response functions	stdoirf
standard error of the cumulative impulse–response functions	stdcirf
standard error of the cumulative, orthogonalized impulse–response functions	stdcoirf
standard error of the Cholesky forecast-error decomposition	stdfevd
standard error of the structural impulse–response functions	stdsirf
standard error of the structural forecast-error decomposition	stdsfevd

In addition to the variables, information is stored in _dta characteristics. Much of the following information is also available in r() following irf describe, where it is often more convenient to obtain the information.

Characteristic _dta[version] contains the version number of the IRF file. That number is currently 1.1.

Characteristic _dta[irfnames] contains a list of all the *irfnames* in the IRF file. For each *irfname*, there are a series of additional characteristics:

Name	Contents
_dta[*irfname*_model]	var, sr var, lr var, or vec
_dta[*irfname*_order]	Cholesky order used in IRF estimates
_dta[*irfname*_exog]	exogenous variables in VAR
_dta[*irfname*_constant]	constant or noconstant, depending on whether noconstant was specified in var or svar
_dta[*irfname*_lags]	lags in model
_dta[*irfname*_tmin]	minimum value of timevar in the estimation sample
_dta[*irfname*_tmax]	maximum value of timevar in the estimation sample
_dta[*irfname*_timevar]	name of tsset timevar
_dta[*irfname*_tsfmt]	format of timevar
_dta[*irfname*_varcns]	constrained or colon-separated list of constraints placed on VAR coefficients
_dta[*irfname*_svarcns]	constrained or colon-separated list of constraints placed on VAR coefficients
_dta[*irfname*_step]	maximum step in IRF estimates
_dta[*irfname*_stderror]	asymptotic, bs, bsp, or none, depending on the type of standard errors requested
_dta[*irfname*_reps]	number of bootstrap replications performed
_dta[*irfname*_version]	version of the IRF file that originally held *irfname* IRF results
_dta[*irfname*_rank]	number of cointegrating equations
_dta[*irfname*_trend]	trend() specified in vec
_dta[*irfname*_veccns]	constraints placed on VECM parameters
_dta[*irfname*_sind]	normalized seasonal indicators included in vec

IRFs and FEVDs

irf create can estimate several types of IRFs and FEVDs for VARs and VECMs. We first discuss IRF results for VAR and SVAR models, and then we discuss them in the context of VECMs. Since the cointegrating VECM is an extension of the stationary VAR framework, the section that discusses the IRF results for VECMs draws on the earlier VAR material.

IRF results for VARs

An introduction to impulse–response functions for VARs

A pth order vector autoregressive model (VAR) with exogenous variables is given by

$$\mathbf{y}_t = \mathbf{v} + \mathbf{A}_1 \mathbf{y}_{t-1} + \cdots + \mathbf{A}_p \mathbf{y}_{t-p} + \mathbf{B}\mathbf{x}_t + \mathbf{u}_t$$

where

$\mathbf{y}_t = (y_{1t}, \dots, y_{Kt})'$ is a $K \times 1$ random vector,
the $\mathbf{A}_i$ are fixed $K \times K$ matrices of parameters,
$\mathbf{x}_t$ is an $M \times 1$ vector of exogenous variables,
$\mathbf{B}$ is a $K \times M$ matrix of coefficients,
$\mathbf{v}$ is a $K \times 1$ vector of fixed parameters, and
$\mathbf{u}_t$ is assumed to be white noise; that is,
$E(\mathbf{u}_t) = \mathbf{0}$
$E(\mathbf{u}_t \mathbf{u}_t') = \mathbf{\Sigma}$
$E(\mathbf{u}_t \mathbf{u}_s') = \mathbf{0}$ for $t \neq s$

As discussed in [TS] **varstable**, a VAR can be rewritten in moving-average form only if it is stable. Any exogenous variables are assumed to be covariance stationary. Since the functions of interest in this section depend only on the exogenous variables through their effect on the estimated $\mathbf{A}_i$, we can simplify the notation by dropping them from the analysis. All the formulas given below still apply, although the $\mathbf{A}_i$ are estimated jointly with $\mathbf{B}$ on the exogenous variables.

Below we discuss conditions under which the impulse–response functions and forecast-error variance decompositions have a causal interpretation. While estimation requires only that the exogenous variables be predetermined, i.e., that $E[\mathbf{x}_{jt}u_{it}] = 0$ for all i, j, and t, assigning a causal interpretation to IRFs and FEVDs requires that the exogenous variables be strictly exogenous, i.e., that $E[\mathbf{x}_{js}u_{it}] = 0$ for all i, j, s, and t.

Impulse–response functions describe how the innovations to one variable affect another variable after a given number of periods. For an example of how impulse–response functions are interpreted, see Stock and Watson (2001). They use impulse–response functions to investigate the effect of surprise shocks to the Federal Funds rate on inflation and unemployment. In another example, Christiano, Eichenbaum, and Evans (1999) use impulse–response functions to investigate how shocks to monetary policy affect other macroeconomic variables.

Consider a VAR without exogenous variables:

$$\mathbf{y}_t = \mathbf{v} + \mathbf{A}_1 \mathbf{y}_{t-1} + \cdots + \mathbf{A}_p \mathbf{y}_{t-p} + \mathbf{u}_t \tag{1}$$

The VAR represents the variables in $\mathbf{y}_t$ as functions of its own lags and serially uncorrelated innovations $\mathbf{u}_t$. All the information about contemporaneous correlations among the K variables in $\mathbf{y}_t$ is contained in $\mathbf{\Sigma}$. In fact, as discussed in [TS] **var svar**, a VAR can be viewed as the reduced form of a dynamic simultaneous-equation model.

To see how the innovations affect the variables in $\mathbf{y}_t$ after, say, i periods, rewrite the model in its moving-average form

$$\mathbf{y}_t = \boldsymbol{\mu} + \sum_{i=0}^{\infty} \boldsymbol{\Phi}_i \mathbf{u}_{t-i} \tag{2}$$

where $\boldsymbol{\mu}$ is the $K \times 1$ time-invariant mean of $\mathbf{y}_t$, and

$$\boldsymbol{\Phi}_i = \begin{cases} \mathbf{I}_K & \text{if } i = 0 \\ \sum_{j=1}^{i} \boldsymbol{\Phi}_{i-j} \mathbf{A}_j & \text{if } i = 1, 2, \ldots \end{cases}$$

We can rewrite a VAR in the moving-average form only if it is stable. Essentially, a VAR is stable if the variables are covariance stationary and none of the autocorrelations are too high (the issue of stability is discussed in greater detail in [TS] **varstable**).

The $\boldsymbol{\Phi}_i$ are the simple impulse–response functions. The j, k element of $\boldsymbol{\Phi}_i$ gives the effect of a one-time unit increase in the kth element of $\mathbf{u}_t$ on the jth element of $\mathbf{y}_t$ after i periods, holding everything else constant. Unfortunately, these effects have no causal interpretation, which would require us to be able to answer the question, "How does an innovation to variable k, holding everything else constant, affect variable j after i periods?" Since the $\mathbf{u}_t$ are contemporaneously correlated, we cannot assume that everything else is held constant. Contemporaneous correlation among the $\mathbf{u}_t$ implies that a shock to one variable is likely to be accompanied by shocks to some of the other variables, so it does not make sense to shock one variable and hold everything else constant. For this reason, (2) cannot provide a causal interpretation.

This shortcoming may be overcome by rewriting (2) in terms of mutually uncorrelated innovations. Suppose that we had a matrix $\mathbf{P}$, such that $\boldsymbol{\Sigma} = \mathbf{P}\mathbf{P}'$. If we had such a $\mathbf{P}$, then $\mathbf{P}^{-1}\boldsymbol{\Sigma}\mathbf{P}'^{-1} = \mathbf{I}_K$, and

$$E\{\mathbf{P}^{-1}\mathbf{u}_t(\mathbf{P}^{-1}\mathbf{u}_t)'\} = \mathbf{P}^{-1}E\{(\mathbf{u}_t\mathbf{u}_t')\mathbf{P}'^{-1}\} = \mathbf{P}^{-1}\boldsymbol{\Sigma}\mathbf{P}'^{-1} = \mathbf{I}_K$$

We can thus use $\mathbf{P}^{-1}$ to orthogonalize the $\mathbf{u}_t$ and rewrite (2) as

$$\mathbf{y}_t = \boldsymbol{\mu} + \sum_{i=0}^{\infty} \boldsymbol{\Phi}_i \mathbf{P}\mathbf{P}^{-1}\mathbf{u}_{t-i}$$

$$= \boldsymbol{\mu} + \sum_{i=0}^{\infty} \boldsymbol{\Theta}_i \mathbf{P}^{-1}\mathbf{u}_{t-i}$$

$$= \boldsymbol{\mu} + \sum_{i=0}^{\infty} \boldsymbol{\Theta}_i \mathbf{w}_{t-i}$$

where $\boldsymbol{\Theta}_i = \boldsymbol{\Phi}_i \mathbf{P}$ and $\mathbf{w}_t = \mathbf{P}^{-1}\mathbf{u}_t$. If we had such a $\mathbf{P}$, the $\mathbf{w}_k$ would be mutually orthogonal, and no information would be lost in the holding-everything-else-constant assumption, implying that the $\boldsymbol{\Theta}_i$ would have the causal interpretation that we seek.

Choosing a $\mathbf{P}$ is very similar to placing identification restrictions on a system of dynamic simultaneous equations. The simple impulse–response functions do not identify the causal relationships that we wish to analyze. Thus we seek at least as many identification restrictions as necessary to identify the causal impulse–response functions.

So, where do we get such a $\mathbf{P}$? Sims (1980) popularized the method of choosing $\mathbf{P}$ to be the Cholesky decomposition of $\widehat{\boldsymbol{\Sigma}}$. The impulse–response functions based on this choice of $\mathbf{P}$ are known as the *orthogonalized impulse–response functions*. Choosing $\mathbf{P}$ to be the Cholesky decomposition of

$\widehat{\Sigma}$ is equivalent to imposing a recursive structure for the corresponding dynamic structural equation model. The ordering of the recursive structure is the same as the ordering imposed in the Cholesky decomposition. Since this choice is arbitrary, some researchers will look at the orthogonalized impulse–response functions with different orderings assumed in the Cholesky decomposition. The order() option available with irf create facilitates this type of analysis.

The SVAR approach integrates the need to identify the causal impulse–response functions into the model specification and estimation process. Sufficient identification restrictions can be obtained by placing either short-run or long-run restrictions on the model. The VAR in (1) can be rewritten as

$$\mathbf{y}_t - \mathbf{v} - \mathbf{A}_1\mathbf{y}_{t-1} - \cdots - \mathbf{A}_p\mathbf{y}_{t-p} = \mathbf{u}_t \tag{3}$$

Similarly, a short-run SVAR model can be written as

$$\mathbf{A}(\mathbf{y}_t - \mathbf{v} - \mathbf{A}_1\mathbf{y}_{t-1} - \cdots - \mathbf{A}_p\mathbf{y}_{t-p}) = \mathbf{A}\mathbf{u}_t = \mathbf{B}\mathbf{e}_t \tag{4}$$

where $\mathbf{A}$ and $\mathbf{B}$ are $K \times K$ nonsingular matrices of parameters to be estimated, $\mathbf{e}_t$ is a $K \times 1$ vector of disturbances with $\mathbf{e}_t \sim N(\mathbf{0}, \mathbf{I}_K)$, and $E[\mathbf{e}_t\mathbf{e}_s'] = \mathbf{0}_K$ for all $s \neq t$. Sufficient constraints must be placed on $\mathbf{A}$ and $\mathbf{B}$ so that $\mathbf{P}$ is identified. One way to see the connection is to draw out the implications of the latter equality in (4). From (4) it can be shown that

$$\Sigma = \mathbf{A}^{-1}\mathbf{B}(\mathbf{A}^{-1}\mathbf{B})'$$

As discussed in [TS] **var svar**, the estimates $\widehat{\mathbf{A}}$ and $\widehat{\mathbf{B}}$ are obtained by maximizing the concentrated log-likelihood function based on the $\widehat{\Sigma}$ obtained from the underlying VAR. The short-run SVAR approach chooses $\mathbf{P} = \widehat{\mathbf{A}}^{-1}\widehat{\mathbf{B}}$ to identify the causal impulse–response functions. The long-run SVAR approach works similarly, with $\mathbf{P} = \widehat{\mathbf{C}} = \widehat{\overline{\mathbf{A}}}^{-1}\widehat{\mathbf{B}}$, where $\widehat{\overline{\mathbf{A}}}^{-1}$ is the matrix of estimated long-run or accumulated effects of the reduced-form VAR shocks.

There is one important difference between long-run and short-run SVAR models. As discussed by Amisano and Gianinni (1997, chapter 6), in the short-run model the constraints are applied directly to the parameters in $\mathbf{A}$ and $\mathbf{B}$. Then $\mathbf{A}$ and $\mathbf{B}$ interact with the estimated parameters of the underlying VAR. In contrast, in a long-run model, the constraints are placed on functions of the estimated VAR parameters. While estimation and inference of the parameters in $\mathbf{C}$ is straightforward, obtaining the asymptotic standard errors of the structural impulse–response functions requires untenable assumptions. For this reason, irf create does not estimate the asymptotic standard errors of the structural impulse–response functions generated by long-run SVAR models. However, bootstrapped standard errors are still available.

An introduction to forecast-error variance decompositions for VARs

Another measure of the effect of the innovations in variable k on variable j is the forecast-error variance decomposition (FEVD). This method, which is also known as *innovation accounting*, measures the fraction of the error in forecasting variable j after h periods that is attributable to the orthogonalized innovations in variable k. Since the derivation of the FEVD requires orthogonalizing the $\mathbf{u}_t$ innovations, the FEVD is always predicated upon a choice of $\mathbf{P}$.

Lütkepohl (1993, section 2.2.2) shows that the h-step forecast error can be written as

$$\mathbf{y}_{t+h} - \widehat{\mathbf{y}}_t(h) = \sum_{i=0}^{h-1} \mathbf{\Phi}_i\mathbf{u}_{t+h-i} \tag{6}$$

where y_{t+h} is the value observed at time $t + h$ and $\widehat{y}_t(h)$ is the h-step-ahead predicted value for y_{t+h} that was made at time t.

Since the $\mathbf{u}_t$ are contemporaneously correlated, their distinct contributions to the forecast error cannot be ascertained. However, if we choose a $\mathbf{P}$ such that $\mathbf{\Sigma} = \mathbf{PP}'$, as above, we can orthogonalize the $\mathbf{u}_t$ into $\mathbf{w}_t = \mathbf{P}^{-1}\mathbf{u}_t$. We can then ascertain the relative contribution of the distinct elements of $\mathbf{w}_t$. Thus we can rewrite (6) as

$$\mathbf{y}_{t+h} - \widehat{\mathbf{y}}_t(h) = \sum_{i=0}^{h-1} \mathbf{\Phi}_i \mathbf{PP}^{-1} \mathbf{u}_{t+h-i}$$

$$= \sum_{i=0}^{h-1} \mathbf{\Theta}_i \mathbf{w}_{t+h-i} \qquad (7)$$

Since the forecast errors can be written in terms of the orthogonalized errors, the forecast-error variance can be written in terms of the orthogonalized error variances. Forecast-error variance decompositions measure the fraction of the total forecast-error variance that is attributable to each of the orthogonalized shocks.

❏ Technical Note

The details in this note are not critical to the discussion that follows. A forecast-error variance decomposition is derived for a given $\mathbf{P}$. Following Lütkepohl (1993, section 2.3.3), letting $\theta_{mn,i}$ be the m, nth element of $\mathbf{\Theta}_i$, we can express the h-step forecast error of the jth component of $\mathbf{y}_t$ as

$$\mathbf{y}_{j,t+h} - \widehat{\mathbf{y}}_j(h) = \sum_{i=0}^{h-1} \theta_{j1,1} \mathbf{w}_{1,t+h-i} + \cdots + \theta_{jK,i} \mathbf{w}_{K,t+h-i}$$

$$= \sum_{k=1}^{K} \theta_{jk,0} \mathbf{w}_{k,t+h} + \cdots + \theta_{jk,h-1} \mathbf{w}_{k,t+1}$$

The $\mathbf{w}_t$, which were constructed using $\mathbf{P}$, are mutually orthogonal with unit variance. This allows us to compute easily the mean squared error (MSE) of the forecast of variable j at horizon h in terms of the contributions of the components of $\mathbf{w}_t$. Specifically,

$$E[\{y_{j,t+h} - y_{j,t}(h)\}^2] = \sum_{k=1}^{K} (\theta_{jk,0}^2 + \cdots + \theta_{jk,h-1}^2)$$

The kth term in the sum above is interpreted as the contribution of the orthogonalized innovations in variable k to the h-step forecast error of variable j. Note that the kth element in the sum above can be rewritten as

$$(\theta_{jk,0}^2 + \cdots + \theta_{jk,h-1}^2) = \sum_{i=0}^{h-1} \left(\mathbf{e}_j' \mathbf{\Theta}_k \mathbf{e}_k\right)^2$$

where $\mathbf{e}_i$ is the ith column of $\mathbf{I}_K$. Normalizing by the forecast error for variable j at horizon h yields

$$\omega_{jk,h} = \frac{\sum_{i=0}^{h-1} \left(\mathbf{e}_j' \mathbf{\Theta}_k \mathbf{e}_k\right)^2}{\text{MSE}[y_{j,t}(h)]}$$

where $\text{MSE}[y_{j,t}(h)] = \sum_{i=0}^{h-1} \sum_{k=1}^{K} \theta_{jk,i}^2$.

❏

Since the FEVD depends on the choice of $\mathbf{P}$, there are different forecast-error variance decompositions associated with each distinct $\mathbf{P}$. irf create can estimate the FEVD for a VAR or an SVAR. For a VAR, $\mathbf{P}$ is the Cholesky decomposition of $\widehat{\boldsymbol{\Sigma}}$. For an SVAR, $\mathbf{P}$ is the estimated structural decomposition, $\mathbf{P} = \widehat{\mathbf{A}}^{-1}\widehat{\mathbf{B}}$ for short-run models and $\mathbf{P} = \widehat{\mathbf{C}}$ for long-run SVAR models. Due to the same complications that arose with the structural impulse–response functions, the asymptotic standard errors of the structural FEVD are not available after long-run SVAR models, but bootstrapped standard errors are still available.

IRF results for VECMs

An introduction to impulse–response functions for VECMs

As discussed in [TS] **vec intro**, the VECM is a reparameterization of the VAR that is especially useful for fitting VARs with cointegrating variables. This implies that the estimated parameters for the corresponding VAR model can be backed out from the estimated parameters of the VECM model. This relationship means we can use the VAR form of the cointegrating VECM to discuss the IRFs for VECMs.

Consider a cointegrating VAR with one lag with no constant or trend,

$$\mathbf{y}_t = \mathbf{A}\mathbf{y}_{t-1} + \mathbf{u}_t \tag{8}$$

where $\mathbf{y}_t$ is a $K \times 1$ vector of endogenous, first-difference stationary variables among which there are $1 \leq r < K$ cointegration equations; $\mathbf{A}$ is $K \times K$ matrix of parameters; and $\mathbf{u}_t$ is a $K \times 1$ vector of i.i.d. disturbances.

We developed intuition for the IRFs from a stationary VAR by rewriting the VAR as an infinite-order vector moving-average (VMA) process. While the Granger representation theorem establishes the existence of a VMA formulation of this model, since the cointegrating VAR is not stable, the inversion is not nearly so intuitive. (See Johansen [1995, chapters 3 and 4] for more details.) For this reason, we use (8) to develop intuition for the IRFs from a cointegrating VAR.

Suppose that K is 3, that $\mathbf{u}_1 = (1, 0, 0)$, and that we want to analyze the time paths of the variables in $\mathbf{y}$ conditional on the initial values $\mathbf{y}_0 = \mathbf{0}$, $\mathbf{A}$, and the condition that there are no more shocks to the system, i.e., $\mathbf{0} = \mathbf{u}_2 = \mathbf{u}_3 = \cdots$. These assumptions and (8) imply that

$$\mathbf{y}_1 = \mathbf{u}_1$$

$$\mathbf{y}_2 = \mathbf{A}\mathbf{y}_1 = \mathbf{A}\mathbf{u}_1$$

$$\mathbf{y}_3 = \mathbf{A}\mathbf{y}_2 = \mathbf{A}^2\mathbf{u}_1$$

and so on. The ith-row element of the first column of $\mathbf{A}^s$ contains the effect of the unit shock to the first variable after s periods. In other words, the first column of $\mathbf{A}^s$ contains the IRF of a unit impulse to the first variable after s periods. We could deduce the IRFs of a unit impulse to any of the other variables by administering the unit shock to one of them instead of to the first variable. Thus we can see that the (i, j)th element of $\mathbf{A}^s$ contains the unit impulse–response function from variable j to variable i after s periods. By starting with orthogonalized shocks of the form $\mathbf{P}^{-1}\mathbf{u}_t$, we can use the same logic to derive the OIRFs to be $\mathbf{A}^s\mathbf{P}$.

In the case of the stationary VAR, stability implies that all the eigenvalues of $\mathbf{A}$ have moduli strictly less than one, which in turn implies that all the elements of $\mathbf{A}^s \to \mathbf{0}$ as $s \to \infty$. This implies that all the IRFs from a stationary VAR taper off to zero as $s \to \infty$. In contrast, in a cointegrating VAR, some of the eigenvalues of $\mathbf{A}$ are 1, while the remaining eigenvalues have moduli strictly less than 1. This implies that in cointegrating VARs some of the elements of $\mathbf{A}^s$ are not going to zero as $s \to \infty$, which in turn implies that some of the IRFs and OIRFs are not going to zero as $s \to \infty$. The fact that the IRFs and OIRFs taper off to zero for stationary VARs but not for cointegrating VARs is one of the key differences between the two models.

When the IRF or OIRF from the innovation in one variable to another tapers off to zero as time goes on, the innovation to the first variable is said to have a transitory effect on the second variable. When the IRF or OIRF does not go to zero, the effect is said to be permanent.

Note that, since some of the IRFs and OIRFs do not taper off to zero, some of the cumulative IRFs and OIRFs diverge over time.

An introduction to forecast-error variance decompositions for VECMs

The results from *An introduction to impulse–response functions for VECMs* can be used to show that the interpretation of FEVDs for a finite number of steps in cointegrating VARs is essentially the same as in the stationary case. Because the MSE of the forecast is diverging, this interpretation is only valid for a finite number of steps. (See [TS] **vec intro** and [TS] **fcast compute** for more information on this point.)

Methods and Formulas

irf create is implemented as an ado-file.

Impulse–response function formulas for VARs

The previous discussion implies that there are three different choices of $\mathbf{P}$ that can be used to obtain distinct $\mathbf{\Theta}_i$. $\mathbf{P}$ is the Cholesky decomposition of $\mathbf{\Sigma}$ for the orthogonalized impulse–response functions. For the structural impulse–response functions, $\mathbf{P} = \mathbf{A}^{-1}\mathbf{B}$ for short-run models, and $\mathbf{P} = \mathbf{C}$ for long-run models. We will distinguish between the three by defining $\mathbf{\Theta}_i^o$ to be the orthogonalized impulse–response functions, $\mathbf{\Theta}_i^{\text{sr}}$ to be the short-run structural impulse–response functions, and $\mathbf{\Theta}_i^{lr}$ to be the long-run structural impulse–response functions.

We also define $\widehat{\mathbf{P}}_c$ to be the Cholesky decomposition of $\widehat{\mathbf{\Sigma}}$, $\widehat{\mathbf{P}}_{\text{sr}} = \widehat{\mathbf{A}}^{-1}\widehat{\mathbf{B}}$ to be the short-run structural decomposition, and $\widehat{\mathbf{P}}_{lr} = \widehat{\mathbf{C}}$ to be the long-run structural decomposition.

Given estimates of the $\widehat{\mathbf{A}}_i$ and $\widehat{\mathbf{\Sigma}}$ from var or svar, the estimates of the simple impulse–response functions and the orthogonalized impulse–response functions are, respectively,

$$\widehat{\mathbf{\Phi}}_i = \sum_{j=1}^{i} \widehat{\mathbf{\Phi}}_{i-j}\widehat{\mathbf{A}}_j$$

and

$$\widehat{\mathbf{\Theta}}_i^o = \widehat{\mathbf{\Phi}}_i\widehat{\mathbf{P}}_c$$

where $\widehat{\mathbf{A}}_j = \mathbf{0}_K$ for $j > p$.

Given the estimates $\widehat{\mathbf{A}}$ and $\widehat{\mathbf{B}}$, or $\widehat{\mathbf{C}}$, from svar, the estimates of the structural impulse–response functions are either

$$\widehat{\boldsymbol{\Theta}}_i^{\text{sr}} = \widehat{\boldsymbol{\Phi}}_i \widehat{\mathbf{P}}_{\text{sr}}$$

or

$$\widehat{\boldsymbol{\Theta}}_i^{lr} = \widehat{\boldsymbol{\Phi}}_i \widehat{\mathbf{P}}_{lr}$$

The estimated structural impulse–response functions stored in an IRF file with the variable name sirf may be from either a short-run model or a long-run model, depending on the estimation results used to create the IRFs. As discussed in [TS] **irf describe**, you can easily determine whether the structural impulse–response functions were generated from a short-run or a long-run SVAR model using irf describe.

Following Lütkepohl (1993, section 3.7), estimates of the cumulative impulse–response functions and the cumulative, orthogonalized impulse–response functions (COIRFs) at period n are, respectively,

$$\widehat{\boldsymbol{\Psi}}_n = \sum_{i=0}^{n} \widehat{\boldsymbol{\Phi}}_i$$

and

$$\widehat{\boldsymbol{\Xi}}_n = \sum_{i=0}^{n} \widehat{\boldsymbol{\Theta}}_i$$

The asymptotic standard errors of the different impulse–response functions are obtained by applications of the delta method. See Lütkepohl (1993, section 3.7) and Amisano and Giannini (1997, chapter 4) for the derivations. See Serfling (1980, section 3.3) for a discussion of the delta method. In presenting the variance–covariance matrix estimators, we make extensive use of the vec() operator, where vec($\mathbf{X}$) is the vector obtained by stacking the columns of $\mathbf{X}$.

Lütkepohl (1993, section 3.7) derives the asymptotic VCEs of vec($\boldsymbol{\Phi}_i$), vec($\boldsymbol{\Theta}_i^o$), vec($\widehat{\boldsymbol{\Psi}}_n$), and vec($\widehat{\boldsymbol{\Xi}}_n$). Since vec($\boldsymbol{\Phi}_i$) is $K^2 \times 1$, the asymptotic VCE of vec($\boldsymbol{\Phi}_i$) is $K^2 \times K^2$, and it is given by

$$\mathbf{G}_i \widehat{\boldsymbol{\Sigma}}_{\widehat{\alpha}} \mathbf{G}_i'$$

where

$$\mathbf{G}_i = \sum_{m=0}^{i-1} \mathbf{J}(\widehat{\mathbf{M}}')^{(i-1-m)} \otimes \widehat{\boldsymbol{\Phi}}_m \qquad \mathbf{G}_i \text{ is } K^2 \times K^2 p$$

$$\mathbf{J} = (\mathbf{I}_K, \mathbf{0}_K, \dots, \mathbf{0}_K) \qquad \mathbf{J} \text{ is } K \times Kp$$

$$\widehat{\mathbf{M}} = \begin{bmatrix} \widehat{\mathbf{A}}_1 & \widehat{\mathbf{A}}_2 & \dots & \widehat{\mathbf{A}}_{p-1} & \widehat{\mathbf{A}}_p \\ \mathbf{I}_K & \mathbf{0}_K & \dots & \mathbf{0}_K & \mathbf{0}_K \\ \mathbf{0}_K & \mathbf{I}_K & & \mathbf{0}_K & \mathbf{0}_K \\ \vdots & & \ddots & \vdots & \vdots \\ \mathbf{0}_K & \mathbf{0}_K & \dots & \mathbf{I}_K & \mathbf{0}_K \end{bmatrix} \qquad \widehat{\mathbf{M}} \text{ is } Kp \times Kp$$

The $\widehat{\mathbf{A}}_i$ are the estimates of the coefficients on the lagged variables in the VAR, and $\widehat{\boldsymbol{\Sigma}}_{\widehat{\alpha}}$ is the VCE matrix of $\widehat{\alpha} = \text{vec}(\widehat{\mathbf{A}}_1, \dots, \widehat{\mathbf{A}}_p)$. $\widehat{\boldsymbol{\Sigma}}_{\widehat{\alpha}}$ is a $K^2 p \times K^2 p$ matrix whose elements come from the VCE of the VAR coefficient estimator. As such, this VCE is the VCE of the constrained estimator if there are any constraints placed on the VAR coefficients.

The $K^2 \times K^2$ asymptotic VCE matrix for vec($\widehat{\boldsymbol{\Psi}}_n$) after n periods is given by

$$\mathbf{F}_n \widehat{\boldsymbol{\Sigma}}_{\widehat{\alpha}} \mathbf{F}_n'$$

where

$$\mathbf{F}_n = \sum_{i=1}^{n} \mathbf{G}_i$$

The $K^2 \times K^2$ asymptotic VCE matrix of the vectorized, orthogonalized, impulse–response functions at horizon i, $\text{vec}(\mathbf{\Theta}_i^o)$, is

$$\mathbf{C}_i \widehat{\mathbf{\Sigma}}_{\widehat{\alpha}} \mathbf{C}_i' + \overline{\mathbf{C}}_i \widehat{\mathbf{\Sigma}}_{\widehat{\sigma}} \overline{\mathbf{C}}_i'$$

where

$\mathbf{C}_0 = \mathbf{0}$		$\mathbf{C}_0$ is $K^2 \times K^2 p$
$\mathbf{C}_i = (\widehat{\mathbf{P}}_c' \otimes \mathbf{I}_K)\mathbf{G}_i, \quad i = 1, 2, \ldots$		$\mathbf{C}_i$ is $K^2 \times K^2 p$
$\overline{\mathbf{C}}_i = (\mathbf{I}_K \otimes \mathbf{\Phi}_i)\mathbf{H}, \quad i = 0, 1, \ldots$		$\overline{\mathbf{C}}_i$ is $K^2 \times K^2$
$\mathbf{H} = \mathbf{L}_K' \left\{ \mathbf{L}_K \mathbf{N}_K (\widehat{\mathbf{P}}_c \otimes \mathbf{I}_K) \mathbf{L}_K' \right\}^{-1}$		$\mathbf{H}$ is $K^2 \times K \frac{(K+1)}{2}$

$\mathbf{L}_K$ solves $\quad \text{vech}(\mathbf{F}) = \mathbf{L}_K \text{ vec}(\mathbf{F}) \qquad\qquad$ $\mathbf{L}_K$ is $K\frac{(K+1)}{2} \times K^2$

$$\text{for } \mathbf{F} \ K \times K \text{ and symmetric}$$

$\mathbf{K}_K$ solves $\quad \mathbf{K}_K \text{vec}(\mathbf{G}) = \text{vec}(\mathbf{G}')$ for any $K \times K$ matrix $\mathbf{G}$ $\qquad$ $\mathbf{K}_K$ is $K^2 \times K^2$

$$\mathbf{N}_K = \tfrac{1}{2}\left(\mathbf{I}_{K^2} + \mathbf{K}_K\right) \qquad\qquad \mathbf{N}_K \text{ is } K^2 \times K^2$$

$$\widehat{\mathbf{\Sigma}}_{\widehat{\sigma}} = 2\mathbf{D}_K^+ (\widehat{\mathbf{\Sigma}} \otimes \widehat{\mathbf{\Sigma}}) \mathbf{D}_K^+ \qquad\qquad \widehat{\mathbf{\Sigma}}_{\widehat{\sigma}} \text{ is } K\frac{(K+1)}{2} \times K\frac{(K+1)}{2}$$

$$\mathbf{D}_K^+ = (\mathbf{D}_K' \mathbf{D}_K)^{-1} \mathbf{D}_K' \qquad\qquad \mathbf{D}_K^+ \text{ is } K\frac{(K+1)}{2} \times K^2$$

$\mathbf{D}_K$ solves $\quad \mathbf{D}_K \text{vech}(\mathbf{F}) = \text{vec}(\mathbf{F}) \quad$ for $\mathbf{F} \ K \times K$ and symmetric $\qquad$ $\mathbf{D}_K$ is $K^2 \times K\frac{(K+1)}{2}$

$$\text{vech}(\mathbf{X}) = \begin{bmatrix} x_{11} \\ x_{21} \\ \vdots \\ x_{K1} \\ x_{22} \\ \vdots \\ x_{K2} \\ \vdots \\ x_{KK} \end{bmatrix} \qquad \text{for } \mathbf{X} \ K \times K \qquad\qquad \text{vech}(\mathbf{X}) \text{ is } K\frac{(K+1)}{2} \times 1$$

Note that $\widehat{\mathbf{\Sigma}}_{\widehat{\sigma}}$ is the VCE of $\text{vech}(\widehat{\mathbf{\Sigma}})$. More details about $\mathbf{L}_K$, $\mathbf{K}_K$, $\mathbf{D}_K$ and vech() are available in Lütkepohl (1993, appendix A.12). Finally, as Lütkepohl (1993, 101) discusses, $\mathbf{D}_K^+$ is the Moore–Penrose inverse of $\mathbf{D}_K$.

As discussed in Amisano and Giannini (1997, chapter 6), the asymptotic standard errors of the structural impulse–response functions are available for short-run SVAR models but not for long-run SVAR models. Following Amisano and Giannini (1997, chapter 5), the asymptotic $K^2 \times K^2$ VCE of the short-run structural impulse–response functions after i periods, when a maximum of h periods are estimated, is the i, i block of

$$\widehat{\mathbf{\Sigma}}(h)_{ij} = \widetilde{\mathbf{G}}_i \widehat{\mathbf{\Sigma}}_{\widehat{\alpha}} \widetilde{\mathbf{G}}_j' + \left\{ \mathbf{I}_K \otimes (\mathbf{J}\widehat{\mathbf{M}}^i \mathbf{J}') \right\} \mathbf{\Sigma}(0) \left\{ \mathbf{I}_K \otimes (\mathbf{J}\widehat{\mathbf{M}}^j \mathbf{J}') \right\}'$$

where

$$\widetilde{\mathbf{G}}_0 = \mathbf{0}_K$$

$\mathbf{G}_0$ is $K^2 \times K^2 p$

$$\widetilde{\mathbf{G}}_i = \sum_{k=0}^{i-1} \left\{ \widehat{\mathbf{P}}'_{\text{sr}} \mathbf{J} (\widehat{\mathbf{M}}')^{i-1-k} \otimes \left(\mathbf{J} \widehat{\mathbf{M}}^k \mathbf{J}' \right) \right\}$$

$\mathbf{G}_i$ is $K^2 \times K^2 p$

$$\widehat{\boldsymbol{\Sigma}}(0) = \mathbf{Q}_2 \widehat{\boldsymbol{\Sigma}}_W \mathbf{Q}'_2$$

$\widehat{\boldsymbol{\Sigma}}(0)$ is $K^2 \times K^2$

$$\widehat{\boldsymbol{\Sigma}}_W = \mathbf{Q}_1 \widehat{\boldsymbol{\Sigma}}_{AB} \mathbf{Q}'_1$$

$\widehat{\boldsymbol{\Sigma}}_W$ is $K^2 \times K^2$

$$\mathbf{Q}_2 = \widehat{\mathbf{P}}'_{\text{sr}} \otimes \widehat{\mathbf{P}}_{\text{sr}}$$

$\mathbf{Q}_2$ is $K^2 \times K^2$

$$\mathbf{Q}_1 = \left\{ (\mathbf{I}_K \otimes \widehat{\mathbf{B}}^{-1}), (-\widehat{\mathbf{P}}'^{-1}_{\text{sr}} \otimes \mathbf{B}^{-1}) \right\}$$

$\mathbf{Q}_1$ is $K^2 \times 2K^2$

and $\widehat{\boldsymbol{\Sigma}}_{AB}$ is the $2K^2 \times 2K^2$ VCE of the estimator of $\text{vec}(\mathbf{A}, \mathbf{B})$.

Forecast-error variance decomposition formulas for VARs

This section provides details of how `irf create` estimates the Cholesky forecast-error variance decompositions, the structural forecast-error variance decompositions, and their standard errors. Beginning with the Cholesky-based forecast-error decompositions, the fraction of the h-step-ahead forecast-error variance of variable j that is attributable to the Cholesky orthogonalized innovations in variable k can be estimated as

$$\widehat{\omega}_{jk,h} = \frac{\sum_{i=0}^{h-1} (\mathbf{e}'_j \widehat{\boldsymbol{\Theta}}_i \mathbf{e}_k)^2}{\widehat{\text{MSE}}_j(h)}$$

where $\text{MSE}_j(h)$ is the jth diagonal element of

$$\sum_{i=0}^{h-1} \widehat{\boldsymbol{\Phi}}_i \widehat{\boldsymbol{\Sigma}} \widehat{\boldsymbol{\Phi}}'_i$$

(See Lütkepohl [1993, 97] for a discussion of this result.) $\widehat{\omega}_{jk,h}$ and $\text{MSE}_j(h)$ are scalars. The square of the standard error of $\widehat{\omega}_{jk,h}$ is

$$\mathbf{d}_{jk,h} \widehat{\boldsymbol{\Sigma}}_\alpha \mathbf{d}'_{jk,h} + \overline{\mathbf{d}}_{jk,h} \widehat{\boldsymbol{\Sigma}}_\sigma \overline{\mathbf{d}}_{jk,h}$$

where

$$\mathbf{d}_{jk,h} = \frac{2}{\text{MSE}_j(h)^2} \sum_{i=0}^{h-1} \left\{ \text{MSE}_j(h) (\mathbf{e}'_j \widehat{\boldsymbol{\Phi}}_i \widehat{\mathbf{P}}_c \mathbf{e}_k) (\mathbf{e}'_k \widehat{\mathbf{P}}'_c \otimes \mathbf{e}'_j) \mathbf{G}_i \right.$$
$$\left. - (\mathbf{e}'_j \boldsymbol{\Phi}_i \widehat{\mathbf{P}}_c \mathbf{e}_k)^2 \sum_{m=0}^{h-1} (\mathbf{e}'_j \widehat{\boldsymbol{\Phi}}_m \widehat{\boldsymbol{\Sigma}} \otimes \mathbf{e}'_j) \mathbf{G}_m \right\}$$

$\mathbf{d}_{jk,h}$ is $1 \times K^2 p$

$$\overline{\mathbf{d}}_{jk,h} = \sum_{i=0}^{h-1} \left\{ \text{MSE}_j(h) (\mathbf{e}'_j \widehat{\boldsymbol{\Phi}}_i \mathbf{P}_c \mathbf{e}_k) (\mathbf{e}'_k \otimes \mathbf{e}'_j \widehat{\boldsymbol{\Phi}}_i) \mathbf{H} \right.$$
$$\left. - (\mathbf{e}'_j \widehat{\boldsymbol{\Phi}}_i \widehat{\mathbf{P}}_c \mathbf{e}_k)^2 \sum_{m=0}^{h-1} (\mathbf{e}'_j \widehat{\boldsymbol{\Phi}}_m \otimes \mathbf{e}_j \widehat{\boldsymbol{\Phi}}_m) \mathbf{D}_K \right\} \frac{1}{\text{MSE}_j(h)^2}$$

$\overline{\mathbf{d}}_{jk,h}$ is $1 \times K \frac{(K+1)}{2}$

$$\mathbf{G}_0 = \mathbf{0}$$

$\mathbf{G}_0$ is $K^2 \times K^2 p$

and $\mathbf{D}_K$ is the $K^2 \times K\{(K+1)/2\}$ duplication matrix defined previously.

For the structural forecast-error decompositions, we follow Amisano and Giannini (1997, section 5.2). They define the matrix of structural forecast-error decompositions at horizon s, when a maximum of h periods are estimated, as

$$\widehat{\mathbf{W}}_s = \widehat{\mathbf{F}}_s^{-1} \widehat{\widetilde{\mathbf{M}}}_s \qquad \text{for } s = 1, \ldots, h + 1$$

$$\widehat{\mathbf{F}}_s = \left(\sum_{i=0}^{s-1} \widehat{\boldsymbol{\Theta}}_i^{\mathrm{sr}} \widehat{\boldsymbol{\Theta}}_i^{\mathrm{sr}\prime} \right) \odot \mathbf{I}_K$$

$$\widehat{\widetilde{\mathbf{M}}}_s = \sum_{i=0}^{s-1} \widehat{\boldsymbol{\Theta}}_i^{\mathrm{sr}} \odot \widehat{\boldsymbol{\Theta}}_i^{\mathrm{sr}}$$

where $\odot$ is the Hadamard, or element-by-element, product.

The $K^2 \times K^2$ asymptotic VCE of $\mathrm{vec}(\widehat{\mathbf{W}}_s)$ is given by

$$\widetilde{\mathbf{Z}}_s \boldsymbol{\Sigma}(h) \widetilde{\mathbf{Z}}_s'$$

where $\widehat{\boldsymbol{\Sigma}}(h)$ is as derived previously, and

$$\widetilde{\mathbf{Z}}_s = \left\{ \frac{\partial \mathrm{vec}(\widehat{\mathbf{W}}_s)}{\partial \mathrm{vec}(\widehat{\boldsymbol{\Theta}}_0^{\mathrm{sr}})}, \frac{\partial \mathrm{vec}(\widehat{\mathbf{W}}_s)}{\partial \mathrm{vec}(\widehat{\boldsymbol{\Theta}}_1^{\mathrm{sr}})}, \cdots, \frac{\partial \mathrm{vec}(\widehat{\mathbf{W}}_s)}{\partial \mathrm{vec}(\widehat{\boldsymbol{\Theta}}_h^{\mathrm{sr}})} \right\}$$

$$\frac{\partial \mathrm{vec}(\widehat{\mathbf{W}}_s)}{\partial \mathrm{vec}(\widehat{\boldsymbol{\Theta}}_j^{\mathrm{sr}})} = 2 \left\{ (\mathbf{I}_K \otimes \widehat{\mathbf{F}}_s^{-1}) \widetilde{\mathbf{D}}(\widehat{\boldsymbol{\Theta}}_j^{\mathrm{sr}}) - (\widehat{\mathbf{W}}_s' \otimes \widehat{\mathbf{F}}_s^{-1}) \widetilde{\mathbf{D}}(\mathbf{I}_K) \mathbf{N}_K (\widehat{\boldsymbol{\Theta}}_j^{\mathrm{sr}} \otimes I_K) \right\}$$

If $\mathbf{X}$ is an $n \times n$ matrix, then $\widetilde{\mathbf{D}}(\mathbf{X})$ is the $n^2 \times n^2$ matrix with $\mathrm{vec}(\mathbf{X})$ on the diagonal and zeros in all the off-diagonal elements, and $\mathbf{N}_K$ is as defined previously.

Impulse–response function formulas for VECMs

We begin by providing the formulas for backing out the estimates of the $\mathbf{A}_i$ from the $\boldsymbol{\Gamma}_i$ estimated by vec. As discussed in [TS] **vec intro**, the VAR in (1) can be rewritten as a VECM:

$$\Delta y_t = v + \boldsymbol{\Pi} y_{t-1} + \boldsymbol{\Gamma}_1 \Delta y_{t-1} + \boldsymbol{\Gamma}_{p-1} \Delta y_{p-2} + \epsilon_t$$

vec estimates $\boldsymbol{\Pi}$ and the $\boldsymbol{\Gamma}_i$. Johansen (1995, 25) notes that

$$\boldsymbol{\Pi} = \sum_{i=1}^{p} \mathbf{A}_i - \mathbf{I}_K \tag{9}$$

where $\mathbf{I}_K$ is the K-dimensional identity matrix, and

$$\boldsymbol{\Gamma}_i = - \sum_{j=i+1}^{p} \mathbf{A}_j \tag{10}$$

Defining

$$\boldsymbol{\Gamma} = \mathbf{I}_K - \sum_{i=1}^{p-1} \boldsymbol{\Gamma}_i$$

and using (9) and (10) allow us to solve for the $\mathbf{A}_i$ as

$$\mathbf{A}_1 = \mathbf{\Pi} + \mathbf{\Gamma}_1 + \mathbf{I}_K$$

$$\mathbf{A}_i = \mathbf{\Gamma}_i - \mathbf{\Gamma}_{i-1} \quad \text{for } i = \{2, \ldots, p-1\}$$

and

$$\mathbf{A}_p = -\mathbf{\Gamma}_{p-1}$$

Using these formulas, we can back out estimates of $\mathbf{A}_i$ from the estimates of the $\mathbf{\Gamma}_i$ and $\mathbf{\Pi}$ produced by `vec`. Then we simply use the formulas for the IRFs and OIRFs presented in *Impulse–response function formulas for VARs*.

The running sums of the IRFs and OIRFs over the steps within each impulse–response pair are the cumulative IRFs and OIRFs.

Algorithms for bootstrapping the VAR IRF and FEVD standard errors

`irf create` offers two bootstrap algorithms for estimating the standard errors of the various impulse–response functions and forecast-error variance decompositions. Both `var` and `svar` contain estimators for the coefficients in a VAR that are conditional on the first p observations. The two bootstrap algorithms are also conditional on the first p observations.

Specifying the `bs` option calculates the standard errors by bootstrapping the residuals. For a bootstrap with R repetitions, this method uses the following algorithm:

1. Fit the model and save the estimated parameters.

2. Use the estimated coefficients to calculate the residuals.

3. Repeat steps 3a to 3c R times.

   3a. Draw a simple random sample of size T with replacement from the residuals. The random samples are drawn over the $K \times 1$ vectors of residuals. When the tth vector is drawn, all K residuals are selected. This preserves the contemporaneous correlations among the residuals.

   3b. Use the p initial observations, the sampled residuals, and the estimated coefficients to construct a new sample dataset.

   3c. Fit the model and calculate the different impulse–response functions and forecast-error variance decompositions.

   3d. Save these estimates as observation r in the bootstrapped dataset.

4. For each impulse–response function and forecast-error variance decomposition, the estimated standard deviation from the R bootstrapped estimates is the estimated standard error of that impulse–response function or forecast-error variance decomposition.

Specifying the `bsp` option estimates the standard errors by a multivariate normal parametric bootstrap. The algorithm for the multivariate normal parametric bootstrap is identical to the one above, with the exception that 3a is replaced by 3a(bsp):

   3a(bsp). Draw T pseudovariates from a multivariate normal distribution with covariance matrix $\widehat{\mathbf{\Sigma}}$.

References

Amisano, G. and C. Giannini. 1997. *Topics in Structural* VAR *Econometrics*. 2nd ed. Heidelberg: Springer.

Christiano, L. J., M. Eichenbaum, and C. L. Evans. 1999. Monetary Policy Shocks: What have we learned and to what end? In *Handbook of Macroeconomics*, vol. 1, ed. J. B. Taylor and M. Woodford. New York: Elsevier Science.

Hamilton, J. D. 1994. *Time Series Analysis*. Princeton: Princeton University Press.

Johansen, S. 1995. *Likelihood-Based Inference in Cointegrated Vector Auto-Regressive Models*. Oxford: Oxford University Press.

Lütkepohl, H. 1993. *Introduction to Multiple Time Series Analysis*. 2nd ed. New York: Springer.

Serfling, R. J. 1980. *Approximation Theorems of Mathematical Statistics*. New York: Wiley.

Sims, C. A. 1980. Macroeconomics and reality. *Econometrica* 48: 1–48.

Stock, J. H. and M. W. Watson. 2001. Vector autoregressions. *Journal of Economic Perspectives* 15: 101–115.

Also See

Complementary:	[TS] **var**, [TS] **var svar**, [TS] **varbasic**, [TS] **vec**; [TS] **irf add**, [TS] **irf cgraph**, [TS] **irf ctable**, [TS] **irf describe**, [TS] **irf drop**, [TS] **irf graph**, [TS] **irf ograph**, [TS] **irf rename**, [TS] **irf set**, [TS] **irf table**
Background:	[U] **11.4.3 Time-series varlists**, [TS] **irf**, [TS] **var intro**, [TS] **vec intro**

Title

> **irf ctable** — Combine tables of impulse–response functions and FEVDs

Syntax

irf <u>ct</u>able $(spec_1)$ $\big[(spec_2) \ \ldots \ \big[(spec_N) \big] \big]$ $\big[, \ options \big]$

where $(spec_k)$ is

   $(irfname \ impulsevar \ responsevar \ stat \ \big[, \ spec\_options \big])$

irfname is the name of a set of IRF results in the active IRF file, *impulsevar* is an endogenous variable name, *responsevar* is an endogenous variable name, and *stat* is one or more statistics from the list below:

stat	description
irf	impulse–response function
oirf	orthogonalized impulse–response function
cirf	cumulative impulse–response function
coirf	cumulative, orthogonalized impulse–response function
fevd	Cholesky forecast-error variance decomposition
sirf	structural impulse–response function
sfevd	structural forecast-error variance decomposition

options	description
set(*filename*)	make *filename* active
noci	do not report confidence intervals
<u>st</u>derror	include standard errors for each statistic
<u>ind</u>ividual	make an individual table for each combination
<u>ti</u>tle("*text*")	use *text* as overall table title
<u>st</u>ep(#)	set common maximum step
<u>l</u>evel(#)	set confidence level; default is level(95)

spec_options	description
noci	do not report confidence intervals
<u>st</u>derror	include standard errors for each statistic
<u>l</u>evel(#)	set confidence level; default is level(95)
[†]<u>it</u>itle("*text*")	use *text* as individual subtitle for specific table

[†] itititle("*text*") is not shown in the dialog box.

spec_options may be specified within a table specification, globally, or both. When specified in a table specification, the *spec_options* affect only the specification in which they are used. When supplied globally, the *spec_options* affect all table specifications. When specified in both places, options for the table specification take precedence.

Description

irf ctable makes a table or a combined table of IRF results. Each block within a pair of matching parentheses—each (*spec_k*)—specifies the information for a specific table. irf ctable combines these tables into one table, unless option individual is specified, in which case separate tables for each block are created.

irf ctable operates on the active IRF file; see [TS] **irf set**.

Options

set(*filename*) specifies the file to be made active; see [TS] **irf set**. If set() is not specified, the active file is used.

noci suppresses reporting of the confidence intervals for each statistic. noci is assumed when the model was fitted by vec because no confidence intervals were estimated.

stderror specifies that standard errors for each statistic also be included in the table.

individual places each block, or (*spec_k*), in its own table. By default, irf ctable combines all the blocks into one table.

title("*text*") specifies a title for the table or the set of tables.

step(#) specifies the maximum number of steps to use for all tables. By default, each table is constructed using all steps available.

level(#) specifies the default confidence level, as a percentage, for confidence intervals, when they are reported. The default is level(95) or as set by set level; see [U] **20.6 Specifying the width of confidence intervals**.

The following option is available with irf ctable but is not shown in the dialog box:

ititle("*text*") specifies an individual subtitle for a specific table. ititle() may be specified only when the individual option is also specified.

Remarks

If you have not read [TS] **irf**, please do so.

Also see [TS] **irf table** for a slightly easier to use, but less powerful, table command.

irf ctable creates a series of tables from IRF results. The information enclosed within each set of parentheses,

(*irfname impulsevar responsevar stat* [, *spec_options*])

forms a request for a specific table.

The first part—*irfname impulsevar responsevar*—identifies a set of impulse–response function estimates or a set of variance decomposition estimates. The next part—*stat*—specifies which statistics are to be included in the table. The last part—*spec_options*—includes options noci, level(), and stderror, and places (or suppresses) additional columns in the table.

Each specific table displays the requested statistics corresponding to the specified combination of *irfname*, *impulsevar*, and *responsevar* over the step horizon. By default, all the individual tables are combined into a single table. Also by default, all the steps, or periods, available are included in the table. You can use the step() option to impose a common maximum for all tables.

▷ Example 1

In the first example of [TS] **irf table**, we fitted a model using var and we saved the impulse–response functions for two different orderings. The commands we used were

```
. use http://www.stata-press.com/data/r9/lutkepohl
. var dlinvestment dlincome dlconsumption
. irf set results4
. irf create ordera, step(8)
. irf create orderb, order(dlincome dlinvestment dlconsumption) step(8)
```

We then formed the desired table by typing

```
. irf table oirf fevd, impulse(dlincome) response(dlconsumption) noci std
> title("Ordera versus orderb")
```

Using irf ctable, we can form the equivalent table by typing

```
. irf ctable (ordera dlincome dlconsumption oirf fevd)
>            (orderb dlincome dlconsumption oirf fevd),
>            noci std title("Ordera versus orderb")
```

Ordera versus orderb

step	(1) oirf	(1) S.E.	(1) fevd	(1) S.E.
0	.005123	.000878	0	0
1	.001635	.000984	.288494	.077483
2	.002948	.000993	.294288	.073722
3	−.000221	.000662	.322454	.075562
4	.000811	.000586	.319227	.074063
5	.000462	.000333	.322579	.075019
6	.000044	.000275	.323552	.075371
7	.000151	.000162	.323383	.075314
8	.000091	.000114	.323499	.075386

step	(2) oirf	(2) S.E.	(2) fevd	(2) S.E.
0	.005461	.000925	0	0
1	.001578	.000988	.327807	.08159
2	.003307	.001042	.328795	.077519
3	−.00019	.000676	.370775	.080604
4	.000846	.000617	.366896	.079019
5	.000491	.000349	.370399	.079941
6	.000069	.000292	.371487	.080323
7	.000158	.000172	.371315	.080287
8	.000096	.000122	.371438	.080366

(1) irfname = ordera, impulse = dlincome, and response = dlconsumption
(2) irfname = orderb, impulse = dlincome, and response = dlconsumption

The output is displayed in a single table. Since the table did not fit horizontally, it automatically wrapped. At the bottom of the table is a list of keys that appear at the top of each column. The results in the table above indicate that the orthogonalized impulse–response functions do not change by much. Since the estimated forecast-error variances do change, we might want to produce two tables that contain the estimated forecast-error variance decompositions and their 95% confidence intervals:

```
. irf ctable (ordera dlincome dlconsumption fevd)
>            (orderb dlincome dlconsumption fevd), individual
```

Table 1

step	(1) fevd	(1) Lower	(1) Upper
0	0	0	0
1	.288494	.13663	.440357
2	.294288	.149797	.43878
3	.322454	.174356	.470552
4	.319227	.174066	.464389
5	.322579	.175544	.469613
6	.323552	.175826	.471277
7	.323383	.17577	.470995
8	.323499	.175744	.471253

95% lower and upper bounds reported
(1) irfname = ordera, impulse = dlincome, and response = dlconsumption

Table 2

step	(2) fevd	(2) Lower	(2) Upper
0	0	0	0
1	.327807	.167893	.487721
2	.328795	.17686	.48073
3	.370775	.212794	.528757
4	.366896	.212022	.52177
5	.370399	.213718	.52708
6	.371487	.214058	.528917
7	.371315	.213956	.528674
8	.371438	.213923	.528953

95% lower and upper bounds reported
(2) irfname = orderb, impulse = dlincome, and response = dlconsumption

Because we specified the `individual` option, the output contains two tables, one for each specific table command. At the bottom of each table is a list of the keys used in that table and a note indicating the level of the confidence intervals that we requested. The results from table 1 and table 2 indicate that each estimated function is well within the confidence interval of the other, so we conclude that the functions are not significantly different.

◁

Saved Results

`irf ctable` saves in `r()`

Scalars
   r(ncols)   number of columns in all tables
   r(k_umax)   number of distinct keys
   r(k)   number of specific table commands
Macros
   r(key*i*)   *i*th key
   r(tnotes)   list of keys applied to each column

Methods and Formulas

`irf ctable` is implemented as an ado-file.

Also See

Complementary:	[TS] **var**, [TS] **var svar**, [TS] **varbasic**, [TS] **vec**; [TS] **irf add**, [TS] **irf create**, [TS] **irf describe**, [TS] **irf drop**, [TS] **irf rename**, [TS] **irf set**
Related:	[TS] **irf cgraph**, [TS] **irf graph**, [TS] **irf ograph**, [TS] **irf table**
Background:	[U] **11.4.3 Time-series varlists**, [TS] **irf**, [TS] **var intro**, [TS] **vec intro**

Title

> **irf describe** — Describe an IRF file

Syntax

irf <u>d</u>escribe [*irf_resultslist*] [, *options*]

options	description
set(*filename*)	make *filename* active
using(*irf_filename*)	describe *irf_filename* without making active
<u>d</u>etail	show additional details of IRF results
<u>v</u>ariables	show underlying structure of the IRF dataset

Description

irf describe describes the IRF results stored in an IRF file.

If set() or using() is not specified, the IRF results of the active IRF file are described.

Options

set(*filename*) specifies the IRF file to be described and set; see [TS] **irf set**. If *filename* is specified without an extension, .irf is assumed.

using(*irf_filename*) specifies the IRF file to be described. The active IRF file, if any, remains unchanged. If *irf_filename* is specified without an extension, .irf is assumed.

detail specifies that irf describe display detailed information about each set of IRF results. detail is implied when *irf_resultslist* is specified.

variables is a programmer's option; additionally displays the output produced by the describe command.

Remarks

If you have not read [TS] **irf**, please do so.

irf describe describes the contents of an IRF file:

```
. irf describe
no irf file active
r(111);

. irf set myfile
(file myfile.irf now active)
```

```
. irf describe
Contains irf results from myfile.irf (dated 10 Jun 2004 13:20)
      irfname | model     endogenous variables and order (*)
--------------+------------------------------------------------------
      wiggins | vec       income investment consumption
        riley | sr svar   dlinvestment dlincome dlconsumption
          poi | lr svar   dlinvestment dlincome dlconsumption
     drukker1 | var       dlinvestment dlincome dlconsumption
     drukker2 | var       dlincome dlinvestment dlconsumption
--------------+------------------------------------------------------
```

(*) order is relevant only when model is var

The output reveals that the `wiggins` model was fitted by `vec`; the `riley` model by a short-run `svar`; the `poi` model by a long-run `svar`; and the `drukker1` and `drukker2` models were fitted by `var` but with their endogenous variables in different orders (and perhaps other differences as well).

We can obtain detailed information about any of the models by specifying the `detail` option. If we specify one or more *irfnames* following `irf describe`, the `detail` option is assumed:

```
. irf describe drukker1
```

irf results for drukker1

```
  Estimation specification
        model:  var
        endog:  dlinvestment dlincome dlconsumption
       sample:  quarterly data from 1960q4 to 1982q4
         lags:  1 2
     constant:  constant
         exog:  none
       varcns:  unconstrained

  IRF specification
         step:  15
        order:  dlinvestment dlincome dlconsumption
    std error:  asymptotic
         reps:  none
```

(Continued on next page)

Saved Results

irf describe saves in r():

Scalars

r(N)	number of observations in the IRF file
r(k)	number of variables in the IRF file
r(width)	width of dataset in the IRF file
r(N_max)	maximum number of observations
r(k_max)	maximum number of variables
r(widthmax)	maximum width of the dataset
r(changed)	flag indicating that data have changed since last saved

Macros

r(version)	version of IRF results file
r(irfnames)	names of IRF results in the IRF file
r(*irfname*_model)	var, sr var, lr var, or vec
r(*irfname*_order)	Cholesky order assumed in IRF estimates
r(*irfname*_exog)	exogenous variables in VAR or underlying VAR
r(*irfname*_constant)	constant or noconstant
r(*irfname*_lags)	lags in model
r(*irfname*_tmin)	minimum value of timevar in the estimation sample
r(*irfname*_tmax)	maximum value of timevar in the estimation sample
r(*irfname*_timevar)	name of tsset timevar
r(*irfname*_tsfmt)	format of timevar in the estimation sample
r(*irfname*_varcns)	unconstrained or colon-separated list of constraints placed on VAR coefficients
r(*irfname*_svarcns)	"." or colon-separated list of constraints placed on SVAR coefficients
r(*irfname*_step)	maximum step in IRF estimates
r(*irfname*_stderror)	asymptotic, bs, bsp, or none, depending on type of standard errors specified to irf create
r(*irfname*_reps)	"." or number of bootstrap replications performed
r(*irfname*_version)	version of IRF file that originally held *irfname* IRF results
r(*irfname*_rank)	"." or number of cointegrating equations
r(*irfname*_trend)	"." or trend() specified in vec
r(*irfname*_veccns)	"." or constraints placed on VECM parameters
r(*irfname*_sind)	"." or normalized seasonal indicators included in vec

Methods and Formulas

irf describe is implemented as an ado-file.

Also See

Complementary:	[TS] **var**, [TS] **var svar**, [TS] **varbasic**, [TS] **vec**; [TS] **irf add**, [TS] **irf cgraph**, [TS] **irf create**, [TS] **irf ctable**, [TS] **irf drop**, [TS] **irf graph**, [TS] **irf ograph**, [TS] **irf rename**, [TS] **irf set**, [TS] **irf table**
Background:	[U] **11.4.3 Time-series varlists**, [TS] **irf**, [TS] **var intro**, [TS] **vec intro**

Title

> **irf drop** — Drop IRF results from the active IRF file

Syntax

irf drop *irf_resultslist* [, set(*filename*)]

Description

irf drop removes IRF results from the active IRF file.

Option

set(*filename*) specifies the file to be made active; see [TS] **irf set**. If set() is not specified, the active file is used.

Remarks

If you have not read [TS] **irf**, please do so.

irf drop drops sets of IRF results from the active IRF file:

```
. irf set myfile
(file myfile.irf now active)
. irf describe
Contains irf results from myfile.irf (dated 10 Jun 2004 13:34)
    irfname |  model      endogenous variables and order (*)
------------+------------------------------------------------------
    wiggins |  vec        income investment consumption
      riley |  sr svar    dlinvestment dlincome dlconsumption
        poi |  lr svar    dlinvestment dlincome dlconsumption
   drukker1 |  var        dlinvestment dlincome dlconsumption
   drukker2 |  var        dlincome dlinvestment dlconsumption

    (*) order is relevant only when model is var
. irf drop drukker2 wiggins
(drukker2 dropped)
(wiggins dropped)
file myfile.irf updated
. irf describe
Contains irf results from myfile.irf (dated 10 Jun 2004 13:34)
    irfname |  model      endogenous variables and order (*)
------------+------------------------------------------------------
      riley |  sr svar    dlinvestment dlincome dlconsumption
        poi |  lr svar    dlinvestment dlincome dlconsumption
   drukker1 |  var        dlinvestment dlincome dlconsumption

    (*) order is relevant only when model is var
```

Methods and Formulas

`irf drop` is implemented as an ado-file.

Also See

Complementary: [TS] **var**, [TS] **var svar**, [TS] **varbasic**, [TS] **vec**; [TS] **irf add**,
[TS] **irf cgraph**, [TS] **irf create**, [TS] **irf ctable**, [TS] **irf describe**,
[TS] **irf graph**, [TS] **irf ograph**, [TS] **irf rename**, [TS] **irf set**,
[TS] **irf table**

Background: [U] **11.4.3 Time-series varlists**,
[TS] **irf**, [TS] **var intro**, [TS] **vec intro**

Title

> **irf graph** — Graph impulse–response functions and FEVDs

Syntax

irf $\underline{g}$raph *stat* [, *options*]

stat	description
irf	impulse–response function
oirf	orthogonalized impulse–response function
cirf	cumulative impulse–response function
coirf	cumulative, orthogonalized impulse–response function
fevd	Cholesky forecast-error variance decomposition
sirf	structural impulse–response function
sfevd	structural forecast-error variance decomposition

Notes: (1) No statistic may appear more than once.
     (2) If confidence intervals are included (the default), only two statistics may be included.
     (3) If confidence intervals are suppressed (option noci), up to four statistics may be included.

options	description
Main	
set(*filename*)	make *filename* active
$\underline{\text{irf}}$(*irfnames*)	use *irfnames* IRF result sets
$\underline{\text{impulse}}$(*endogvars*)	use *endogvars* as impulse variables
$\underline{\text{response}}$(*endogvars*)	use *endogvars* as response variables
Options	
noci	suppress confidence bands
$\underline{\text{level}}$(#)	set confidence level; default is level(95)
$\underline{\text{lstep}}$(#)	use # for first step
$\underline{\text{ustep}}$(#)	use # for maximum step
$\underline{\text{ind}}$ividual	graph each combination individually
iname(*namestub* [, replace])	stub for naming the individual graphs
$\underline{\text{isaving}}$(*filenamestub* [, replace])	stub for saving the individual graphs to files
Plot 1, Plot 2, Plot 3, Plot 4	
plot#opts(*cline_options*)	affect rendition of the line plotting the # *stat*
CI 1, CI 2	
$\underline{\text{ci}}$#opts(*area_options*)	affect rendition of the confidence interval for the # *stat*

Y-Axis, X-Axis, Title, Caption, Legend, Overall

twoway_options	any options other than by() documented in [G] *twoway_options*

By

byopts(*by_option*)	how subgraphs are combined, labeled, etc.

Description

irf graph graphs impulse–response functions and forecast-error variance decompositions over time.

Options

_____ Main _____

set(*filename*) specifies the file to be made active; see [TS] **irf set**. If set() is not specified, the active file is used.

irf(*irfnames*) specifies the IRF result sets to be used. If irf() is not specified, each of the results in the active IRF file is used. (Files often contain just one set of IRF results stored under a single *irfname*; in that case, those results are used.)

impulse(*endogvars*) and response(*endogvars*) specify the impulse and response variables. Usually one of each is specified, and one graph is drawn. If multiple variables are specified, a separate subgraph is drawn for each impulse–response combination. If impulse() and response() are not specified, subgraphs are drawn for all combination of impulse and response variables.

_____ Options _____

noci suppresses graphing the confidence interval for each statistic. noci is assumed when the model was fitted by vec because no confidence intervals were estimated.

level(*#*) specifies the default confidence level, as a percentage, for confidence intervals, when they are reported. The default is level(95) or as set by set level; see [U] **20.6 Specifying the width of confidence intervals**. Also see [TS] **irf cgraph** for a graph command that allows the confidence level to vary over the graphs.

lstep(*#*) specifies the first step, or period, to be included in the graphs. lstep(0) is the default.

ustep(*#*), *# ≥ 1*, specifies the maximum step, or period, to be included in the graphs.

individual specifies that each graph be displayed individually. By default, irf graph combines the subgraphs into a single image. When individual is specified, byopts() may not be specified, but options isaving() and iname() may be specified.

iname(*namestub* [, replace]) specifies that the *i*th individual graph be saved in memory under the name *namestubi*, which must be a valid Stata name of 24 characters or fewer. iname() may only be specified with the individual option.

isaving(*filenamestub* [, replace]) specifies that the *i*th individual graph should be saved to disk in the current working directory under the name *filenamestubi*.gph. isaving() may only be specified when option individual is also specified.

> **Plot 1, Plot 2, Plot 3, Plot 4**

plot1opts(*cline_options*), ..., plot4opts(*cline_options*) affect the rendition of the plotted statistics (the *stat*). plot1opts() affects the rendition of the first statistic; plot2opts(), the second; and so on. *cline_options* are as described in [G] *cline_options*.

> **CI 1, CI 2**

ci1opts(*area_options*) and ci2opts(*area_options*) affect the rendition of the confidence intervals for the first (ci1opts()) and second (ci2opts()) statistics in *stat*.

> **Y-Axis, X-Axis, Title, Caption, Legend, Overall**

twoway_options are any of the options documented in [G] *twoway_options*, excluding by(). These include options for titling the graph (see [G] *title_options*) and options for saving the graph to disk (see [G] *saving_option*). Note that the saving() and name() options may not be combined with the individual option.

> **By**

byopts(*by_option*) is as documented in [G] *by_option* and may not be specified when individual is also specified. byopts() affects how the subgraphs are combined, labeled, etc.

Remarks

If you have not read [TS] **irf**, please do so.

Also see [TS] **irf cgraph**, which produces combined graphs; [TS] **irf ograph**, which produces overlayed graphs; and [TS] **irf table**, which displays results in tabular form.

irf graph produces one or more graphs and displays them arrayed into a single image unless option individual is specified, in which case the individual graphs are displayed separately. Each individual graph consists of all the specified *stat* and represents a single impulse–response combination.

Since all the specified *stat* appear on the same graph, putting together statistics with very different scales is not recommended. For instance, sometimes sirf and oirf are on similar scales while irf is on a different scale. In such cases, combining sirf and oirf on the same graph looks fine, but combining either with irf produces an uninformative graph.

▷ Example 1

Suppose that we have results generated from two different SVAR models. We want to know whether the shapes of the structural impulse–response functions and the structural forecast-error variance decompositions are similar in the two models. We are also interested in knowing whether the structural impulse–response functions and the structural forecast-error variance decompositions differ significantly from their Cholesky counterparts.

Filling in the background, we have previously issued the commands

```
. use http://www.stata-press.com/data/r9/lutkepohl
. mat a = (., 0, 0\0,.,0\.,.,.)
. mat b = I(3)
. svar dlinvestment dlincome dlconsumption, aeq(a) beq(b)
. irf create modela, set(results3) step(8)
. svar dlincome dlinvestment dlconsumption, aeq(a) beq(b)
. irf create modelb, step(8)
```

To see whether the shapes of the structural impulse–response functions and the structural forecast-error variance decompositions are similar in the two models, we type

```
. irf graph oirf sirf, impulse(dlincome) response(dlconsumption)
```

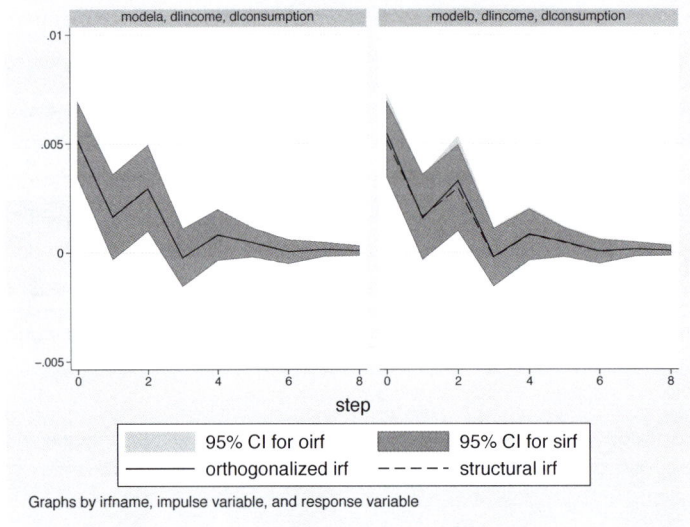

Graphs by irfname, impulse variable, and response variable

The graph reveals that the `oirf` and the `sirf` estimates are essentially the same for both models and that the shapes of the functions are very similar for the two models.

To see whether the structural impulse–response functions and the structural forecast-error variance decompositions differ significantly from their Cholesky counterparts, we type

```
. irf graph fevd sfevd, impulse(dlincome) response(dlconsumption) lstep(1)
> legend(cols(1))
```

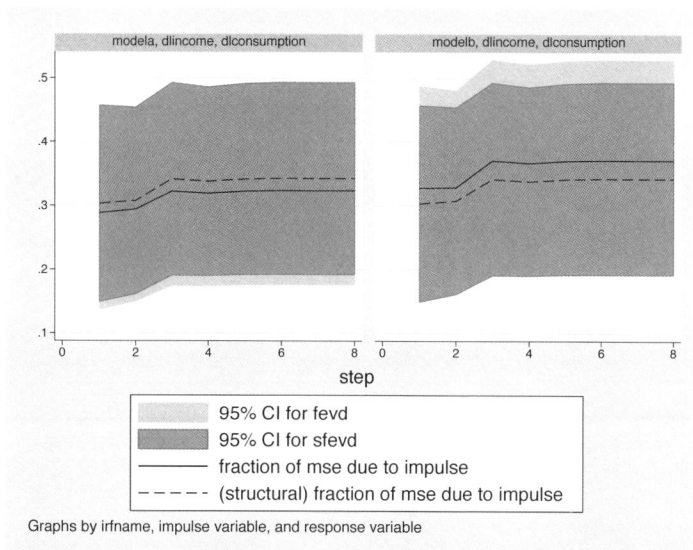

This combined graph reveals that the shapes of these functions are also similar for the two models. However, the graph illuminates one minor difference between them: In modela, the estimated structural forecast-error variance is slightly larger than the Cholesky-based estimates, whereas in modelb the Cholesky-based estimates are slightly larger than the structural estimates. For both models, however, the structural estimates are close to the center of the wide confidence intervals for the two estimates.

◁

▷ Example 2

Let us focus on the results from modela. Suppose that we were interested in examining how dlconsumption responded to impulses in its own structural innovations, structural innovations to dlincome, and structural innovations to dlinvestment. We type

(Continued on next page)

. irf graph sirf, irf(modela) response(dlconsumption)

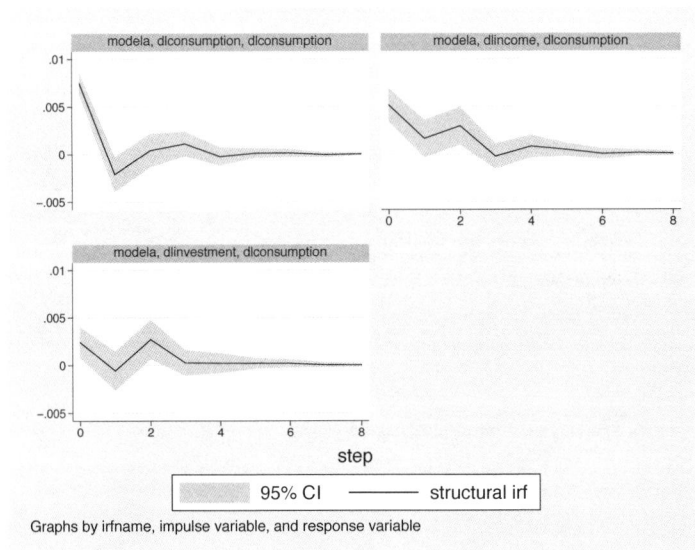

Graphs by irfname, impulse variable, and response variable

The upper left graph shows the structural impulse–response function of an innovation in dl-consumption on dlconsumption. It indicates that the identification restrictions used in modela imply that a positive shock to dlconsumption causes an increase in dlconsumption, followed by a decrease, followed by an increase, and so on, until the effect dies out after roughly 5 periods.

The upper right graph shows the structural impulse–response function of an innovation in dlin-come on dlconsumption, indicating that a positive shock to dlincome causes an increase in dlconsumption, which dies out after 4 or 5 periods.

◁

❏ Technical Note

[TS] **irf table** contains a technical note warning you to be careful in naming variables when you fit models. What is said there applies equally here.

❏

Saved Results

irf graph saves in r():

Scalars
   r(k)   number of graphs
Macros

r(stats)	*statlist*	r(byopts)	contents of byopts()
r(irfname)	*resultslist*	r(saving)	supplied saving() option
r(impulse)	*impulselist*	r(name)	supplied name() option
r(response)	*responselist*	r(individual)	individual or blank
r(plot*i*)	contents of plot*i*()	r(isaving)	contents of saving()
r(ci)	level applied to confidence	r(iname)	contents of name()
	intervals or noci	r(subtitle*j*)	subtitle for individual graph *j*
r(ciopts*i*)	contents of ciopts*i*()		

Methods and Formulas

irf graph is implemented as an ado-file.

Also See

Complementary:	[TS] **var**, [TS] **var svar**, [TS] **varbasic**, [TS] **vec**; [TS] **irf add**, [TS] **irf create**, [TS] **irf describe**, [TS] **irf drop**, [TS] **irf rename**, [TS] **irf set**
Related:	[TS] **irf cgraph**, [TS] **irf ctable**, [TS] **irf ograph**, [TS] **irf table**
Background:	[U] **11.4.3 Time-series varlists**, [TS] **irf**, [TS] **var intro**, [TS] **vec intro**

Title

> **irf ograph** — Graph overlaid impulse–response functions and FEVDs

Syntax

irf <u>og</u>raph $(spec_1)$ $\big[\,(spec_2)\,\ldots\big[\,(spec_{15})\,\big]\big]$ $\big[\,,\ options\,\big]$

where $(spec_k)$ is

($irfname$ $impulsevar$ $responsevar$ $stat$ $\big[\,,\ spec\_options\,\big]$)

irfname may be specified as the name of a set of IRF results or as ".", which means the first named result in the active `irf` file; *impulsevar* is an endogenous variable name, *response* is an endogenous variable name, and *stat* is one of the following:

stat	description
irf	impulse–response function
oirf	orthogonalized impulse–response function
cirf	cumulative impulse–response function
coirf	cumulative, orthogonalized impulse–response function
fevd	Cholesky forecast-error variance decomposition
sirf	structural impulse–response function
sfevd	structural forecast-error variance decomposition

options	description
Main	
set(*filename*)	make *filename* active
common_options	level, steps, and CIs
Y-Axis, X-Axis, Title, Caption, Legend, Overall	
twoway_options	any options other than by() documented in [G] ***twoway_options***

spec_options	description
Main	
common_options	level, steps, and CIs
Plot 1, Plot 2, Plot 3	
cline_options	affect rendition of the plotted lines
CI 1, CI 2, CI 3	
<u>ci</u>opts(*area_options*)	affect rendition of the confidence intervals

common_options	description
Main	
ci	add confidence bands to the graph
<u>level</u>(#)	set confidence level; default is level(95)
<u>lstep</u>(#)	use # for first step
<u>ust</u>ep(#)	use # for maximum step

spec_options may be specified within a plot specification, globally, or in both. When specified in a plot specification, the *spec_options* affect only the specification in which they are used. When supplied globally, the *spec_options* affect all plot specifications. When supplied in both places, options in the plot specification take precedence.

Description

irf ograph displays plots of irf results on a single graph (single pair of axes).

To become familiar with this command, we recommend that you type db irf ograph.

Options

⌐ Main ⌐

set(*filename*) specifies the file to be made active; see [TS] **irf set**. If set() is not specified, the active file is used.

ci adds confidence bands to the graph. Option noci may be used within a plot specification to suppress its confidence bands when the ci option is supplied globally.

level(#) specifies the confidence level, as a percentage, for confidence bands; see [U] **20.6 Specifying the width of confidence intervals**.

lstep(#) specifies the first step, or period, to be included in the graph. lstep(0) is the default.

ustep(#), $# \geq 1$, specifies the maximum step, or period, to be included.

⌐ Plot 1, Plot 2, Plot 3 ⌐

cline_options affect the rendition of the plotted lines; see [G] *cline_options*.

⌐ CI 1, CI 2, CI 3 ⌐

ciopts(*area_options*) affect the rendition of the confidence bands for the plotted statistic; see [G] *area_options*. ciopts() implies ci.

⌐ Y-Axis, X-Axis, Title, Caption, Legend, Overall ⌐

twoway_options are any of the options documented in [G] *twoway_options*, excluding by(). These include options for titling the graph (see [G] *title_options*) and options for saving the graph to disk (see [G] *saving_option*).

Remarks

If you have not read [TS] **irf**, please do so.

`irf ograph` overlays plots of impulse–response functions and forecast-error variance decompositions on a single graph.

▷ Example 1

We have previously issued the commands

```
. use http://www.stata-press.com/data/r9/lutkepohl
. var dlinvestment dlincome dlconsumption if qtr <= q(1978q4), lags(1/2) dfk
. irf create order1, step(10) set(myirf1, new)
. irf create order2, step(10) order(dlincome dlinvestment dlconsumption)
```

We now wish to compare the `oirf` for impulse `dlincome` and response `dlinvestment` for two different Cholesky orderings:

```
. irf ograph (order1 dlincome dlconsumption oirf)
>             (order2 dlincome dlconsumption oirf)
```

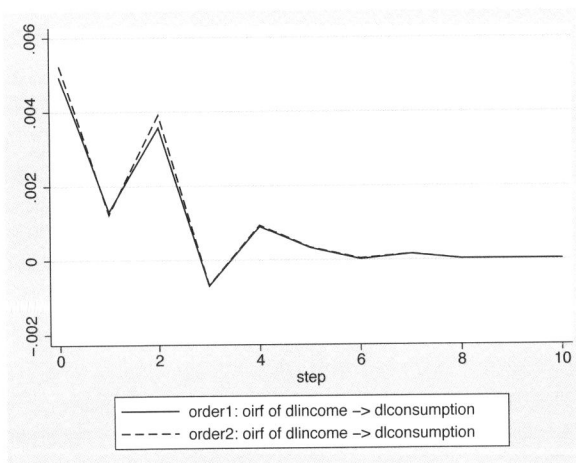

◁

❑ Technical Note

Graph options allow you to change the appearance of each plot. The following graph contains the plots of the forecast-error variance decompositions (FEVDs) for impulse `dlincome` and each response using the results from the first collection of results in the active IRF file (using the "." shortcut). In the second plot, we supply the `clpat(dash)` option (an abbreviation for `clpattern(dash)`) to give the line a dashed pattern. In the third plot, we supply the `m(o) clpat(dashdot) recast(connected)` options to get small circles connected by a line with a dash–dot pattern; the `cilines` option plots the confidence bands using lines instead of areas. We use the `title()` option to add a descriptive title to the graph and supply the `ci` option globally to add confidence bands to all the plots.

```
. irf ograph (. dlincome dlincome      fevd)
>            (. dlincome dlconsumption fevd, clpat(dash))
>            (. dlincome dlinvestment  fevd, cilines m(o) clpat(dashdot)
>                                      recast(connected))
>      , ci title("Comparison of forecast-error variance decomposition")
```

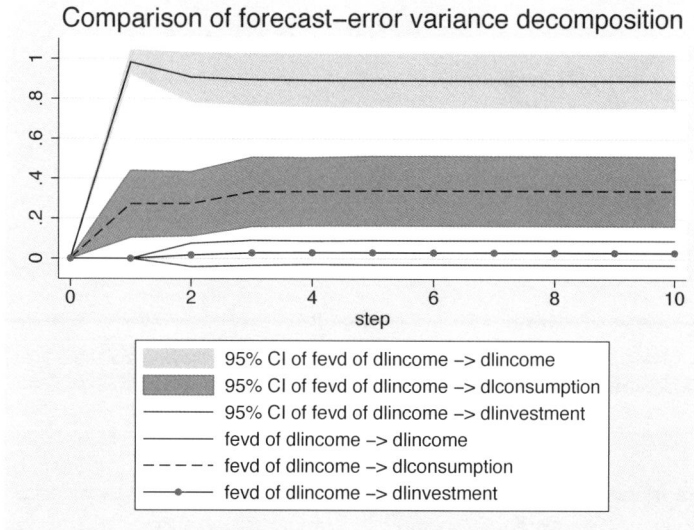

The clpattern() option is described in [G] **connect_options**, msymbol() is described in [G] **marker_options**, title() is described in [G] **title_options**, and recast() is described in [G] **advanced_options**.

Saved Results

irf ograph saves in r():

Scalars
  r(plots)   number of plot specifications
  r(ciplots)   number of plotted confidence bands
Macros
  r(irfname$k$)   *irfname* from ($spec_k$)
  r(impulse$k$)   impulse from ($spec_k$)
  r(response$k$) response from ($spec_k$)
  r(stat$k$)   stat from ($spec_k$)
  r(ci$k$)   level from ($spec_k$) or noci

Methods and Formulas

irf ograph is implemented as an ado-file.

Also See

Complementary:	[TS] **var**, [TS] **var svar**, [TS] **varbasic**, [TS] **vec**; [TS] **irf add**, [TS] **irf create**, [TS] **irf describe**, [TS] **irf drop**, [TS] **irf rename**, [TS] **irf set**
Related:	[TS] **irf cgraph**, [TS] **irf ctable**, [TS] **irf graph**, [TS] **irf table**
Background:	[U] **11.4.3 Time-series varlists**, [TS] **irf**, [TS] **var intro**, [TS] **vec intro**

Title

irf rename — Rename an IRF result in an IRF file

Syntax

irf <u>ren</u>ame *oldname* *newname* $\left[\,, \text{set}(filename)\,\right]$

Description

irf rename changes the name of a set of IRF results stored in the active IRF file.

Option

set(*filename*) specifies the file to be made active; see [TS] **irf set**. If set() is not specified, the active file is used.

Remarks

If you have not read [TS] **irf**, please do so.

irf rename changes the name associated with a set of IRF results:

```
. irf describe, set(myfile)
(file myfile.irf now active)
Contains irf results from myfile.irf (dated 10 Jun 2004 13:39)
       irfname | model     endogenous variables and order (*)
-----------------------------------------------------------------------------
       wiggins | vec       income investment consumption
         riley | sr svar   dlinvestment dlincome dlconsumption
           poi | lr svar   dlinvestment dlincome dlconsumption
      drukker1 | var       dlinvestment dlincome dlconsumption
      drukker2 | var       dlincome dlinvestment dlconsumption
-----------------------------------------------------------------------------
  (*) order is relevant only when model is var
. irf rename drukker2 preferred
(144 real changes made)
drukker2 renamed to preferred
. irf describe
Contains irf results from myfile.irf (dated 10 Jun 2004 13:39)
       irfname | model     endogenous variables and order (*)
-----------------------------------------------------------------------------
       wiggins | vec       income investment consumption
         riley | sr svar   dlinvestment dlincome dlconsumption
           poi | lr svar   dlinvestment dlincome dlconsumption
      drukker1 | var       dlinvestment dlincome dlconsumption
      preferred| var       dlincome dlinvestment dlconsumption
-----------------------------------------------------------------------------
  (*) order is relevant only when model is var
```

Saved Results

irf rename saves in r():

Macros
        r(irfnames)   irfnames after rename
        r(oldnew)   *oldname newname*

Methods and Formulas

irf rename is implemented as an ado-file.

Also See

Complementary: [TS] **var**, [TS] **var svar**, [TS] **varbasic**, [TS] **vec**; [TS] **irf add**,
                      [TS] **irf cgraph**, [TS] **irf create**, [TS] **irf ctable**, [TS] **irf describe**,
                      [TS] **irf drop**, [TS] **irf graph**, [TS] **irf ograph**, [TS] **irf set**, [TS] **irf table**

Background: [U] **11.4.3 Time-series varlists**,
                      [TS] **irf**, [TS] **var intro**, [TS] **vec intro**

Title

irf set — Set the active IRF file

Syntax

Report identity of active file

    irf set

Set, and if necessary create, active file

    irf set *irf_filename*

Create, and if necessary replace, active file

    irf set *irf_filename*, replace

Clear any active IRF file

    irf set, clear

If *irf_filename* is specified without an extension, .irf is assumed.

Description

In the first syntax, irf set reports the identity of the active file, if there is one. Also see [TS] **irf describe** for obtaining reports on the contents of an IRF file.

In the second syntax, irf set *irf_filename* specifies that the file be set as the active file and, if the file does not exist, that it be created, as well.

In the third syntax, irf set *irf_filename*, replace specifies that even if file *irf_filename* exists, a new, empty file is to be created and set.

In the rarely used fourth syntax, irf set, clear specifies that, if any IRF file is set, it be unset and that there be no active IRF file.

IRF files are just files: they can be erased by erase, listed by dir, and copied by copy; see [D] **erase**, [D] **dir**, and [D] **copy**.

Options

replace specifies that if *irf_filename* already exists, the file is to be erased and a new, empty IRF file is to be created in its place. If it does not already exist, a new, empty file is created.

clear unsets the active IRF file.

163

Remarks

If you have not read [TS] **irf**, please do so.

`irf set` reports the identity of the active IRF file:

```
. irf set
no irf file active
```

`irf set` *irf_filename* creates and sets an IRF file:

```
. irf set results1
(file results1.irf now active)
```

Note that we specified the name `results1`, and `results1.irf` became the active file. The suffix `.irf` was added for us.

`irf set` *irf_filename* can also be used to create a new file:

```
. use http://www.stata-press.com/data/r9/lutkepohl
(Quarterly SA West German macro data, Bil DM, from Lutkepohl 1993 Table E.1)
. var dlincome dlconsumption, exog(l.dlinvestment)
  (output omitted )
. irf set results2
(file results2.irf created)
(file results2.irf now active)
. irf create order1
(file results2.irf updated)
```

Saved Results

`irf set` saves in `r()`:

Macros
    `r(irffile)`   name of active IRF file, if there is an active IRF

Methods and Formulas

`irf set` is implemented as an ado-file.

Also See

Complementary: [TS] **var**, [TS] **var svar**, [TS] **varbasic**, [TS] **vec**; [TS] **irf add**,
                       [TS] **irf cgraph**, [TS] **irf create**, [TS] **irf ctable**, [TS] **irf describe**,
                       [TS] **irf drop**, [TS] **irf graph**, [TS] **irf ograph**, [TS] **irf rename**,
                       [TS] **irf table**

Background: [U] **11.4.3 Time-series varlists**,
                       [TS] **irf**, [TS] **var intro**, [TS] **vec intro**

Title

> **irf table** — Create tables of impulse–response functions and FEVDs

Syntax

irf <u>t</u>able [*stat*] [, *options*]

stat	description
Main	
irf	impulse–response function
oirf	orthogonalized impulse–response function
cirf	cumulative impulse–response function
coirf	cumulative, orthogonalized impulse–response function
fevd	Cholesky forecast-error variance decomposition
sirf	structural impulse–response function
sfevd	structural forecast-error variance decomposition

If *stat* are not specified, all statistics are included, unless option nostructural is also specified, in which case sirf and sfevd are excluded.

options	description
Main	
<u>set</u>(*filename*)	make *filename* active
<u>irf</u>(*irfnames*)	use *irfnames* IRF result sets
<u>impulse</u>(*endogvars*)	use *endogvars* as impulse variables
<u>response</u>(*endogvars*)	use *endogvars* as response variables
<u>individual</u>	make an individual table for each result set
<u>title</u>("*text*")	use *text* for overall table title
Options	
<u>level</u>(#)	set confidence level; default is level(95)
noci	suppress confidence intervals
<u>stderror</u>	include standard errors in the tables
<u>nostructural</u>	suppress sirf and sfevd from the default list of statistics
<u>step</u>(#)	use common maximum step horizon # for all tables

Description

irf table makes a table from the specified IRF results.

The rows of the tables are the time since impulse. Each column represents a combination of impulse() variable and response() variable for a *stat* from the irf() results.

Options

```
        Main
```

set(*filename*) specifies the file to be made active; see [TS] **irf set**. If set() is not specified, the active file is used.

> Note that all results are obtained from a single IRF file. If you have results in different files that you want in a single table, use `irf add` to copy results into a single file; see [TS] **irf add**.

irf(*irfnames*) specifies the IRF result sets to be used. If irf() is not specified, all the results in the active IRF file are used. (Files often contain just one set of IRF results, stored under a single *irfname*; in that case, those results are used. When there are multiple IRF results, you may also wish to specify option individual.)

impulse(*endogvars*) specifies the impulse variables for which the statistics are to be reported. If impulse() is not specified, each endogenous variable, in turn, is used.

response(*endogvars*) specifies the response variables for which the statistics are to be reported. If response() is not specified, each endogenous variable, in turn, is used.

individual specifies that each set of IRF results be placed in its own table, with its own title and footer. By default, `irf table` places all the IRF results in a single table with one title and one footer. individual may not be combined with title().

title("*text*") specifies a title for the overall table.

```
        Options
```

level(*#*) specifies the default confidence level, as a percentage, for confidence intervals, when they are reported. The default is level(95) or as set by set level; see [U] **20.6 Specifying the width of confidence intervals**.

noci suppresses reporting of the confidence intervals for each statistic. noci is assumed when the model was fitted by vec because no confidence intervals were estimated.

stderror specifies that standard errors for each statistic also be included in the table.

nostructural specifies that *stat*, when not specified, exclude sirf and sfevd.

step(*#*) specifies the maximum step horizon for all tables. If step() is not specified, each table is constructed using all steps available.

Remarks

If you have not read [TS] **irf**, please do so.

Also see [TS] **irf graph**, which produces output in graphical form, and see [TS] **irf ctable**, which also produces tabular output. `irf ctable` is more difficult to use but provides more control over how tables are formed.

▷ Example 1

We have fitted a model using var, and we saved the impulse–response functions from two different orderings. The commands we previously used were

```
. use http://www.stata-press.com/data/r9/lutkepohl
. var dlinvestment dlincome dlconsumption
. irf set results4
. irf create ordera, step(8)
. irf create orderb, order(dlincome dlinvestment dlconsumption) step(8)
```

We now wish to compare the two orderings:

```
. irf table oirf fevd, impulse(dlincome) response(dlconsumption)
> noci std title("Ordera versus orderb")
```
                         Ordera versus orderb

step	(1) oirf	(1) S.E.	(1) fevd	(1) S.E.
0	.005123	.000878	0	0
1	.001635	.000984	.288494	.077483
2	.002948	.000993	.294288	.073722
3	-.000221	.000662	.322454	.075562
4	.000811	.000586	.319227	.074063
5	.000462	.000333	.322579	.075019
6	.000044	.000275	.323552	.075371
7	.000151	.000162	.323383	.075314
8	.000091	.000114	.323499	.075386

step	(2) oirf	(2) S.E.	(2) fevd	(2) S.E.
0	.005461	.000925	0	0
1	.001578	.000988	.327807	.08159
2	.003307	.001042	.328795	.077519
3	-.00019	.000676	.370775	.080604
4	.000846	.000617	.366896	.079019
5	.000491	.000349	.370399	.079941
6	.000069	.000292	.371487	.080323
7	.000158	.000172	.371315	.080287
8	.000096	.000122	.371438	.080366

(1) irfname = ordera, impulse = dlincome, and response = dlconsumption
(2) irfname = orderb, impulse = dlincome, and response = dlconsumption

The output is displayed as a "single" table; since the table did not fit horizontally, it wrapped automatically. At the bottom of the table is a definition of the keys that appear at the top of each column. The results in the table above indicate that the orthogonalized impulse–response functions do not change by much.

◁

▷ Example 2

Because the estimated forecast-error variances do change significantly, we might want to produce two tables that contain the estimated forecast-error variance decompositions and their 95% confidence intervals:

(Continued on next page)

```
. irf table fevd, impulse(dlincome) response(dlconsumption) individual
```

Results from ordera

step	(1) fevd	(1) Lower	(1) Upper
0	0	0	0
1	.288494	.13663	.440357
2	.294288	.149797	.43878
3	.322454	.174356	.470552
4	.319227	.174066	.464389
5	.322579	.175544	.469613
6	.323552	.175826	.471277
7	.323383	.17577	.470995
8	.323499	.175744	.471253

95% lower and upper bounds reported
(1) irfname = ordera, impulse = dlincome, and response = dlconsumption

Results from orderb

step	(1) fevd	(1) Lower	(1) Upper
0	0	0	0
1	.327807	.167893	.487721
2	.328795	.17686	.48073
3	.370775	.212794	.528757
4	.366896	.212022	.52177
5	.370399	.213718	.52708
6	.371487	.214058	.528917
7	.371315	.213956	.528674
8	.371438	.213923	.528953

95% lower and upper bounds reported
(1) irfname = orderb, impulse = dlincome, and response = dlconsumption

Since we specified the `individual` option, the output contains two tables, one for each set of IRF results. Examining the results in the tables indicates that each of the estimated functions is well within the confidence interval of the other, so we conclude that the functions are not significantly different.

◁

❑ Technical Note

Be careful in how you name variables when you fit models. For instance, say that you fitted one model using `var` and used time-series operators to form one of the endogenous variables

```
. var d.linvestment ...
```

and in another model, you created a new variable:

```
. gen dlinvestment = d.linvestment
. var dlinvestment ...
```

Say that you saved IRF results from both (perhaps they differ in the number of lags). Now you wish to use `irf table` to compare them. You would not be able to specify `response(d.linvestment)` or `response(dlinvestment)` because neither variable is in both models. Similarly, you could not specify `impulse(d.linvestment)` or `impulse(dlinvestment)` for the same reason.

All is not lost; if `impulse()` is not specified, all endogenous variables are used, and similarly if `response()` is not specified, so you could obtain the result you desired by simply not specifying the options, but you will also obtain a lot more, besides. If you want to specify the `impulse()` or `response()` options, be sure to name variables consistently.

Also, you may forget how the endogenous variables were named. If so, `irf describe, detail` can provide the answer. In `irf describe`'s output, the endogenous variables are listed next to `endog`.

❏

Saved Results

If the `individual` option is not specified, `irf table` saves in `r()`:

Scalars
  `r(ncols)` number of columns in table
  `r(k_umax)` number of distinct keys
  `r(k)`   number of specific table commands
Macros
  `r(key`*i*`)` *i*th key
  `r(tnotes)` list of keys applied to each column

If the `individual` option is specified, then for each `irfname`, `irf table` saves in `r()`:

Scalars
 `r(`*irfname*`_ncols)`   number of columns in table for *irfname*
 `r(`*irfname*`_k_umax)` number of distinct keys in table for *irfname*
 `r(`*irfname*`_k)`   number of specific table commands used to create table for *irfname*
Macros
 `r(`*irfname*`_key`*i*`)`   *i*th key for *irfname* table
 `r(`*irfname*`_tnotes)` list of keys applied to each column in table for *irfname*

Methods and Formulas

`irf table` is implemented as an ado-file.

Also See

Complementary:	[TS] **var**, [TS] **var svar**, [TS] **varbasic**, [TS] **vec**; [TS] **irf add**, [TS] **irf create**, [TS] **irf describe**, [TS] **irf drop**, [TS] **irf rename**, [TS] **irf set**
Related:	[TS] **irf cgraph**, [TS] **irf ctable**, [TS] **irf graph**, [TS] **irf ograph**
Background:	[U] **11.4.3 Time-series varlists**, [TS] **irf**, [TS] **var intro**, [TS] **vec intro**

Title

---

newey — Regression with Newey–West standard errors

Syntax

newey *depvar* $\left[\textit{indepvars}\right]$ $\left[\textit{if}\right]$ $\left[\textit{in}\right]$ $\left[\textit{weight}\right]$, lag(#) $\left[\textit{options}\right]$

options	description
Model	
*lag(#)	set maximum lag order of autocorrelation
<u>nocon</u>stant	suppress constant term
Reporting	
<u>level</u>(#)	set confidence level; default is level(95)

*lag(#) is required.

You must tsset your data before using newey; see [TS] **tsset**.

depvar and *indepvars* may contain time-series operators; see [U] **11.4.3 Time-series varlists**.

by, rolling, statsby, and xi may be used with newey; see [U] **11.1.10 Prefix commands**.

aweights are allowed; see [U] **11.1.6 weight**.

See [U] **20 Estimation and postestimation commands** for additional capabilities of estimation commands.

Description

newey produces Newey–West standard errors for coefficients estimated by OLS regression. The error structure is assumed to be heteroskedastic and possibly autocorrelated up to some lag.

Options

⌐ **Model** ⌐

lag(#) specifies the maximum lag to be considered in the autocorrelation structure. If you specify lag(0), the output is exactly the same as regress, robust. lag() is required.

noconstant; see [TS] **estimation options**.

⌐ **Reporting** ⌐

level(#); see [TS] **estimation options**.

Remarks

The Huber/White/sandwich robust variance estimator (see White 1980) produces consistent standard errors for OLS regression coefficient estimates in the presence of heteroskedasticity. The Newey–West (1987) variance estimator is an extension that produces consistent estimates when there is autocorrelation in addition to possible heteroskedasticity.

The Newey–West variance estimator handles autocorrelation up to and including a lag of m, where m is specified by stipulating the `lag()` option. Thus it assumes that any autocorrelation at lags greater than m can be ignored.

Note that if `lag(0)` is specified, the variance estimates produced by `newey` are simply the Huber/White/sandwich robust variances estimates calculated by `regress, robust`; see [R] **regress**.

▷ Example 1

`newey, lag(0)` is equivalent to `regress, robust`:

```
. use http://www.stata-press.com/data/r9/auto
(1978 Automobile Data)

. regress price weight displ, robust
```

Linear regression

```
                                              Number of obs =      74
                                              F(  2,     71) =   14.44
                                              Prob > F       =  0.0000
                                              R-squared      =  0.2909
                                              Root MSE       =  2518.4
```

price	Coef.	Robust Std. Err.	t	P>|t|	[95% Conf. Interval]	
weight	1.823366	.7808755	2.34	0.022	.2663445	3.380387
displacement	2.087054	7.436967	0.28	0.780	-12.74184	16.91595
_cons	247.907	1129.602	0.22	0.827	-2004.455	2500.269

```
. generate t = _n

. tsset t
        time variable:  t, 1 to 74

. newey price weight displ, lag(0)
```

Regression with Newey-West standard errors

maximum lag: 0

```
                                        Number of obs  =      74
                                        F(  2,     71) =   14.44
                                        Prob > F       =  0.0000
```

price	Coef.	Newey-West Std. Err.	t	P>|t|	[95% Conf. Interval]	
weight	1.823366	.7808755	2.34	0.022	.2663445	3.380387
displacement	2.087054	7.436967	0.28	0.780	-12.74184	16.91595
_cons	247.907	1129.602	0.22	0.827	-2004.455	2500.269

Since `newey` requires the dataset to be `tsset`, we generated a dummy time variable `t`, which, in this example, played no role in the estimation.

◁

▷ Example 2

Say that we have time-series measurements on variables `usr` and `idle` and now wish to fit an OLS model but obtain Newey–West standard errors allowing for a lag of up to 3:

```
. use http://www.stata-press.com/data/r9/idle2, clear

. tsset time
        time variable:  time, 1 to 30
```

```
. newey usr idle, lag(3)
Regression with Newey-West standard errors          Number of obs  =        30
maximum lag: 3                                       F(  1,    28)  =     10.90
                                                     Prob > F       =    0.0026
```

usr	Coef.	Newey-West Std. Err.	t	P>|t|	[95% Conf. Interval]
idle	-.2281501	.0690927	-3.30	0.003	-.3696801 -.08662
_cons	23.13483	6.327031	3.66	0.001	10.17449 36.09516

◁

Saved Results

newey saves in e():

Scalars

e(N)	number of observations	e(F)	F statistic
e(df_m)	model degrees of freedom	e(lag)	maximum lag
e(df_r)	residual degrees of freedom		

Macros

e(cmd)	newey	e(title)	title in estimation output
e(depvar)	name of dependent variable	e(vcetype)	title used to label Std. Err.
e(wtype)	weight type	e(properties)	b V
e(wexp)	weight expression	e(predict)	program used to implement predict

Matrices

e(b)	coefficient vector	e(V)	variance–covariance matrix of the estimators

Functions

e(sample)	marks estimation sample

Methods and Formulas

newey is implemented as an ado-file.

newey calculates the estimates

$$\widehat{\beta}_{\text{OLS}} = (\mathbf{X}'\mathbf{X})^{-1}\mathbf{X}'\mathbf{y}$$

$$\widehat{\text{Var}}(\widehat{\beta}_{\text{OLS}}) = (\mathbf{X}'\mathbf{X})^{-1}\mathbf{X}'\widehat{\mathbf{\Omega}}\mathbf{X}(\mathbf{X}'\mathbf{X})^{-1}$$

That is, the coefficient estimates are simply those of OLS linear regression.

For the case of lag(0) (no autocorrelation), the variance estimates are calculated using the White formulation:

$$\mathbf{X}'\widehat{\mathbf{\Omega}}\mathbf{X} = \mathbf{X}'\widehat{\mathbf{\Omega}}_0\mathbf{X} = \frac{n}{n-k}\sum_i \widehat{e}_i^2 \mathbf{x}_i'\mathbf{x}_i$$

Here $\widehat{e}_i = y_i - \mathbf{x}_i\widehat{\beta}_{\text{OLS}}$, where $\mathbf{x}_i$ is the ith row of the $\mathbf{X}$ matrix, n is the number of observations, and k is the number of predictors in the model, including the constant if there is one. Note that the above formula is exactly the same as that used by regress, robust with the regression-like formula (the default) for the multiplier q_c; see *Methods and Formulas* of [R] **regress**.

For the case of `lag(m)`, $m > 0$, the variance estimates are calculated using the Newey–West (1987) formulation

$$\mathbf{X}'\widehat{\mathbf{\Omega}}\mathbf{X} = \mathbf{X}'\widehat{\mathbf{\Omega}}_0\mathbf{X} + \frac{n}{n-k}\sum_{l=1}^{m}\left(1 - \frac{l}{m+1}\right)\sum_{t=l+1}^{n}\widehat{e}_t\widehat{e}_{t-l}(\mathbf{x}'_t\mathbf{x}_{t-l} + \mathbf{x}'_{t-l}\mathbf{x}_t)$$

where $\mathbf{x}_t$ is the row of the $\mathbf{X}$ matrix observed at time t.

Whitney K. Newey (1954–) earned degrees in economics at Brigham Young University and MIT. After a period at Princeton, he returned to MIT as Professor in 1990. His interests in theoretical and applied econometrics include bootstrapping, nonparametric estimation of models, semiparametric models and choosing the number of instrumental variables.

Kenneth D. West (1953–) earned a Bachelor's degree in economics and mathematics at Wesleyan University and then a Ph.D. in economics at MIT. After a period at Princeton, he joined the University of Wisconsin in 1988. His interests include empirical macroeconomics and time-series econometrics.

References

Hardin, J. W. 1997. sg72: Newey–West standard errors for probit, logit, and poisson models. *Stata Technical Bulletin* 39: 32–35. Reprinted in *Stata Technical Bulletin Reprints*, vol. 7, pp. 182–186.

Newey, W. K. and K. D. West. 1987. A simple, positive semi-definite, heteroskedasticity and autocorrelation consistent covariance matrix. *Econometrica* 55: 703–708.

White, H. 1980. A heteroskedasticity-consistent covariance matrix estimator and a direct test for heteroskedasticity. *Econometrica* 48: 817–838.

Also See

Complementary:	[TS] **newey postestimation**, [TS] **tsset**
Related:	[R] **regress**, [SVY] **variance estimation**, [XT] **xtgls**, [XT] **xtpcse**
Background:	[U] **11.1.10 Prefix commands**,
	[U] **20 Estimation and postestimation commands**,
	[TS] **estimation options**

Title

newey postestimation — Postestimation tools for newey

Description

The following postestimation commands are available for `newey`:

command	description
estat	VCE and estimation sample summary
estimates	cataloging estimation results
lincom	point estimates, standard errors, testing, and inference for linear combinations of coefficients
linktest	link test for model specification
mfx	marginal effects or elasticities
nlcom	point estimates, standard errors, testing, and inference for nonlinear combinations of coefficients
predict	predictions, residuals, influence statistics, and other diagnostic measures
predictnl	point estimates, standard errors, testing, and inference for generalized predictions
test	Wald tests for simple and composite linear hypotheses
testnl	Wald tests of nonlinear hypotheses

See the corresponding entries in the *Stata Base Reference Manual* for details.

Syntax for predict

predict [*type*] *newvar* [*if*] [*in*] [, *statistic*]

statistic	description
xb	linear prediction; the default
stdp	standard error of the linear prediction
<u>resid</u>uals	residuals

These statistics are available both in and out of sample; type `predict ... if e(sample) ...` if wanted only for the estimation sample.

Options for predict

xb, the default, calculates the linear prediction.

stdp calculates the standard error of the linear prediction.

residuals calculates the residuals.

174

Methods and Formulas

All postestimation commands listed above are implemented as ado-files.

Also See

Complementary:	[TS] **newey**,
	[R] **estimates**, [R] **lincom**, [R] **linktest**,
	[R] **mfx**, [R] **nlcom**, [R] **predictnl**, [R] **test**, [R] **testnl**
Background:	[U] **13.5 Accessing coefficients and standard errors**,
	[U] **20 Estimation and postestimation commands**,
	[R] **estat**, [R] **predict**

Title

> **pergram** — Periodogram

Syntax

> pergram *varname* $\left[\,if\,\right]$ $\left[\,in\,\right]$ $\left[\,,\ options\,\right]$

options	description
Main	
generate(*newvar*)	generate *newvar* to contain the raw periodogram values
Plot	
cline_options	affect rendition of the plotted points connected by lines
Add plot	
addplot(*plot*)	add other plots to the generated graph
Y-Axis, X-Axis, Title, Caption, Legend, Overall	
twoway_options	any options other than by() documented in [G] ***twoway_options***
†nograph	suppress the graph

† nograph is not shown in the dialog box.

You must tsset your data before using pergram; see [TS] **tsset**. In addition, the time series must be dense (nonmissing with no gaps in the time variable) in the specified sample.

varname may contain time-series operators; see [U] **11.4.3 Time-series varlists**.

Description

pergram plots the log-standardized periodogram for a dense time series.

Options

```
┌─ Main ──────────────────────────────────────────────────────────────
```

generate(*newvar*) specifies a new variable to contain the raw periodogram values. Note that the generated graph log-transforms and scales the values by the sample variance and then truncates them to the $\left[\,-6,6\,\right]$ interval before graphing them.

```
┌─ Plot ──────────────────────────────────────────────────────────────
```

cline_options affect the rendition of the plotted points connected by lines; see [G] ***cline_options***.

```
┌─ Add plot ──────────────────────────────────────────────────────────
```

addplot(*plot*) adds specified plots to the generated graph; see [G] ***addplot_option***.

⌐ Y-Axis, X-Axis, Title, Caption, Legend, Overall ⌐

twoway_options are any of the options documented in [G] *twoway_options*, excluding by(). These include options for titling the graph (see [G] *title_options*) and saving the graph to disk (see [G] *saving_option*).

The following option is available with pergram but is not shown in the dialog box:

nograph prevents pergram from constructing a graph.

Remarks

A good discussion of the periodogram is provided in Chatfield (1996), Hamilton (1994), and Newton (1988). Chatfield is also a very good introductory reference for time-series analysis. Another classic reference is Box, Jenkins, and Reinsel (1994). pergram produces a scatterplot in which the points of the scatterplot are connected. The points themselves represent the log-standardized periodogram, and the connections between points represent the (continuous) log-standardized sample spectral density.

In the following examples, we present the periodograms together with an interpretation of the main features of the plots.

▷ Example 1

We have time-series data consisting of 144 observations on the monthly number of international airline passengers (in thousands) between 1949 and 1960 (Box, Jenkins, and Reinsel 1994, Series G). We can graph the raw series and the log periodogram for these data by typing

```
. use http://www.stata-press.com/data/r9/air2
(TIMESLAB: Airline passengers)

. scatter air time, m(o) c(l)
```

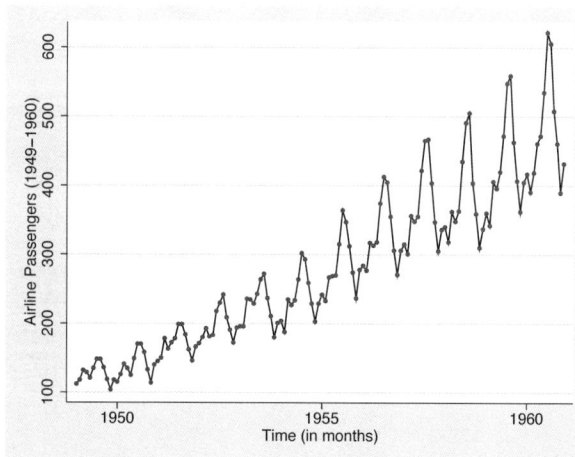

```
. pergram air
```

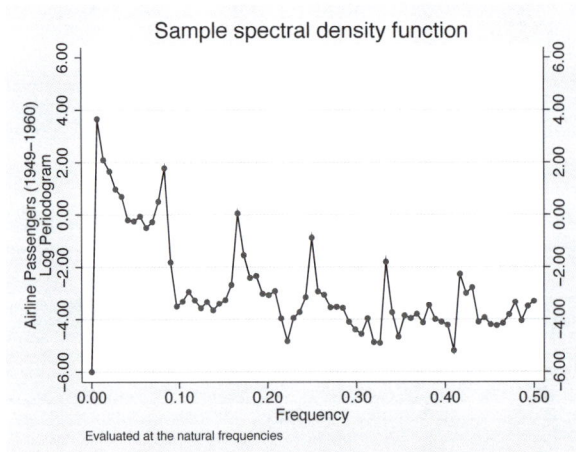

The periodogram highlights the annual cycle together with the harmonics. Notice the peak at a frequency of about 0.08 cycles per month (cpm). The period is the reciprocal of frequency, and the reciprocal of 0.08 cpm is approximately 12 months per cycle. The similarity in shape of each group of twelve observations reveals the annual cycle. The magnitude of the cycle is increasing, resulting in the peaks in the periodogram at the harmonics of the principal annual cycle.

◁

▷ Example 2

This example uses 215 observations on the annual number of sunspots from 1749 to 1963 (Box and Jenkins 1976, Series E). The graph of the raw series and the log periodogram for these data are given as

```
. use http://www.stata-press.com/data/r9/sunspot
(TIMESLAB: Wolfer sunspot data)
. scatter spot time, m(o) c(l)
```

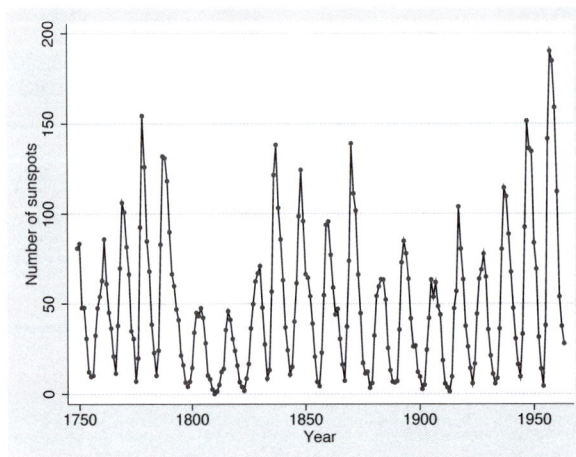

. pergram spot

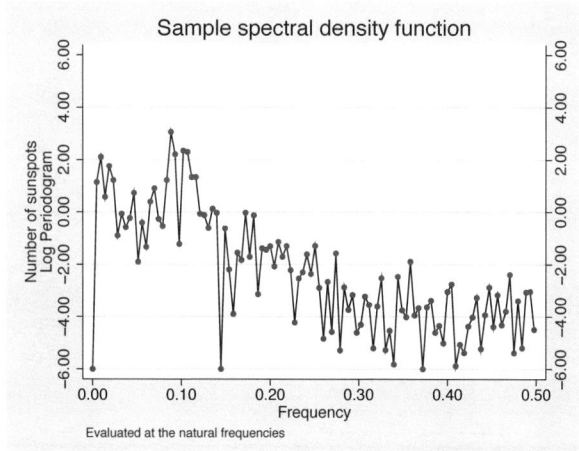

The periodogram peaks at a frequency of slightly under 0.10 cycles per year, indicating a ten- to twelve-year cycle in sunspot activity.

◁

▷ Example 3

Here we examine the number of trapped Canadian lynx from 1821 through 1934 (Newton 1988, 587). The raw series and the log periodogram are given as

```
. use http://www.stata-press.com/data/r9/lynx2
(TIMESLAB: Canadian lynx)
. scatter lynx time, m(o) c(l)
```

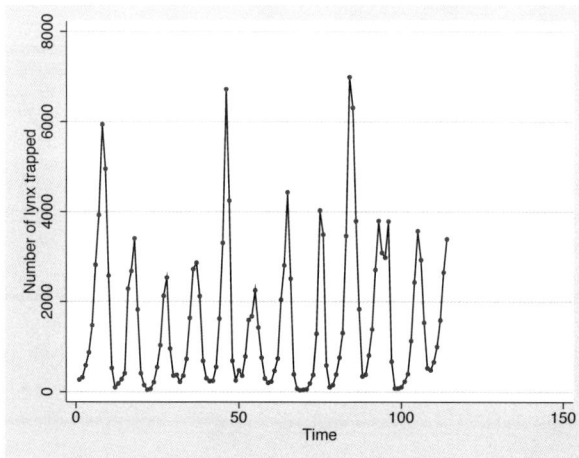

```
. pergram lynx
```

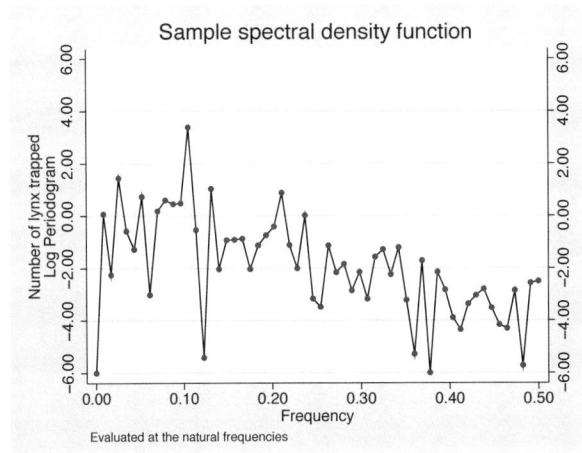

Sample spectral density function

Evaluated at the natural frequencies

The periodogram indicates that there is a cycle with a duration of about 10 years for these data but that it is otherwise random in nature.

◁

▷ Example 4

To more clearly highlight what the periodogram depicts, we present the result of analyzing a time series of the sum of four sinusoids (of different periods). The periodogram should be able to decompose the time series into four different sinusoids whose periods may be determined from the plot.

```
. use http://www.stata-press.com/data/r9/cos4
(TIMESLAB: Sum of 4 Cosines)

. scatter sumfc time, m(o) c(l)
```

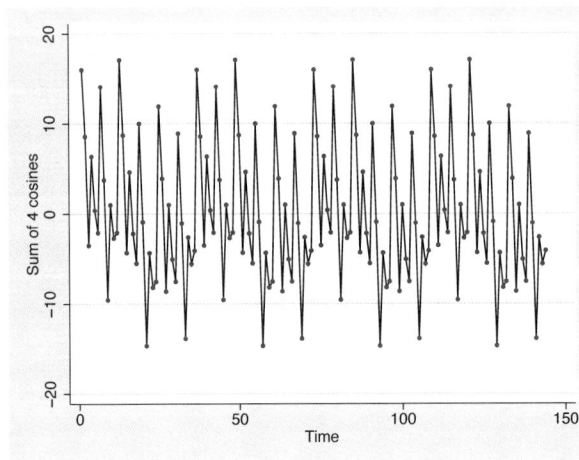

```
. pergram sumfc, gen(ordinate)
```

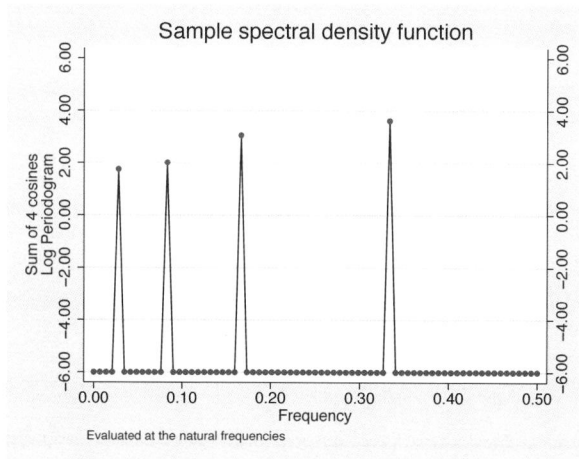

The periodogram clearly shows the four contributions to the original time series. From the plot, we can see that the periods of the summands were 3, 6, 12, and 36, although you can confirm this by using

```
. generate double omega = (_n-1)/144
. generate double period = 1/omega
(1 missing value generated)
. list period omega if ordinate> 1e-5 & omega <=.5
```

	period	omega
5.	36	.02777778
13.	12	.08333333
25.	6	.16666667
49.	3	.33333333

◁

Methods and Formulas

pergram is implemented as an ado-file.

We use the notation of Newton (1988) in the following discussion.

A time series of interest is decomposed into a unique set of sinusoids of various frequencies and amplitudes.

A plot of the sinusoidal amplitudes (ordinates) versus the frequencies for the sinusoidal decomposition of a time series gives us the spectral density of the time series. If we calculate the sinusoidal amplitudes for a discrete set of "natural" frequencies $(1/n, 2/n, \ldots, q/n)$, we obtain the periodogram.

Let $x(1), \ldots, x(n)$ be a time series, and let $\omega_k = (k-1)/n$ denote the natural frequencies for $k = 1, \ldots, (n/2) + 1$. Define

$$C_k^2 = \frac{1}{n^2} \left| \sum_{t=1}^{n} x(t) e^{2\pi i(t-1)\omega_k} \right|^2$$

A plot of nC_k^2 versus ω_k is then called the periodogram.

The sample spectral density is defined for a continuous frequency ω as

$$\widehat{f}(\omega) = \begin{cases} \dfrac{1}{n} \left| \displaystyle\sum_{t=1}^{n} x(t) e^{2\pi i (t-1)\omega} \right|^2 & \text{if } \omega \in [0, .5] \\ \widehat{f}(1-\omega) & \text{if } \omega \in [.5, 1] \end{cases}$$

Note that the periodogram (and sample spectral density) is symmetric about $\omega = .5$. Further standardize the periodogram such that

$$\frac{1}{n} \sum_{k=2}^{n} \frac{nC_k^2}{\widehat{\sigma}^2} = 1$$

where $\widehat{\sigma}^2$ is the sample variance of the time series so that the average value of the ordinate is one.

Once the amplitudes are standardized, we may then take the natural log of the values and produce the log periodogram. In doing so, we truncate the graph at ± 6. Note that we drop the word "log" and simply refer to the "log periodogram" as the "periodogram" in text.

References

Box, G. E. P. and G. M. Jenkins. 1976. *Time Series Analysis: Forecasting and Control*. Oakland, CA: Holden–Day.

Box, G. E. P., G. M. Jenkins, and G. C. Reinsel. 1994. *Time Series Analysis: Forecasting and Control*. 3rd ed. Englewood Cliffs, NJ: Prentice Hall.

Chatfield, C. 1996. *The Analysis of Time Series: An Introduction*. 5th ed. London: Chapman & Hall.

Hamilton, J. D. 1994. *Time Series Analysis*. Princeton: Princeton University Press.

Newton, H. J. 1988. *TIMESLAB: A Time Series Analysis Laboratory*. Pacific Grove, CA: Wadsworth & Brooks/Cole.

Also See

Complementary:	[TS] **tsset**
Related:	[TS] **corrgram**, [TS] **cumsp**, [TS] **wntestb**
Background:	*Stata Graphics Reference Manual*

Title

---

pperron — Phillips–Perron unit-root test

Syntax

pperron *varname* $\begin{bmatrix} if \end{bmatrix}$ $\begin{bmatrix} in \end{bmatrix}$ $\begin{bmatrix} , & options \end{bmatrix}$

options	description
Main	
<u>noc</u>onstant	suppress constant term
<u>tr</u>end	include trend term in regression
<u>reg</u>ress	display regression table
<u>lag</u>s(#)	use # Newey–West lags

You must tsset your data before using pperron; see [TS] **tsset**.

varname may contain time-series operators; see [U] **11.4.3 Time-series varlists**.

Description

pperron performs the Phillips–Perron (1987) test that a variable has a unit root. The null hypothesis is that the variable contains a unit root, and the alternative is that the variable was generated by a stationary process. pperron uses Newey–West (1987) standard errors to account for serial correlation, while the augmented Dickey–Fuller test implemented in dfuller uses additional lags of the first-differenced variable.

Options

> Main

noconstant suppresses the constant term (intercept) in the model.

trend specifies that a trend term be included in the associated regression. This option may not be specified if noconstant is specified.

regress specifies that the associated regression table appear in the output. By default, the regression table is not produced.

lags(#) specifies the number of Newey–West lags to use in calculating the standard error. The default is to use int $\left\{ 4(T/100)^{2/9} \right\}$ lags.

Remarks

As noted in [TS] **dfuller**, the Dickey–Fuller test involves fitting the regression model

$$y_t = \alpha + \rho y_{t-1} + \delta t + u_t \tag{1}$$

by ordinary least squares (OLS), but serial correlation will present a problem. To account for this, the augmented Dickey–Fuller test's regression includes lags of the first-differences of y_t.

The Phillips–Perron test involves fitting (1), and the results are used to calculate the test statistics. Phillips and Perron (1988) proposed two alternative statistics, which `pperron` presents. Phillips and Perron's test statistics can be viewed as Dickey–Fuller statistics that have been made robust to serial correlation by using the Newey–West (1987) heteroskedasticity and autocorrelation consistent covariance matrix estimator.

Hamilton (1994, chapter 17) and [TS] **dfuller** discuss four different cases into which unit-root tests can be classified. The Phillips–Perron test applies to cases one, two, and four but not to case three. Cases one and two assume that the variable has a unit root without drift under the null hypothesis, the only difference being whether the constant term α is included in regression (1). Case four assumes that the variable has a random walk, with or without drift, under the null hypothesis. Case three, which assumes the variable has a random walk with drift under the null hypothesis, is just a special case of case four, so the fact that the Phillips–Perron test does not apply is not restrictive. The table below summarizes the relevant cases:

Case	Process under null hypothesis	Regression restrictions	`dfuller` option
1	Random walk without drift	$\alpha = 0, \delta = 0$	`noconstant`
2	Random walk without drift	$\delta = 0$	(default)
4	Random walk with or without drift	(none)	`trend`

The critical values for the Phillips–Perron test are the same as those for the augmented Dickey–Fuller test. See Hamilton (1994, chapter 17) for more information.

▷ Example 1

Here we use the international airline passengers dataset (Box, Jenkins, and Reinsel 1994, Series G). This dataset has 144 observations on the monthly number of international airline passengers from 1949 through 1960. Because the data exhibit a clear upward trend over time, we will use the `trend` option.

```
. use http://www.stata-press.com/data/r9/air2
(TIMESLAB: Airline passengers)

. pperron air, lags(4) trend regress

Phillips-Perron test for unit root          Number of obs   =      143
                                            Newey-West lags =        4

                        ------------- Interpolated Dickey-Fuller -------------
                 Test         1% Critical       5% Critical      10% Critical
             Statistic           Value             Value             Value
------------------------------------------------------------------------------
 Z(rho)        -46.405          -27.687           -20.872           -17.643
 Z(t)           -5.049           -4.026            -3.444            -3.144
------------------------------------------------------------------------------
MacKinnon approximate p-value for Z(t) = 0.0002

------------------------------------------------------------------------------
         air |      Coef.   Std. Err.      t    P>|t|     [95% Conf. Interval]
-------------+----------------------------------------------------------------
         air |
         L1. |   .7318116   .0578092    12.66   0.000     .6175196    .8461035
      _trend |   .7107559   .1670563     4.25   0.000     .3804767    1.041035
       _cons |   25.95168   7.325951     3.54   0.001     11.46788    40.43547
------------------------------------------------------------------------------
```

Just as in the example in [TS] **dfuller**, we reject the null hypothesis of a unit root at all common significance levels. The interpolated critical values for Z_t differ slightly from those shown in the example in [TS] **dfuller** because the sample sizes are different: with the augmented Dickey–Fuller regression we lose observations because of the inclusion of lagged difference terms as regressors.

◁

Saved Results

pperron saves in r():

Scalars

r(N)	number of observations	r(Zt)	Phillips–Perron τ test statistic
r(lags)	number of lagged differences used	r(Zrho)	Phillips–Perron ρ test statistic
r(pval)	MacKinnon approximate p-value		
	(not included if noconstant specified)		

Methods and Formulas

pperron is implemented as an ado-file.

In the OLS estimation of an AR(1) process with Gaussian errors,

$$y_i = \rho y_{i-1} + \epsilon_i$$

where ϵ_i are independently and identically distributed as $N(0, \sigma^2)$ and $y_0 = 0$, the OLS estimate (based on an n-observation time series) of the autocorrelation parameter ρ is given by

$$\widehat{\rho}_n = \frac{\sum_{i=1}^{n} y_{i-1} y_i}{\sum_{i=1}^{n} y_i^2}$$

If $|\rho| < 1$, then $\sqrt{n}(\widehat{\rho}_n - \rho) \to N(0, 1 - \rho^2)$. If this result were valid for when $\rho = 1$, then the resulting distribution would have a variance of zero. When $\rho = 1$, the OLS estimate $\widehat{\rho}$ still converges to one, though we need to find a nondegenerate distribution so that we can test $H_0 : \rho = 1$. See Hamilton (1994, chapter 17).

The Phillips–Perron test involves fitting the regression

$$y_i = \alpha + \rho y_{i-1} + \epsilon_i$$

where we may exclude the constant or include a trend term. There are two statistics, Z_ρ and Z_τ, calculated as

$$Z_\rho = n(\widehat{\rho}_n - 1) - \frac{1}{2}\frac{n^2\widehat{\sigma}^2}{s_n^2}\left(\widehat{\lambda}_n^2 - \widehat{\gamma}_{0,n}\right)$$

$$Z_\tau = \sqrt{\frac{\widehat{\gamma}_{0,n}}{\widehat{\lambda}_n^2}}\frac{\widehat{\rho}_n - 1}{\widehat{\sigma}} - \frac{1}{2}\left(\widehat{\lambda}_n^2 - \widehat{\gamma}_{0,n}\right)\frac{1}{\widehat{\lambda}_n}\frac{n\widehat{\sigma}}{s_n}$$

$$\widehat{\gamma}_{j,n} = \frac{1}{n}\sum_{i=j+1}^{n}\widehat{u}_i\widehat{u}_{i-j}$$

$$\widehat{\lambda}_n^2 = \widehat{\gamma}_{0,n} + 2\sum_{j=1}^{q}\left(1 - \frac{j}{q+1}\right)\widehat{\gamma}_{j,n}$$

$$s_n^2 = \frac{1}{n-k}\sum_{i=1}^{n}\widehat{u}_i^2$$

where u_i is the OLS residual, k is the number of covariates in the regression, q is the number of Newey–West lags to use in calculating $\widehat{\lambda}_n^2$, and $\widehat{\sigma}$ is the OLS standard error of $\widehat{\rho}$.

The critical values, which have the same distribution as the Dickey–Fuller statistic (see Dickey and Fuller 1979) included in the output, are linearly interpolated from the table of values that appear in Fuller (1976), and the MacKinnon approximate p-values use the regression surface published in MacKinnon (1994).

Peter Charles Bonest Phillips (1948–) was born in Weymouth, England, and earned degrees in economics at the University of Auckland, New Zealand, and the London School of Economics. After periods at the Universities of Essex and Birmingham, Phillips moved to Yale in 1979. He also holds appointments at the University of Auckland and the University of York. His main research interests are in econometric theory, financial econometrics, time-series and panel-data econometrics, and applied macroeconomics.

Pierre Perron (1959–) was born in Québec, Canada and earned degrees at McGill, Queen's, and Yale in economics. After posts at Princeton and the Université de Montréal, he joined Boston University in 1997. His research interests include time-series analysis, econometrics, and applied macroeconomics.

References

Box, G. E. P., G. M. Jenkins, and G. C. Reinsel. 1994. *Time Series Analysis: Forecasting and Control.* 3rd ed. Englewood Cliffs, NJ: Prentice Hall.

Dickey, D. A. and W. A. Fuller. 1979. Distribution of the estimators for autoregressive time series with a unit root. *Journal of the American Statistical Association* 74: 427–431.

Fuller, W. A. 1976. *Introduction to Statistical Time Series.* New York: Wiley.

Hamilton, J. D. 1994. *Time Series Analysis.* Princeton: Princeton University Press.

MacKinnon, J. G. 1994. Approximate asymptotic distribution functions for unit-root and cointegration tests. *Journal of Business and Economic Statistics* 12: 167–176.

Newey, W. K. and K. D. West. 1987. A simple, positive semi-definite, heteroskedasticity and autocorrelation consistent covariance matrix. *Econometrica* 55: 703–708.

Phillips, P. C. B. and P. Perron. 1988. Testing for a unit root in time series regression. *Biometrika* 75: 335–346.

Also See

Complementary: [TS] **tsset**

Related: [TS] **dfgls**, [TS] **dfuller**

Title

> **prais** — Prais–Winsten and Cochrane–Orcutt regression

Syntax

> prais *depvar* [*indepvars*] [*if*] [*in*] [, *options*]

options	description
Model	
<u>rho</u>type(<u>regress</u>)	base ρ on single-lag OLS of residuals; the default
<u>rho</u>type(freg)	base ρ on single-lead OLS of residuals
<u>rho</u>type(<u>ts</u>corr)	base ρ on autocorrelation of residuals
<u>rho</u>type(dw)	base ρ on autocorrelation based on Durbin–Watson
<u>rho</u>type(<u>theil</u>)	base ρ on adjusted autocorrelation
<u>rho</u>type(<u>nagar</u>)	base ρ on adjusted Durbin–Watson
corc	use Cochrane–Orcutt transformation
<u>sse</u>search	search for ρ that minimizes SSE
<u>two</u>step	stop after the first iteration
<u>nocon</u>stant	suppress constant term
<u>has</u>cons	has user-defined constant
<u>save</u>space	conserve memory during estimation
SE/Robust	
<u>r</u>obust	compute standard errors using the robust/sandwich estimator
<u>cl</u>uster(*varname*)	adjust standard errors for intragroup correlation
hc2	use $1/(1-h)$ bias correction for robust
hc3	use $1/(1-h)^2$ bias correction for robust
Reporting	
<u>l</u>evel(#)	set confidence level; default is level(95)
nodw	do not report the Durbin–Watson statistic
Max options	
maximize_options	control the maximization process; seldom used

You must tsset your data before using prais; see [TS] **tsset**.

depvar and *indepvars* may contain time-series operators; see [U] **11.4.3 Time-series varlists**.

by, rolling, statsby, and xi may be used with prais; see [U] **11.1.10 Prefix commands**.

See [U] **20 Estimation and postestimation commands** for additional capabilities of estimation commands.

Description

prais uses the generalized least-squares method to estimate the parameters in a linear regression model in which the errors are serially correlated. Specifically, the errors are assumed to follow a first-order autoregressive process.

Options

___Model___

rhotype(*rhomethod*) selects a specific computation for the autocorrelation parameter ρ, where *rhomethod* can be

<u>regr</u>ess	$\rho_{\text{reg}} = \beta$ from the residual regression $\epsilon_t = \beta\epsilon_{t-1}$
freg	$\rho_{\text{freg}} = \beta$ from the residual regression $\epsilon_t = \beta\epsilon_{t+1}$
<u>ts</u>corr	$\rho_{\text{tscorr}} = \epsilon'\epsilon_{t-1}/\epsilon'\epsilon$, where ϵ is the vector of residuals
dw	$\rho_{\text{dw}} = 1 - \text{dw}/2$, where dw is the Durbin–Watson d statistic
<u>t</u>heil	$\rho_{\text{theil}} = \rho_{\text{tscorr}}(N - k)/N$
<u>na</u>gar	$\rho_{\text{nagar}} = (\rho_{\text{dw}} * N^2 + k^2)/(N^2 - k^2)$

The prais estimator can use any consistent estimate of ρ to transform the equation, and each of these estimates meets that requirement. The default is regress, which produces the minimum sum-of-squares solution (ssesearch option) for the Cochrane–Orcutt transformation—none of these computations will produce the minimum sum-of-squares solution for the full Prais–Winsten transformation. See Judge, Griffiths, Hill, Lütkepohl, and Lee (1985) for a discussion of each of the estimates of ρ.

corc specifies that the Cochrane–Orcutt transformation be used to estimate the equation. With this option, the Prais–Winsten transformation of the first observation is not performed, and the first observation is dropped when estimating the transformed equation; see *Methods and Formulas* below.

ssesearch specifies that a search be performed for the value of ρ that minimizes the sum-of-squared errors of the transformed equation (Cochrane–Orcutt or Prais–Winsten transformation). The search method is a combination of quadratic and modified bisection searches using golden sections.

twostep specifies that prais stop on the first iteration after the equation is transformed by ρ—the two-step efficient estimator. Although it is customary to iterate these estimators to convergence, they are efficient at each step.

noconstant; see [TS] **estimation options**.

hascons indicates that a user-defined constant, or a set of variables that in linear combination forms a constant, has been included in the regression. For some computational concerns, see the discussion in [R] **regress**.

savespace specifies that prais attempt to save as much space as possible by retaining only those variables required for estimation. The original data are restored after estimation. This option is rarely used and should be used only if there is insufficient space to fit a model without the option.

___SE/Robust___

robust, cluster(*varname*); see [TS] **estimation options** for details.

Note that all estimates from prais are conditional on the estimated value of ρ. This means that robust variance estimates in this case are only robust to heteroskedasticity and are not generally robust to misspecification of the functional form or omitted variables. The estimation of the functional form is intertwined with the estimation of ρ, and all estimates are conditional on ρ. Thus estimates cannot be robust to misspecification of functional form. For these reasons, it is probably best to interpret robust in the spirit of White's (1980) original paper on estimation of heteroskedastic-consistent covariance matrices.

hc2 and hc3 specify an alternative bias correction for the robust variance calculation; for more information, see [R] **regress**. hc2 and hc3 may not be specified with cluster(). Specifying hc2 or hc3 implies robust.

| Reporting |

level(*#*); see [TS] **estimation options**.

nodw suppresses reporting of the Durbin–Watson statistic.

| Max options |

maximize_options: $\boxed{\text{no}}$log, tolerance(*#*); see [R] **maximize**. These options are seldom used.

Remarks

prais fits a linear regression of *depvar* on *indepvars* that is corrected for first-order serially correlated residuals using the Prais–Winsten (1954) transformed regression estimator, the Cochrane–Orcutt (1949) transformed regression estimator, or a version of the search method suggested by Hildreth and Lu (1960).

The most common autocorrelated error process is the first-order autoregressive process. Under this assumption, the linear regression model can be written as

$$y_t = \mathbf{x}_t \boldsymbol{\beta} + u_t$$

where the errors satisfy

$$u_t = \rho\, u_{t-1} + e_t$$

and the e_t are independently and identically distributed as $N(0, \sigma^2)$. The covariance matrix $\boldsymbol{\Psi}$ of the error term e can then be written as

$$\boldsymbol{\Psi} = \frac{1}{1 - \rho^2} \begin{bmatrix} 1 & \rho & \rho^2 & \cdots & \rho^{T-1} \\ \rho & 1 & \rho & \cdots & \rho^{T-2} \\ \rho^2 & \rho & 1 & \cdots & \rho^{T-3} \\ \vdots & \vdots & \vdots & \ddots & \vdots \\ \rho^{T-1} & \rho^{T-2} & \rho^{T-3} & \cdots & 1 \end{bmatrix}$$

The Prais–Winsten estimator is a generalized least squares (GLS) estimator. The Prais–Winsten method (as described in Judge et al. 1985) is derived from the AR(1) model for the error term described above. Whereas the Cochrane–Orcutt method uses a lag definition and loses the first observation in the iterative method, the Prais–Winsten method preserves that first observation. In small samples, this can be a significant advantage.

❑ Technical Note

To fit a model with autocorrelated errors, you must specify your data as time series and have (or create) a variable denoting the time at which an observation was collected. The data for the regression should be equally spaced in time.

❑

▷ Example 1

Say that we wish to fit a time-series model of `usr` on `idle` but are concerned that the residuals may be serially correlated. We will declare the variable `t` to represent time by typing

```
. use http://www.stata-press.com/data/r9/idle
. tsset t
        time variable:  t, 1 to 30
```

We can obtain Cochrane–Orcutt estimates by specifying the `corc` option:

```
. prais usr idle, corc
Iteration 0:  rho = 0.0000
Iteration 1:  rho = 0.3518
 (output omitted )
Iteration 13:  rho = 0.5708
```

Cochrane-Orcutt AR(1) regression -- iterated estimates

Source	SS	df	MS		
				Number of obs =	29
				F(1, 27) =	6.49
Model	40.1309584	1	40.1309584	Prob > F =	0.0168
Residual	166.898474	27	6.18142498	R-squared =	0.1938
				Adj R-squared =	0.1640
Total	207.029433	28	7.39390831	Root MSE =	2.4862

usr	Coef.	Std. Err.	t	P>|t|	[95% Conf. Interval]	
idle	-.1254511	.0492356	-2.55	0.017	-.2264742	-.024428
_cons	14.54641	4.272299	3.40	0.002	5.78038	23.31245
rho	.5707918					

```
Durbin-Watson statistic (original)    1.295766
Durbin-Watson statistic (transformed) 1.466222
```

The fitted model is

$$\text{usr}_t = -.1254 \, \text{idle}_t + 14.55 + u_t \quad \text{and} \quad u_t = .5708 \, u_{t-1} + e_t$$

We can also fit the model with the Prais–Winsten method,

(Continued on next page)

```
. prais usr idle

Iteration 0:  rho = 0.0000
Iteration 1:  rho = 0.3518
 (output omitted )
Iteration 14:  rho = 0.5535
```

Prais-Winsten AR(1) regression -- iterated estimates

Source	SS	df	MS		Number of obs	=	30
					F(1, 28)	=	7.12
Model	43.0076941	1	43.0076941		Prob > F	=	0.0125
Residual	169.165739	28	6.04163354		R-squared	=	0.2027
					Adj R-squared	=	0.1742
Total	212.173433	29	7.31632528		Root MSE	=	2.458

usr	Coef.	Std. Err.	t	P>|t|	[95% Conf. Interval]	
idle	-.1356522	.0472195	-2.87	0.008	-.2323769	-.0389275
_cons	15.20415	4.160391	3.65	0.001	6.681978	23.72633
rho	.5535476					

```
Durbin-Watson statistic (original)     1.295766
Durbin-Watson statistic (transformed) 1.476004
```

where the Prais–Winsten fitted model is

$$\text{usr}_t = -.1357\,\text{idle}_t + 15.20 + u_t \quad \text{and} \quad u_t = .5535\,u_{t-1} + e_t$$

As the results indicate, for these data there is little difference between the Cochrane–Orcutt and Prais–Winsten estimators, whereas the OLS estimate of the slope parameter is substantially different.

◁

▷ Example 2

We have data on quarterly sales, in millions of dollars, for five years, and we would like to use this information to model sales for company X. First, we fit a linear model by OLS and obtain the Durbin–Watson statistic using `estat dwatson`; see [R] **regress postestimation time series**.

```
. use http://www.stata-press.com/data/r9/qsales

. regress csales isales
```

Source	SS	df	MS		Number of obs	=	20
					F(1, 18)	=	14888.15
Model	110.256901	1	110.256901		Prob > F	=	0.0000
Residual	.133302302	18	.007405683		R-squared	=	0.9988
					Adj R-squared	=	0.9987
Total	110.390204	19	5.81001072		Root MSE	=	.08606

csales	Coef.	Std. Err.	t	P>|t|	[95% Conf. Interval]	
isales	.1762828	.0014447	122.02	0.000	.1732475	.1793181
_cons	-1.454753	.2141461	-6.79	0.000	-1.904657	-1.004849

```
. estat dwatson
Durbin-Watson d-statistic(  2,     20) =  .7347276
```

Noting that the Durbin–Watson statistic is far from 2 (the expected value under the null hypothesis of no serial correlation) and well below the 5% lower limit of 1.2, we conclude that the disturbances are serially correlated. (Upper and lower bounds for the d statistic can be found in most econometrics texts; e.g., Harvey, 1993. The bounds have been derived for only a limited combination of regressors and observations.) To reinforce this conclusion, we use two other tests to test for serial correlation in the error distribution.

```
. estat bgodfrey, lags(1)
```

Breusch-Godfrey LM test for autocorrelation

lags(p)	chi2	df	Prob > chi2
1	7.998	1	0.0047

HO: no serial correlation

```
. estat durbinalt
```

Durbin's alternative test for autocorrelation

lags(p)	chi2	df	Prob > chi2
1	11.329	1	0.0008

HO: no serial correlation

`estat bgodfrey` reports the Breusch–Godfrey Lagrange multiplier test statistic, and `estat durbinalt` reports the Durbin's alternative test statistic. Both tests give a small p-value and thus reject the null hypothesis of no serial correlation. These two tests are asymptotically equivalent when testing for AR(1) process. See [R] **regress postestimation time series** if you are not familiar with these two tests.

We correct for autocorrelation using the `ssesearch` option of `prais` to search for the value of ρ that minimizes the sum-of-squared residuals of the Cochrane–Orcutt transformed equation. Normally, the default Prais–Winsten transformations is used with such a small dataset, but the less-efficient Cochrane–Orcutt transformation allows us to demonstrate an aspect of the estimator's convergence.

```
. prais csales isales, corc ssesearch
Iteration 1:   rho = 0.8944, criterion =  -.07298558
Iteration 2:   rho = 0.8944, criterion =  -.07298558
  (output omitted )
Iteration 15:  rho = 0.9588, criterion =  -.07167037
```

Cochrane-Orcutt AR(1) regression -- SSE search estimates

Source	SS	df	MS		
Model	2.33199178	1	2.33199178	Number of obs =	19
Residual	.071670369	17	.004215904	F(1, 17) =	553.14
				Prob > F =	0.0000
				R-squared =	0.9702
				Adj R-squared =	0.9684
Total	2.40366215	18	.133536786	Root MSE =	.06493

csales	Coef.	Std. Err.	t	P>|t|	[95% Conf. Interval]	
isales	.1605233	.0068253	23.52	0.000	.1461233	.1749234
_cons	1.738946	1.432674	1.21	0.241	-1.283732	4.761624
rho	.9588209					

```
Durbin-Watson statistic (original)     0.734728
Durbin-Watson statistic (transformed) 1.724419
```

We noted in *Options* that, with the default computation of ρ, the Cochrane–Orcutt method produces an estimate of ρ that minimizes the sum-of-squared residuals—the same criterion as the `ssesearch` option. Given that the two methods produce the same results, why would the search method ever be preferred? It turns out that the back-and-forth iterations employed by Cochrane–Orcutt may have difficulty converging if the value of ρ is large. Using the same data, the Cochrane–Orcutt iterative procedure requires over 350 iterations to converge, and a higher tolerance must be specified to prevent premature convergence:

```
. prais csales isales, corc tol(1e-9) iterate(500)
Iteration 0:  rho = 0.0000
Iteration 1:  rho = 0.6312
Iteration 2:  rho = 0.6866
 (output omitted )
Iteration 377:  rho = 0.9588
Iteration 378:  rho = 0.9588
Iteration 379:  rho = 0.9588
```

Cochrane-Orcutt AR(1) regression -- iterated estimates

Source	SS	df	MS
Model	2.33199171	1	2.33199171
Residual	.071670369	17	.004215904
Total	2.40366208	18	.133536782

Number of obs =	19
F(1, 17) =	553.14
Prob > F =	0.0000
R-squared =	0.9702
Adj R-squared =	0.9684
Root MSE =	.06493

csales	Coef.	Std. Err.	t	P>|t|	[95% Conf. Interval]
isales	.1605233	.0068253	23.52	0.000	.1461233 .1749234
_cons	1.738946	1.432674	1.21	0.241	-1.283732 4.761625
rho	.9588209				

```
Durbin-Watson statistic (original)     0.734728
Durbin-Watson statistic (transformed) 1.724419
```

Once convergence is achieved, the two methods produce identical results.

◁

Saved Results

prais saves in e():

Scalars

e(N)	number of observations	e(N_gaps)	number of gaps
e(mss)	model sum of squares	e(N_clust)	number of clusters
e(df_m)	model degrees of freedom	e(rho)	autocorrelation parameter ρ
e(rss)	residual sum of squares	e(dw)	Durbin–Watson d statistic for
e(df_r)	residual degrees of freedom		transformed regression
e(r2)	R-squared	e(dw_0)	Durbin–Watson d statistic of
e(r2_a)	adjusted R-squared		untransformed regression
e(F)	F statistic	e(tol)	target tolerance
e(rmse)	root mean squared error	e(max_ic)	maximum number of iterations
e(ll)	log likelihood	e(ic)	number of iterations

Macros

e(cmd)	prais	e(method)	twostep, iterated, or SSE search
e(depvar)	name of dependent variable	e(tranmeth)	corc or prais
e(title)	title in estimation output	e(rhotype)	method specified in rhotype option
e(clustvar)	name of cluster variable	e(properties)	b V
e(cons)	noconstant or not reported	e(predict)	program used to implement predict

Matrices

e(b)	coefficient vector	e(V)	variance–covariance matrix
			of the estimators

Functions

e(sample)	estimation sample

Methods and Formulas

prais is implemented as an ado-file.

Consider the command 'prais y x z'. The 0th iteration is obtained by estimating a, b, and c from the standard linear regression:

$$y_t = ax_t + bz_t + c + u_t$$

An estimate of the correlation in the residuals is then obtained. By default, prais uses the auxiliary regression:

$$u_t = \rho u_{t-1} + e_t$$

This can be changed to any of the computations noted in the rhotype() option.

Next, we apply a Cochrane–Orcutt transformation (1) for observations $t = 2, \ldots, n$

$$y_t - \rho y_{t-1} = a(x_t - \rho x_{t-1}) + b(z_t - \rho z_{t-1}) + c(1 - \rho) + v_t \tag{1}$$

and the transformation ($1'$) for $t = 1$

$$\sqrt{1 - \rho^2} y_1 = a(\sqrt{1 - \rho^2} x_1) + b(\sqrt{1 - \rho^2} z_1) + c\sqrt{1 - \rho^2} + \sqrt{1 - \rho^2} v_1 \tag{$1'$}$$

Thus the differences between the Cochrane–Orcutt and the Prais–Winsten methods are that the latter uses $(1')$ in addition to (1), whereas the former uses only (1), necessarily decreasing the sample size by one.

Equations (1) and $(1')$ are used to transform the data and obtain new estimates of a, b, and c.

When the `twostep` option is specified, the estimation process stops at this point and reports these estimates. Under the default behavior of iterating to convergence, this process is repeated until the change in the estimate of ρ is within a specified tolerance.

The new estimates are used to produce fitted values

$$\widehat{y}_t = \widehat{a}x_t + \widehat{b}z_t + \widehat{c}$$

and then ρ is re-estimated using, by default, the regression defined by

$$y_t - \widehat{y}_t = \rho(y_{t-1} - \widehat{y}_{t-1}) + u_t \tag{2}$$

We then re-estimate (1) using the new estimate of ρ and continue to iterate between (1) and (2) until the estimate of ρ converges.

Convergence is declared after `iterate()` iterations or when the absolute difference in the estimated correlation between two iterations is less than `tol()`; see [R] **maximize**. Sargan (1964) has shown that this process will always converge.

Under the `ssesearch` option, a combined quadratic and bisection search using golden sections searches for the value of ρ that minimizes the sum-of-squared residuals from the transformed equation. The transformation may be either the Cochrane–Orcutt (1 only) or the Prais–Winsten (1 and $1'$).

All reported statistics are based on the ρ-transformed variables, and ρ is assumed to be estimated without error. See Judge et al. (1985) for details.

The Durbin–Watson d statistic reported by `prais` and `estat dwatson` is

$$d = \frac{\sum_{j=1}^{n-1} (u_{j+1} - u_j)^2}{\sum_{j=1}^{n} u_j^2}$$

where u_j represents the residual of the jth observation.

Sigbert Jon Prais (1928–) was born in Frankfurt and moved to Britain in 1934 as a refugee. After earning degrees at the Universities of Birmingham and Cambridge and serving in various posts in research and industry, he settled at the National Institute of Economic and Social Research. Prais's interests extend widely across economics, including studies of the influence of education on economic progress.

Donald Cochrane (1917–1983) was an Australian economist and econometrician. He was born in Melbourne and earned degrees at Melbourne and Cambridge. After war-time service in the Royal Australian Air Force, he held chairs at Melbourne and Monash, being active also in work for various international organizations and national committees.

Guy Henderson Orcutt (1917–) was born in Michigan and earned degrees in physics and economics at the University of Michigan. He worked at Harvard, the University of Wisconsin, and Yale. He has contributed to econometrics and economics in several fields, most distinctively in developing micro-analytical models of economic behavior.

Acknowledgment

We thank Richard Dickens of the Centre for Economic Performance at the London School of Economics and Political Science for testing and assistance with an early version of this command.

References

Chatterjee, S., A. S. Hadi, and B. Price. 2000. *Regression Analysis by Example.* 3rd ed. New York: Wiley.

Cochrane, D. and G. H. Orcutt. 1949. Application of least-squares regression to relationships containing autocorrelated error terms. *Journal of the American Statistical Association* 44: 32–61.

Durbin, J. and G. S. Watson. 1950 and 1951. Testing for serial correlation in least-squares regression. *Biometrika* 37: 409–428 and 38: 159–178.

Hardin, J. W. 1995. sts10: Prais–Winsten regression. *Stata Technical Bulletin* 25: 26–29. Reprinted in *Stata Technical Bulletin Reprints*, vol. 5, pp. 234–237.

Harvey, A. C. 1993. *The Econometric Analysis of Time Series.* Cambridge, MA: MIT Press.

Hildreth, C. and J. Y. Lu. 1960. Demand relations with autocorrelated disturbances. *Agricultural Experiment Station Technical Bulletin* 276. East Lansing, MI: Michigan State University.

Johnston, J. and J. DiNardo. 1997. *Econometric Methods.* 4th ed. New York: McGraw–Hill.

Judge, G. G., W. E. Griffiths, R. C. Hill, H. Lütkepohl, and T.-C. Lee. 1985. *The Theory and Practice of Econometrics.* 2nd ed. New York: Wiley.

King, M. L. and D. E. A. Giles (ed.). 1987. *Specification analysis in the linear model (in honor of Donald Cochrane).* London: Routledge and Kegan Paul.

Kmenta, J. 1997. *Elements of Econometrics.* 2nd ed. Ann Arbor: University of Michigan Press.

Prais, S. J. and C. B. Winsten. 1954. Trend Estimators and Serial Correlation. *Cowles Commission Discussion Paper No. 383*, Chicago.

Sargan, J. D. 1964. Wages and prices in the United Kingdom: A study in econometric methodology. In *Econometric Analysis for National Economic Planning*, ed. P. E. Hart, G. Mills, J. K. Whitaker, 25–64. London: Butterworths.

Theil, H. 1971. *Principles of Econometrics.* New York: Wiley.

White, H. 1980. A heteroskedasticity-consistent covariance matrix estimator and a direct test for heteroskedasticity. *Econometrica* 48: 817–838.

Zellner, A. 1990. Guy H. Orcutt: Contributions to economic statistics. *Journal of Economic Behavior and Organization* 14: 43–51.

Also See

Complementary:	[TS] **prais postestimation**, [TS] **tsset**
Related:	[TS] **arima**,
	[R] **regress**, [R] **regress postestimation time series**
Background:	[U] **11.1.10 Prefix commands**,
	[U] **20 Estimation and postestimation commands**,
	[TS] **estimation options**,
	[R] **maximize**

Title

> **prais postestimation** — Postestimation tools for prais

Description

The following postestimation commands are available for `prais`:

command	description
estat	AIC, BIC, VCE, and estimation sample summary
estimates	cataloging estimation results
lincom	point estimates, standard errors, testing, and inference for linear combinations of coefficients
linktest	link test for model specification
mfx	marginal effects or elasticities
nlcom	point estimates, standard errors, testing, and inference for nonlinear combinations of coefficients
predict	predictions, residuals, influence statistics, and other diagnostic measures
predictnl	point estimates, standard errors, testing, and inference for generalized predictions
test	Wald tests for simple and composite linear hypotheses
testnl	Wald tests of nonlinear hypotheses

See the corresponding entries in the *Stata Base Reference Manual* for details.

Syntax for predict

> predict [*type*] *newvar* [*if*] [*in*] [, *statistic*]

statistic	description
xb	linear prediction; the default
stdp	standard error of the linear prediction
<u>res</u>iduals	residuals

These statistics are available both in and out of sample; type `predict ... if e(sample) ...` if wanted only for the estimation sample.

Options for predict

xb, the default, calculates the fitted values—the prediction of $x_j b$ for the specified equation. This is the linear predictor from the fitted regression model; it does not apply the estimate of ρ to prior residuals.

stdp calculates the standard error of the prediction for the specified equation, in other words, the standard error of the predicted expected value or mean for the observation's covariate pattern. This is also referred to as the standard error of the fitted value.

As computed for `prais`, this is strictly the standard error from the variance in the estimates of the parameters of the linear model and assumes that ρ is estimated without error.

residuals calculates the residuals from the linear prediction.

Methods and Formulas

All postestimation commands listed above are implemented as ado-files.

Also See

Complementary:	[TS] **prais**,
	[R] **estimates**, [R] **lincom**, [R] **linktest**,
	[R] **mfx**, [R] **nlcom**, [R] **predictnl**, [R] **test**, [R] **testnl**
Background:	[U] **13.5 Accessing coefficients and standard errors**,
	[U] **20 Estimation and postestimation commands**,
	[R] **estat**, [R] **predict**

Title

> **rolling** — Rolling window and recursive estimation

Syntax

rolling [*exp_list*] [*if*] [*in*] [, *options*] : *command*

options	description
Main	
* window(*#*)	number of consecutive data points in each sample
recursive	use recursive samples
rrecursive	use reverse recursive samples
Options	
clear	replace data in memory with results
saving(*filename*, ...)	save results to *filename*; save statistics in double precision; save results to *filename* every # replications
stepsize(*#*)	number of time periods to advance window
start(*time_constant*)	period at which rolling is to start
end(*time_constant*)	period at which rolling is to end
keep(*varname*[, start])	save *varname* along with results; optionally, use value at left edge of window
Reporting	
nodots	suppress the replication dots
noisily	display any output from *command*
trace	trace command's execution
Advanced	
reject(*exp*)	identify invalid results

* window(*#*) is required.

You must tsset your data before using rolling; see [TS] **tsset**.

aweights are allowed in *command*; see [U] **11.1.6 weight**.

exp_list contains	(*name*: *elist*)
	elist
	eexp
elist contains	*newvar* = (*exp*)
	(*exp*)
eexp is	*specname*
	[*eqno*]*specname*
specname is	_b
	_b[]
	_se
	_se[]

eqno is   # #

                           name

exp is a standard Stata expression; see [U] **13 Functions and expressions**.

Distinguish between [ ], which are to be typed, and [ ], which indicate optional arguments.

Description

rolling is a moving sampler that collects statistics from *command* after executing *command* on subsets of the data in memory. Typing

    . rolling *exp_list*, window(50) clear: *command*

executes *command* on sample windows of span 50. That is, rolling will first execute *command* using time periods 1 through 50 of the dataset, then using time periods 2 through 51, 3 through 52, and so on. rolling can also perform recursive and reverse recursive analyses, in which the starting or ending time period is held fixed and the window size grows.

command defines the statistical command to be executed. Most Stata commands and user-written programs can be used with rolling, as long as they follow standard Stata syntax and allow the *if* qualifier; see [U] **11 Language syntax**. The by prefix cannot be part of *command*.

exp_list specifies the statistics to be collected from the execution of *command*. If no expressions are given, *exp_list* assumes a default of _b if *command* stores results in e() and of all the scalars if *command* stores results in r() and not in e(). Otherwise, not specifying an expression in *exp_list* is an error.

Options

        Main

window(#) defines the window size used each time *command* is executed. The window size refers to calendar time periods, not the number of observations. If there is missing data (for example, because of weekends), the actual number of observations used by *command* may be less than window(#). window(#) is required.

recursive specifies that a recursive analysis be done. The starting time period is held fixed, the ending time period advances, and the window size grows.

rrecursive specifies that a reverse recursive analysis be done. Here the ending time period is held fixed, the starting time period advances, and the window size shrinks.

        Options

clear specifies that Stata replace the data in memory with the collected statistics even though the current data in memory has not been saved to disk.

saving(*filename*[, *suboptions*]) creates a Stata data file (.dta file) consisting of, for each statistic in *exp_list*, a variable containing the window replicates.

   double specifies that the results for each replication be stored as doubles, meaning 8-byte reals. By default, they are stored as floats, meaning 4-byte reals.

   every(#) specifies that results be written to disk every #th replication. every() should only be specified in conjunction with saving() when *command* takes a long time for each replication. This will allow recovery of partial results should your computer crash. See [P] **postfile**.

stepsize(*#*) specifies the number of periods the window is to be advanced each time *command* is executed.

start(*time_constant*) specifies the date on which rolling is to start. start() may be specified as an integer or as a date literal.

end(*time_constant*) specifies the date on which rolling is to end. end() may be specified as an integer or as a date literal.

keep(*varname*[, start]) specifies a variable to be posted along with the results. The value posted is the value that corresponds to the right edge of the window. Specifying the start option requests that the value corresponding to the left edge of the window be posted instead. This option is frequently used to record calendar dates.

___

⌐ Reporting ⌐_____

nodots suppresses display of the replication dot for each window on which *command* is executed. By default, a single dot character is printed for each window. A single red 'x' is printed if *command* returns with an error or if any of the values in *exp_list* is missing.

noisily causes the output of *command* to be displayed for each window on which *command* is executed. This option implies the nodots option.

trace causes a trace of the execution of *command* to be displayed. This option implies the noisily and nodots options.

___

⌐ Advanced ⌐_____

reject(*exp*) identifies an expression that indicates when results should be rejected. When *exp* is true, the saved statistics are set to missing values.

Remarks

rolling executes a command on each of a series of windows of observations and stores the results. rolling can perform what are commonly called rolling regressions, recursive regressions, and reverse recursive regressions. However, rolling is not limited to just linear regression analysis: any command that saves results in e() or r() can be used with rolling.

Suppose that you have data collected at 100 consecutive points in time, numbered one through 100, and you wish to perform a rolling regression with a window size of 20 time periods. Typing

```
. rolling _b, window(20) clear: regress depvar indepvar
```

causes Stata to regress *depvar* on *indepvar* using periods 1 through 20, store the regression coefficients (_b), run the regression using periods 2 through 21, and so on, finishing with a regression using periods 81 through 100 (the last 20 periods).

The stepsize() option specifies how far ahead the window is moved each time. For example, if you specify step(2), then *command* is executed on time periods 1 through 20, then 3 through 22, 5 through 24, etc. By default, rolling replaces the dataset in memory with the computed statistics unless the saving() option is specified, in which case the computed statistics are stored in the filename specified. If the dataset in memory has been changed since it was last saved and you do not specify saving(), you must use clear.

Additionally, rolling can perform recursive and reverse recursive analyses. In a recursive analysis, the starting date is held fixed, and the window size grows as the ending date is advanced. In a reverse recursive analysis, the ending date is held fixed, and the window size shrinks as the starting date is advanced.

▷ Example 1

We have data on the daily returns to IBM stock (`ibm`), the S&P 500 (`spx`), and short-term interest rates (`irx`), and we want to create a series containing the beta of IBM using the previous 200 trading days at each date. We will also record the standard errors, so that we can obtain 95% confidence intervals for the betas. See, for example, Stock and Watson (2003, 102) for more information on estimating betas. We type

```
. use http://www.stata-press.com/data/r9/ibm, clear
(Source: Yahoo! Finance)
. tsset t
        time variable:  t, 1 to 494
. generate ibmadj = ibm - irx
(1 missing value generated)
. generate spxadj = spx - irx
(1 missing value generated)
. rolling _b _se, window(200) saving(betas, replace) keep(date): regress ibmadj
> spxadj
(running regress on estimation sample)
(note: file betas.dta not found)
Rolling replications (295)
 ───┼─── 1 ───┼─── 2 ───┼─── 3 ───┼─── 4 ───┼─── 5
.................................................        50
.................................................       100
.................................................       150
.................................................       200
.................................................       250
...........................................
file betas.dta saved
```

Our dataset has both a time variable `t` that runs consecutively and a date variable `date` that measures the calendar date and therefore has gaps at weekends and holidays. Had we used the `date` variable as our time variable, `rolling` would have used windows consisting of 200 calendar days instead of 200 trading days, and each window would not have exactly 200 observations. We used the `keep(date)` option so that we could refer to the `date` variable when working with the results dataset.

We can `list` a portion of the dataset created by `rolling` to see what it contains:

```
. use betas, clear
(rolling: regress)
. sort date
. list in 1/3
```

	start	end	date	_b_spx~j	_b_cons	_se_sp~j	_se_cons
1.	1	200	16oct2003	1.043422	-.0181504	.0658531	.0748295
2.	2	201	17oct2003	1.039024	-.0126876	.0656893	.074609
3.	3	202	20oct2003	1.038371	-.0235616	.0654591	.0743851

The variables `start` and `end` indicate the first and last observations used each time `rolling` called `regress`, and the `date` variable contains the calendar date corresponding the time period represented by `end`. The remaining variables are the estimated coefficients and standard errors from the regression. In our example , `_b_spxadj` contains the estimated betas, and `_b_cons` contains the estimated alphas. The variables `_se_spxadj` and `_se_cons` have the corresponding standard errors.

Finally, we compute the confidence intervals for the betas and examine how they have changed over time:

```
. generate lower = _b_spxadj - 1.96*_se_spxadj
. generate upper = _b_spxadj + 1.96*_se_spxadj
. twoway (line _b_spxadj date) (rline lower upper date) if date>=d(1oct2003),
> ytitle("Beta")
```

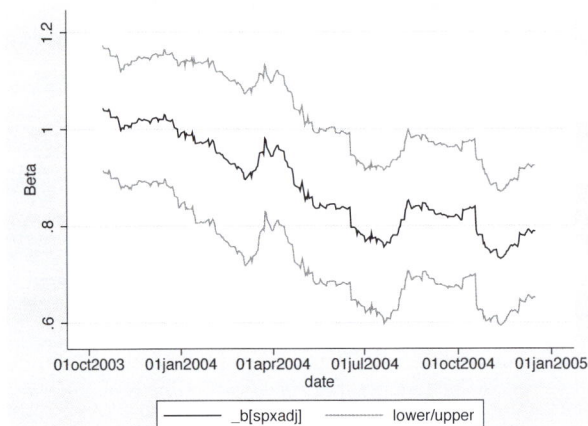

As 2004 progressed, IBM's stock returns were less influenced by returns in the broader market. Beginning in June of 2004, IBM's beta became significantly different from unity at the 95% confidence level, as indicated by the fact that the confidence interval does not contain one from then onwards.

◁

In addition to rolling window analyses, rolling can also perform recursive ones. Suppose again that you have data collected at 100 consecutive points in time, and now you type

```
. rolling _b, window(20) recursive clear: regress depvar indepvar
```

Stata will first regress *depvar* on *indepvar* using observations 1 through 20, store the coefficients, run the regression using observations 1 through 21, observations 1 through 22, and so on, finishing with a regression using all 100 observations. Unlike a rolling regression, in which case the number of observations is held constant and the starting and ending points are shifted, a recursive regression holds the starting point fixed and increases the number of observations. Recursive analyses are often used in forecasting situations. As time goes by, more information becomes available that can be used in making forecasts. See, for example, Kmenta (1997, 423–424).

▷ Example 2

Using the same dataset, we type

```
. use http://www.stata-press.com/data/r9/ibm, clear
(Source: Yahoo! Finance)
. tsset t
        time variable:  t, 1 to 494
. generate ibmadj = ibm - irx
(1 missing value generated)
. generate spxadj = spx - irx
(1 missing value generated)
```

```
. rolling _b _se, recursive window(200) clear: regress ibmadj spxadj
(output omitted)
. list in 1/3
```

	start	end	_b_spx~j	_b_cons	_se_sp~j	_se_cons
1.	1	200	1.043422	-.0181504	.0658531	.0748295
2.	1	201	1.039024	-.0126876	.0656893	.074609
3.	1	202	1.037687	-.016475	.0655896	.0743481

Notice that here the starting time period remains fixed and the window grows larger.

◁

In a reverse recursive analysis, the ending date is held fixed, and the window size becomes smaller as the starting date is advanced. For example, with a dataset that has observations numbered one through 100, typing

```
. rolling _b, window(20) reverse recursive clear: regress depvar indepvar
```

creates a dataset in which the first observation has the results based on periods 1 through 100, the second observation has the results based on 2 through 100, the third having 3 through 100, and so on up to the last observation having results based on periods 81 through 100 (the last 20 observations).

▷ Example 3

Using the data on stock returns, we want to build a model in which we predict today's IBM stock return on the basis of the yesterday's returns on IBM and the S&P 500. That is, letting i_t and s_t denote the returns to IBM and the S&P 500 on date t, we want to fit the regression model

$$i_t = \beta_0 + \beta_1 i_{t-1} + \beta_2 s_{t-1} + \epsilon_t$$

where ϵ_t is a regression error term, and then compute

$$\widehat{i_{t+1}} = \widehat{\beta_0} + \widehat{\beta_1} i_t + \widehat{\beta_2} s_t$$

We will use recursive regression because we suspect that the more data we have to fit the regression model, the better the model will predict returns. We will use at least 20 time periods in fitting the regression.

One alternative would be to use `rolling` with the `recursive` option to fit the regressions, collect the coefficients, and then compute the predicted values afterward. However, we will instead write a short program that computes the forecasts automatically and then use `rolling, recursive` on that program. The program must accept an `if` expression so that `rolling` can indicate to the program which observations are to be used. Our program is

```
program myforecast, rclass
        syntax [if]
        regress ibm L.ibm L.spx 'if'
        // Find last time period of estimation sample and
        // make forecast for period just after that
        summ t if e(sample)
        local last = r(max)
        local fcast = _b[_cons] + _b[L.ibm]*ibm['last'] + ///
                        _b[L.spx]*spx['last']
        return scalar forecast = 'fcast'
        // Next period's actual return
        // Will return missing value for final period
        return scalar actual = ibm['last'+1]
end
```

Now we call `rolling`:

```
. rolling actual=r(actual) forecast=r(forecast), recursive window(20): myforecast
  (output omitted)

. corr actual forecast
(obs=474)
```

	actual	forecast
actual	1.0000	
forecast	-0.0957	1.0000

Clearly, our model does not work too well—the correlation between actual returns and our forecasts is negative!

◁

Saved Results

`rolling` does not set any r- or e-class macros. The results from the command used with `rolling`, depending on the last window of data used, are available after `rolling` has finished.

Methods and Formulas

`rolling` is implemented as an ado-file.

Acknowledgment

We would like to thank Christopher Baum for an earlier rolling regression command.

References

Kmenta, J. 1997. *Elements of Econometrics*. 2nd ed. Ann Arbor: University of Michigan Press.

Stock, J. H. and M. W. Watson. 2003. *Introduction to Econometrics*. Boston: Addison–Wesley.

Also See

Related:	[D] **statsby**
Background:	[U] **13.6 Accessing results from Stata commands**,
	[R] **saved results**

Title

tsappend — Add observations to a time-series dataset

Syntax

tsappend , add(*#*) last(*date*) tsfmt(*string*) [*options*]

options	description
add(#)	add # observations
last(date)	add observations at *date*
tsfmt(string)	use time-series function *string* with last(*date*)
panel(*panel_id*)	add observations to panel *panel_id*

* Either add(*#*) is required, or last(*date*) and tsfmt(*string*) are required.

You must tsset your data before using tsappend; see [TS] **tsset**.

Description

tsappend appends observations to a time-series dataset or to a panel dataset. tsappend uses and updates the information set by tsset.

Options

add(*#*) specifies the number of observations to add.

last(*date*) and tsfmt(*string*) must be specified together and are an alternative to add().

last(*date*) specifies the date of the last observation to add.

tsfmt(*string*) specifies the name of the Stata time-series function to use in converting the date specified in last() to an integer. The function names are d (daily), w (weekly), m (monthly), q (quarterly), y (yearly), and h (half-yearly).

For instance, you might specify last(17may2004) tsfmt(d) or last(2001m1) tsfmt(m).

panel(*panel_id*) specifies that observations be added only to panels with the ID specified in panel().

Remarks

Remarks are presented under the headings

Introduction
Use of tsappend with time-series data
Use of tsappend with panel data

Introduction

tsappend adds observations to a time-series dataset or to a panel dataset. You must tsset your data before using tsappend. tsappend simultaneously removes any gaps from the dataset.

There are two ways to use tsappend: you can specify the add(#) option to request that # observations be added, or you can specify the last(*date*) option to request that observations be appended until the date specified is reached. If you specify last(), you must also specify tsfmt(). tsfmt() specifies the Stata time-series date function that converts the date held in last() to an integer.

tsappend works with time-series of panel data. With panel data, tsappend adds the requested observations to all the panels, unless the panel() option is also specified.

Use of tsappend with time-series data

tsappend can be useful for appending observations when dynamically predicting a time series. Consider an example in which tsappend adds the extra observations before dynamically predicting from an AR(1) regression:

```
. use http://www.stata-press.com/data/r9/tsappend1
. regress y l.y
```

Source	SS	df	MS		Number of obs =	479
					F(1, 477) =	119.29
Model	115.349555	1	115.349555		Prob > F =	0.0000
Residual	461.241577	477	.966963473		R-squared =	0.2001
					Adj R-squared =	0.1984
Total	576.591132	478	1.2062576		Root MSE =	.98334

y	Coef.	Std. Err.	t	P>|t|	[95% Conf. Interval]
y					
L1.	.4493507	.0411417	10.92	0.000	.3685093 .5301921
_cons	11.11877	.8314581	13.37	0.000	9.484993 12.75254

```
. mat b = e(b)
. mat colnames b = L.xb one
. tsset
        time variable:  t2, 1960m2 to 2000m1
. tsappend , add(12)
. tsset
        time variable:  t2, 1960m2 to 2001m1
. predict xb if t2<=m(2000m2)
(option xb assumed; fitted values)
(12 missing values generated)
. gen one=1
. mat score xb=b if t2 >= m(2000m2), replace
```

The calls to tsset before and after tsappend were unnecessary. Their output reveals that tsappend added another year of observations. We then used predict and matrix score to obtain the dynamic predictions, which allows us to produce the following graph:

```
. line y xb t2 if t2>=m(1995m1), ytitle("") xtitle("time")
```

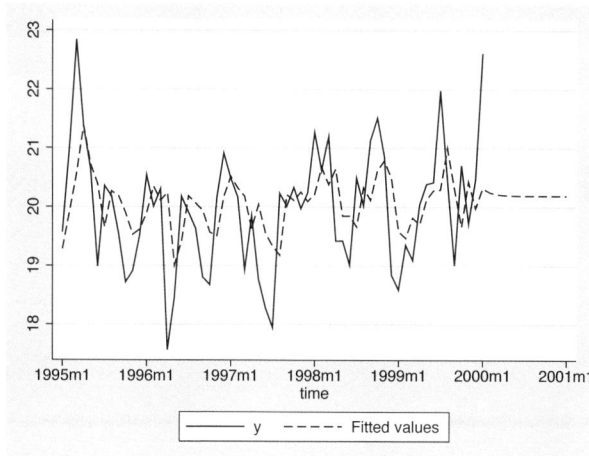

In the call to tsappend, instead of saying that we wanted to add 12 observations, we could have specified that we wanted to fill in observations through the first month of 2001:

```
. use http://www.stata-press.com/data/r9/tsappend1, clear
. tsset
        time variable:  t2, 1960m2 to 2000m1
. tsappend, last(2001m1) tsfmt(m)
. tsset
        time variable:  t2, 1960m2 to 2001m1
```

Note that we specified the m() function in the tsfmt() option. [D] **functions** contains a list of time-series functions for translating date literals to integers. Since we have monthly data, and since [D] **functions** tells us that we want to use the m() function, we specified the option tsfmt(m). The following table shows the most common types of time-series data, their formats, the appropriate translation functions, and the corresponding options for tsappend:

Description	Format	Function	Option
daily	%td	d()	tsfmt(d)
weekly	%tw	w()	tsfmt(w)
monthly	%tm	m()	tsfmt(m)
quarterly	%tq	q()	tsfmt(q)
yearly	%ty	y()	tsfmt(y)
half-yearly	%th	h()	tsfmt(h)

Use of tsappend with panel data

tsappend's actions on panel data are very similar to its action on time-series data, except that tsappend preforms those actions on each of the time series within the panels.

If the end dates vary over panels, last() and add() will produce different results. add(#) always adds # observations to each panel. If the data end at different time periods before tsappend, add() is used, the data will still end at different periods after tsappend, add(). In contrast, tsappend, last() tsfmt() will cause all the panels to end on the specified last date. If the beginning dates differ across panels, using `tsappend, last() tsfmt()` to provide a uniform ending date will not create balanced panels because the number of observations per panel will still differ.

Consider the panel data summarized in the output below:

```
. use http://www.stata-press.com/data/r9/tsappend3, clear

. xtdes
      id:  1, 2, ..., 3                                    n =          3
      t2:  456, 457, ..., 480                              T =         25
           Delta(t2) = 1; (480-456)+1 = 25
           (id*t2 uniquely identifies each observation)
Distribution of T_i:   min      5%     25%      50%     75%    95%     max
                        13      13      13       20      24     24      24

   Freq.  Percent    Cum. |  Pattern
   ------------------------+---------------------------
       1    33.33   33.33  |  ............1111111111111
       1    33.33   66.67  |  1111.11111111111111111111
       1    33.33  100.00  |  111111111111111111111.....
   ------------------------+---------------------------
       3   100.00         |  XXXXXXXXXXXXXXXXXXXXXXXXX

. by id: sum t2
```

```
-> id = 1
    Variable |       Obs        Mean    Std. Dev.       Min        Max
-------------+--------------------------------------------------------
          t2 |        13         474     3.89444        468        480
```

```
-> id = 2
    Variable |       Obs        Mean    Std. Dev.       Min        Max
-------------+--------------------------------------------------------
          t2 |        20       465.5     5.91608        456        475
```

```
-> id = 3
    Variable |       Obs        Mean    Std. Dev.       Min        Max
-------------+--------------------------------------------------------
          t2 |        24     468.3333    7.322786       456        480
```

The output from `xtdes` and `summarize` on these data tells us that one panel starts later than the other, that another panel ends before the other two, and that the remaining panel has a gap in the time variable but otherwise spans the entire time frame.

Now consider the data after a call to `tsappend, add(6)`:

```
. tsappend, add(6)

. xtdes
      id:  1, 2, ..., 3                                    n =          3
      t2:  456, 457, ..., 486                              T =         31
           Delta(t2) = 1; (486-456)+1 = 31
           (id*t2 uniquely identifies each observation)
Distribution of T_i:   min      5%     25%      50%     75%    95%     max
                        19      19      19       26      31     31      31

   Freq.  Percent    Cum. |  Pattern
   ------------------------+---------------------------------
       1    33.33   33.33  |  ............1111111111111111111
       1    33.33   66.67  |  1111111111111111111111111.....
       1    33.33  100.00  |  1111111111111111111111111111111
   ------------------------+---------------------------------
       3   100.00         |  XXXXXXXXXXXXXXXXXXXXXXXXXXXXXXX
```

```
. by id: sum t2
```

```
-> id = 1
    Variable |     Obs       Mean   Std. Dev.       Min        Max
-------------+--------------------------------------------------------
          t2 |      19        477    5.627314        468        486
```

```
-> id = 2
    Variable |     Obs       Mean   Std. Dev.       Min        Max
-------------+--------------------------------------------------------
          t2 |      26      468.5    7.648529        456        481
```

```
-> id = 3
    Variable |     Obs       Mean   Std. Dev.       Min        Max
-------------+--------------------------------------------------------
          t2 |      31        471    9.092121        456        486
```

This output from xtdes and summarize after the call to tsappend shows that the call to tsappend, add(6) added six observations to each panel and filled in the gap in the time variable in the second panel. As noted above, tsappend, add() did not cause a uniform end date over the panels.

The following output illustrates the contrast between tsappend, add() and tsappend, last() tsfmt() with panel data that end at different dates. The output from xtdes and summarize shows that the call to tsappend, last() tsfmt() filled in the gap in t2 and caused all the panels to end at the specified end date. The output also shows that the panels remain unbalanced because one panel has a later entry date than the other two.

```
. use http://www.stata-press.com/data/r9/tsappend2

. tsappend, last(2000m7) tsfmt(m)

. xtdes
   id:  1, 2, ..., 3                                      n =          3
   t2:  456, 457, ..., 486                                T =         31
        Delta(t2) = 1; (486-456)+1 = 31
        (id*t2 uniquely identifies each observation)
Distribution of T_i:    min      5%     25%      50%      75%     95%    max
                         19      19      19       31       31      31     31

   Freq.  Percent    Cum. |  Pattern
---------------------------+----------------------------------------
      2    66.67   66.67 |  1111111111111111111111111111111
      1    33.33  100.00 |  ............1111111111111111111
---------------------------+----------------------------------------
      3   100.00         |  XXXXXXXXXXXXXXXXXXXXXXXXXXXXXXX
```

(Continued on next page)

```
. by id: sum t2
```

```
-> id = 1
```

Variable	Obs	Mean	Std. Dev.	Min	Max
t2	19	477	5.627314	468	486

```
-> id = 2
```

Variable	Obs	Mean	Std. Dev.	Min	Max
t2	31	471	9.092121	456	486

```
-> id = 3
```

Variable	Obs	Mean	Std. Dev.	Min	Max
t2	31	471	9.092121	456	486

Saved Results

tsappend saves in r():

Scalars
   r(add)   number of observations added

Methods and Formulas

tsappend is implemented as an ado-file.

Also See

Complementary:	[TS] **tsset**
Related:	[D] **egen**, [D] **generate**
Background:	[U] **11.4.3 Time-series varlists**

Title

> **tsfill** — Fill in missing times with missing observations in time-series data

Syntax

tsfill [, <u>f</u>ull]

You must tsset your data before using tsfill; see [TS] **tsset**.

Description

tsfill is used after tsset to fill in gaps in time-series data and gaps in panel data with new observations, which contain missing values. For instance, perhaps observations for *timevar* = 1, 3, 5, 6, ..., 22 exist. tsfill would create observations for *timevar* = 2 and *timevar* = 4 containing all missing values. There is seldom reason to do this because Stata's time-series operators consider *timevar* not the observation number. Referring to L.gnp to obtain lagged gnp values would correctly produce a missing value for *timevar* = 3, even if the data were not filled in. Referring to L2.gnp would correctly return the value of gnp in the first observation for *timevar* = 3, even if the data were not filled in.

Option

full is for use with panel data only. With panel data, tsfill by default fills in observations for each panel according to the minimum and maximum values of *timevar* for the panel. Thus if the first panel spanned the times 5–20 and the second panel the times 1–15, after tsfill they would still span the same time periods; observations would be created to fill in any missing times from 5–20 in the first panel and from 1–15 in the second.

If full is specified, observations are created so that both panels span the time 1–20, the overall minimum and maximum of *timevar* across panels.

Remarks

Remarks are presented under the headings

> *Use of tsfill with time-series data*
> *Use of tsfill with panel data*

(Continued on next page)

Use of tsfill with time-series data

You have monthly data, with gaps:

```
. list mdate income
```

	mdate	income
1.	1995m7	1153
2.	1995m8	1181
3.	1995m11	1236
4.	1995m12	1297
5.	1996m1	1265
6.	1996m3	1282

You can fill in the gaps by interpolation easily with `tsfill` and `ipolate`. `tsfill` creates the missing observations:

```
. tsfill
. list mdate income ipinc
```

	mdate	income	
1.	1995m7	1153	
2.	1995m8	1181	
3.	1995m9	.	← new
4.	1995m10	.	← new
5.	1995m11	1236	
6.	1995m12	1297	
7.	1996m1	1265	
8.	1996m2	.	← new
9.	1996m3	1282	

We can now use `ipolate` (see [D] **ipolate**) to fill them in:

```
. ipolate income mdate, gen(ipinc)
. list mdate income ipinc
```

	mdate	income	ipinc
1.	1995m7	1153	1153
2.	1995m8	1181	1181
3.	1995m9	.	1199.333
4.	1995m10	.	1217.667
5.	1995m11	1236	1236
6.	1995m12	1297	1297
7.	1996m1	1265	1265
8.	1996m2	.	1273.5
9.	1996m3	1282	1282

Use of tsfill with panel data

You have the following panel dataset:

```
. list edlevel year income
```

	edlevel	year	income
1.	1	1988	14500
2.	1	1989	14750
3.	1	1990	14950
4.	1	1991	15100
5.	2	1989	22100
6.	2	1990	22200
7.	2	1992	22800

Just as with nonpanel time-series datasets, you can use tsfill to fill in the gaps:

```
. tsfill
. list edlevel year income
```

	edlevel	year	income	
1.	1	1988	14500	
2.	1	1989	14750	
3.	1	1990	14950	
4.	1	1991	15100	
5.	2	1989	22100	
6.	2	1990	22200	
7.	2	1991	.	← new
8.	2	1992	22800	

You could instead use tsfill to produce fully balanced panels using the full option:

```
. tsfill, full
. list edlevel year income, sep(0)
```

	edlevel	year	income	
1.	1	1988	14500	
2.	1	1989	14750	
3.	1	1990	14950	
4.	1	1991	15100	
5.	1	1992	.	← new
6.	2	1988	.	← new
7.	2	1989	22100	
8.	2	1990	22200	
9.	2	1991	.	← new
10.	2	1992	22800	

Methods and Formulas

tsfill is implemented as an ado-file.

Also See

Complementary:	[TS] **tsset**
Related:	[TS] **tsappend**
Background:	[U] **11.4.3 Time-series varlists**,
	[U] **12.5.4 Time-series formats**,
	[U] **24.3 Time-series dates**,
	[U] **26.13 Models with time-series data**

Title

> **tsline** — Plot time-series data

Syntax

Time-series line plot

> [<u>tw</u>oway] tsline *varlist* [*if*] [*in*] [, *scatter_options twoway_options*]

Time-series range plot with lines

> [<u>tw</u>oway] tsrline y_1 y_2 [*if*] [*in*] [, *rline_options twoway_options*]

where the time variable is assumed set by **tsset**, *varlist* has the interpretation $y_1 [y_2 \ldots y_k]$.

Description

tsline draws line plots for time-series data.

tsrline draws a range plot with lines for time-series data.

tsline and **tsrline** are both commands and *plottypes* as defined in [G] **graph twoway**. Thus the syntax for **tsline** is

> . graph twoway tsline ...

> . twoway tsline ...

> . tsline ...

and similarly for **tsrline**. Being plottypes, these commands may be combined with other plottypes in the **twoway** family, as in,

> . twoway (tsrline ...) (tsline ...) (lfit ...) ...

which can equivalently be written

> . tsrline ... || tsline ... || lfit ... || ...

Options

┌─ Plot 1, Plot 2, Plot 3, Plot 4 ─────────────────────────────────

scatter_options are any of the options allowed by the **graph twoway scatter** command except that *marker_options*, *marker_placement_option*, and *marker_label_options* will be ignored if specified; see [G] **graph twoway scatter**.

rline_options are any of the options allowed by the **graph twoway rline** command; see [G] **graph twoway rline**.

Y-Axis, T-Axis, R-Axis, Title, Caption, Legend, Overall, By

twoway_options are any of the options documented in [G] ***twoway_options***. These include options for titling the graph (see [G] ***title_options***), options for saving the graph to disk (see [G] ***saving_option***), and the `by()` option, which will allow you to simultaneously plot different subsets of the data (see [G] ***by_option***).

Also see the `recast()` option discussed in [G] ***advanced_options*** for information on how to plot spikes, bars, etc., instead of lines.

Remarks

▷ Example 1

We simulated two separate time series (each of 200 observations) and placed them in a Stata dataset, `tsline1.dta`. The first series simulates an AR(2) process with $\phi_1 = 0.8$ and $\phi_2 = 0.2$; the second series simulates an MA(2) process with $\theta_1 = 0.8$ and $\theta_2 = 0.2$. We use `tsline` to graph these two series.

```
. use http://www.stata-press.com/data/r9/tsline1, clear
. tsset lags
        time variable:  lags, 0 to 199
. tsline ar ma
```

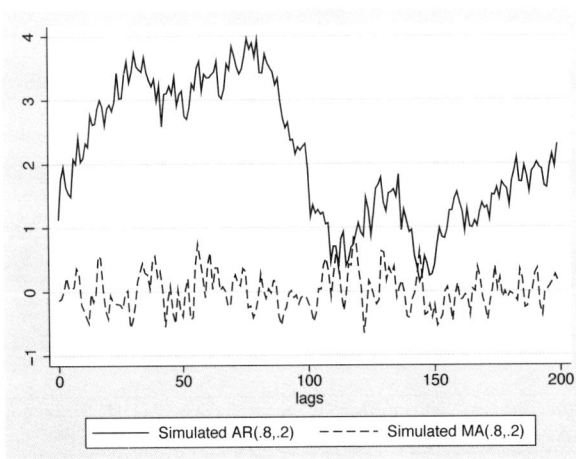

◁

▷ Example 2

Suppose that we kept a calorie log for an entire calendar year. At the end of the year, we would have a dataset (e.g., `tsline2.dta`) that contains the amount of calories consumed for 365 days. We could then use `tsset` to identify the date variable and `tsline` to plot calories versus time. Knowing that we tend to eat a little more food on Thanksgiving and Christmas day, we use the `ttick()` and `ttext()` options to point these days out on the time axis.

```
. use http://www.stata-press.com/data/r9/tsline2, clear
. tsset day
        time variable:  day, 01jan2002 to 31dec2002
. tsline calories, ttick(28nov2002 25dec2002, tpos(in))
> ttext(3470 28nov2002 "thanks" 3470 25dec2002 "x-mas", orient(vert))
```

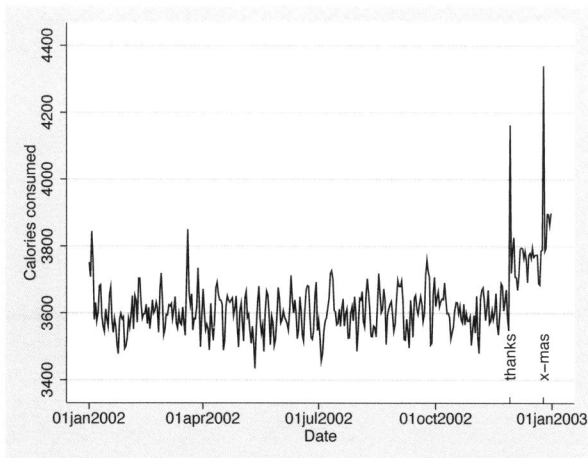

We were uncertain of the exact values we logged, so we also gave a range for each day. Here is a plot of the summer months.

```
. use http://www.stata-press.com/data/r9/tsline2, clear
. tsset day
        time variable:  day, 01jan2002 to 31dec2002
. tsrline lcal ucal if tin(1may2002,31aug2002) || tsline cal ||
> if tin(1may2002,31aug2002), ytitle(Calories)
```

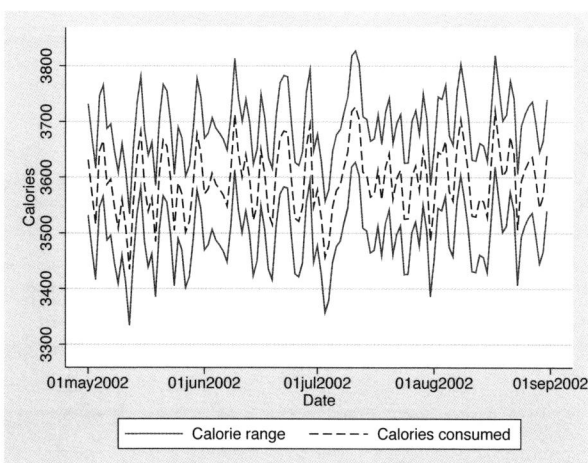

Options associated with the time axis allow dates to be specified in place of elapsed time values. For instance, we used

```
ttick(28nov2002 25dec2002 , tpos(in))
```

to place tick marks at the specified dates. This works similarly for `tlabel`, `tmlabel`, and `tmtick`.

Suppose that we wanted to place vertical lines for the previously mentioned holidays. We could specify the dates in the `tline()` option as follows:

```
. use http://www.stata-press.com/data/r9/tsline2, clear
. tsset day
        time variable:  day, 01jan2002 to 31dec2002
. tsline calories, tline(28nov2002 25dec2002)
```

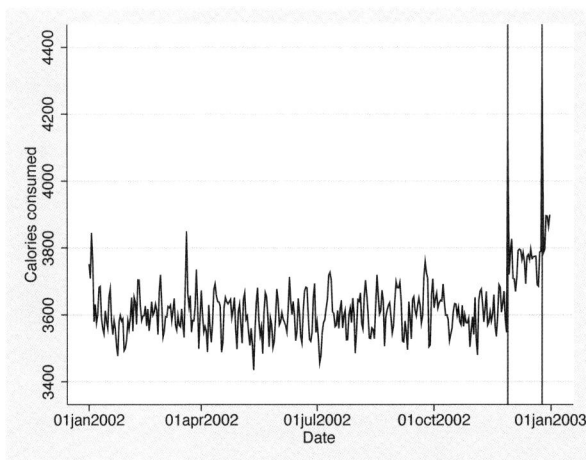

(Continued on next page)

We could also modify the format of the time axis so that only the day in the year is displayed in the labeled ticks:

```
. use http://www.stata-press.com/data/r9/tsline2, clear
. tsset day
        time variable:  day, 01jan2002 to 31dec2002
. tsline calories, tlabel(, format(%tdmd)) ttitle("Date (2002)")
```

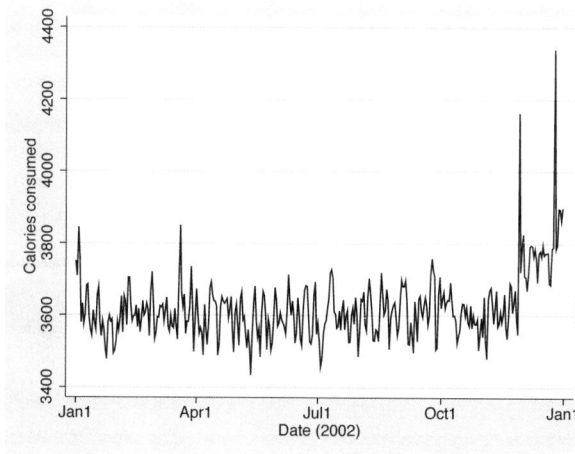

◁

Methods and Formulas

tsline is implemented as an ado-file.

Also See

Complementary:	[TS] **tsset**,
	[G] **graph twoway**
Related:	[XT] **xtline**

Title

tsreport — Report time-series aspects of a dataset or estimation sample

Syntax

tsreport $[if]$ $[in]$ $[, options]$

options	description
Main	
<u>r</u>eport	display number of gaps in the time series
report0	display count of gaps even if no gaps
<u>l</u>ist	display gaps data in tabular list
<u>p</u>anel	ignore panel changes

tsreport typed without options produces no output but provides its standard saved results.

Description

tsreport reports time gaps in a sample of observations. The report option displays a one-line statement showing the count of the gaps, and list displays a complete list of records that follow gaps. A return value r(N_gaps) indicates the number of gaps in the sample.

Options

> Main

report specifies that the number of gaps in the time series be reported, if any gaps exist.

report0 specifies that the count of gaps be reported, even if there are no gaps.

list specifies that a tabular list of gaps be displayed.

panel specifies that panel changes not be counted as gaps. Whether panel changes are counted as gaps usually depends on how the calling command handles panels.

Remarks

Time-series commands sometimes require that observations be on a fixed time interval with no gaps, or the command may not function properly. tsreport provides a tool for reporting the gaps in a sample.

▷ Example 1

The following monthly panel data have two panels and a missing month (March) in the second panel:

```
. list edlevel month income in 1/6, sep(0)
```

	edlevel	month	income
1.	1	1998m1	687
2.	1	1998m2	783
3.	1	1998m3	790
4.	2	1998m1	1435
5.	2	1998m2	1522
6.	2	1998m4	1532

Invoking `tsreport` without the `panel` option gives us the following report:

```
. tsreport, report
Number of gaps in sample:  2   (gap count includes panel changes)
```

We could get a list of gaps and better see what has been counted as a gap by using the `list` option:

```
. tsreport, report list
Number of gaps in sample:  2   (gap count includes panel changes)
Observations with preceding time gaps
(gaps include panel changes)
```

Record	edlevel	month
4	2	1998m1
6	2	1998m4

We now see why `tsreport` is reporting two gaps. It is counting the known gap in March of the second panel and also counting the change from the first to the second panel. (If we are programmers writing a procedure that does not account for panels, a change from one panel to the next represents a break in the time series just as a gap in the data does.)

We may prefer that the changes in panels not be counted as gaps. We can obtain a count without the panel change by using the `panel` option:

```
. tsreport, report panel
Number of gaps in sample:  1
```

To obtain a fuller report, we type

```
. tsreport, report list panel
Number of gaps in sample:  1
Observations with preceding time gaps
```

Record	edlevel	month
6	2	1998m4

◁

Saved Results

tsreport saves in r():

Scalars
      r(N_gaps)   number of gaps in sample

Methods and Formulas

tsreport is implemented as an ado-file.

Also See

Complementary: [TS] **tsset**

Background: [U] **11.4.3 Time-series varlists**,

                      [U] **12.5.4 Time-series formats**,

                      [U] **24.3 Time-series dates**,

                      [U] **26.13 Models with time-series data**

Title

tsrevar — Time-series operator programming command

Syntax

tsrevar [*varlist*] [*if*] [*in*] [, substitute list]

You must tsset your data before using tsrevar; see [TS] **tsset**.

Description

tsrevar, substitute takes a *varlist* that might contain *op.varname* combinations and substitutes equivalent temporary variables for the combinations.

tsrevar, list creates no new variables. It returns in r(varlist) the list of base variables corresponding to *varlist*.

Options

substitute specifies that tsrevar resolve *op.varname* combinations by creating temporary variables as described above. substitute is the default action taken by tsrevar; you do not need to specify the option.

list specifies that tsrevar return a list of base variable names.

Remarks

tsrevar substitutes temporary variables for any *op.varname* combinations in a variable list. For instance, the original *varlist* might be "gnp L.gnp r", and tsrevar, substitute would create *newvar* = L.gnp and create the equivalent varlist "gnp *newvar* r". This new varlist could then be used with commands that do not otherwise support time-series operators, or it could be used in a program to make execution faster at the expense of using more memory.

tsrevar, substitute might create no new variables, one new variable, or many new variables, depending on the number of *op.varname* combinations appearing in *varlist*. Any new variables created are temporary. The new, equivalent varlist is returned in r(varlist). Note that the new varlist corresponds one to one with the original *varlist*.

tsrevar, list returns in r(varlist) the list of base variable names of *varlist* with the time-series operators removed. tsrevar, list creates no new variables. For instance, if the original *varlist* were "gnp l.gnp l2.gnp r l.cd", then r(varlist) would contain "gnp r cd". This is useful for programmers who might want to create programs to keep only the variables corresponding to *varlist*.

▷ Example 1

        . tsrevar l.gnp d.gnp r

creates two temporary variables containing the values for l.gnp and d.gnp. The variable r appears in the new variable list but does not require a temporary variable.

The resulting variable list is

```
. display "'r(varlist)'"
__00014P __00014Q r
```

We can see the results by listing the new variables alongside the original value of gnp.

```
. list gnp 'r(varlist)'  in 1/5
```

	gnp	__00014P	__00014Q	r
1.	128	.	.	3.2
2.	135	128	7	3.8
3.	132	135	-3	2.6
4.	138	132	6	3.9
5.	145	138	7	4.2

Remember that temporary variables automatically vanish when the program concludes.

If we had needed only the base variable names, we could have specified

```
. tsrevar l.gnp d.gnp r, list
. display "'r(varlist)'"
gnp r
```

The order of the list will probably differ from the original list; base variables are listed only once and are listed in the order they appear in the dataset.

◁

❑ Technical Note

tsrevar, substitute avoids creating duplicate variables. Consider

```
. tsrevar gnp l.gnp r cd l.cd l.gnp
```

Note that l.gnp appears twice in the varlist. tsrevar will create only one new variable for l.gnp and use that new variable twice in the resulting r(varlist). Moreover, tsrevar will even do this across multiple calls:

```
. tsrevar gnp l.gnp cd l.cd
. tsrevar cpi l.gnp
```

Note that l.gnp appears in two separate calls. At the first call, tsrevar creates a temporary variable corresponding to l.gnp. At the second call, tsrevar remembers what it has done and uses that same temporary variable for l.gnp again.

❑

Saved Results

tsrevar saves in r():

Macros

    r(varlist)   the modified variable list or list of base variable names

Also See

Related: [P] **syntax**, [P] **unab**

Background: [U] **11 Language syntax**,
                  [U] **11.4.3 Time-series varlists**,
                  [U] **18 Programming Stata**

Title

> **tsset** — Declare a dataset to be time-series data

Syntax

Declare data to be time series and specify the time variable

> tsset [*panelvar*] *timevar* [, *options*]

Display how dataset is currently tsset

> tsset

Clear time-series settings

> tsset, clear

where *panelvar* is a variable that identifies the panels and *timevar* is a variable that identifies the time periods.

options	description
d̲aily	display time scales as daily (%td, 0 = 1jan1960)
w̲eekly	display time scales as weekly (%tw, 0 = 1960w1)
m̲onthly	display time scales as monthly (%tm, 0 = 1960m1)
quarterly	display time scales as quarterly (%tq, 0 = 1960q1)
h̲alfyearly	display time scales as half-yearly (%th, 0 = 1960h1)
y̲early	display time scales as yearly (%ty, 1960 = 1960)
generic	display time scales as generic (%tg, 0 = ?)
f̲ormat(%*fmt*)	indicate how *timevar* will be displayed

Description

tsset declares the data to be a time series and designates that *timevar* represents time. Stata's time-series operators then use *timevar* in computing operations on variables.

tsset without arguments displays how the dataset is currently tsset and sorts the data on *timevar* or *panelvar timevar* if it is sorted differently from that.

tsset, clear is a rarely used programmer's command to declare that the data are no longer a time series.

Options

daily, weekly, monthly, quarterly, halfyearly, yearly, generic, and format(%*fmt*), indicate how *timevar* will be displayed—which %t format, if any, will be placed on *timevar*. Whether *timevar* is formatted is optional; all tsset requires is that *timevar* take on integer values.

The optional format states how *timevar* is to be measured (for instance, *timevar* = 5 might mean the fifth day of January 1960, or the fifth week of January 1960, or the fifth something else) and how you want it displayed (*timevar* = 5 might be displayed as 05jan1960, 1960–5, or in a host of other ways). In addition, the format makes the `tin()` and `twithin()` selection functions work so that later, after `tsset`ing the data, you can type things like `regress` ... `if tin(1jan1998,1apr1998)` to run the regression on the subsample 1jan1998 ≤ *timevar* ≤ 1apr1998.

You set the format by placing a `%t` format on *timevar*. You can do it yourself, or `tsset` can do it.

The time scales `%t` understands are daily (`%td`, 0 = 1jan1960), weekly (`%tw`, 0 = 1960w1), monthly (`%tm`, 0 = 1960m1), quarterly (`%tq`, 0 = 1960q1), half-yearly (`%th`, 0 = 1960h1), yearly (`%ty`, 1960 = 1960), and generic (`%tg`, 0 = ?).

Say that *timevar* is recorded in Stata quarterly units, meaning 0 = 1960q1, 1 = 1960q2, etc. (Perhaps your data start in 1990q1; then the first observation has *timevar* = 120.) You could format *timevar* and then `tsset` your data,

> . format *timevar* `%tq`
> . tsset *timevar*

or you could `tsset` your data and then format *timevar*,

> . tsset *timevar*
> . format *timevar* `%tq`

or you could `tsset` your data specifying `tsset`'s `format()` option,

> . tsset *timevar*, `format(%tq)`

or you could `tsset` your data specifying `tsset`'s `quarterly` option,

> . tsset *timevar*, `quarterly`

These alternatives yield the same result; use whichever appeals to you. See [U] **24.3 Time-series dates** and [U] **12.5.4 Time-series formats** for more information on the `%t` format and the advantages of setting it.

`clear` declares that the data are no longer a time series; this is a rarely used programmer's option.

Remarks

`tsset` sets *timevar* so that Stata's time-series operators are understood in varlists and expressions. The time-series operators are

operator	meaning
L.	lag x_{t-1}
L2.	2-period lag x_{t-2}
...	
F.	lead x_{t+1}
F2.	2-period lead x_{t+2}
...	
D.	difference $x_t - x_{t-1}$
D2.	difference of difference $x_t - x_{t-1} - (x_{t-1} - x_{t-2}) = x_t - 2x_{t-1} + x_{t-2}$
...	
S.	"seasonal" difference $x_t - x_{t-1}$
S2.	lag-2 (seasonal) difference $x_t - x_{t-2}$
...	

Time-series operators may be repeated and combined. L3.gnp refers to the third lag of variable gnp, and so do LLL.gnp, LL2.gnp, and L2L.gnp. LF.gnp is the same as gnp. DS12.gnp refers to the one-period difference of the 12-period difference. LDS12.gnp refers to the same concept, lagged once.

Note that D1. = S1. but D2. ≠ S2., D3. ≠ S3., and so on. D2. refers to the difference of the difference. S2. refers to the two-period difference. If you wanted the difference of the difference of the 12-period difference of gnp, you would write D2S12.gnp.

Operators may be typed in uppercase or lowercase. Most users would type d2s12.gnp instead of D2S12.gnp.

You may type operators however you wish; Stata internally converts operators to their canonical form. If you typed ld2ls12d.gnp, Stata would present the operated variable as L2D3S12.gnp.

In addition to *operator#*, Stata understands *operator*(*numlist*) to mean a set of operated variables. For instance, typing L(1/3).gnp in a varlist is the same as typing 'L.gnp L2.gnp L3.gnp'. The operators can also be applied to a list of variables by enclosing the variables in parentheses; e.g.,

```
. list year L(1/3).(gnp cpi)
```

	year	L.gnp	L2.gnp	L3.gnp	L.cpi	L2.cpi	L3.cpi
1.	1989	.	.	.	.	.	.
2.	1990	5452.8	.	.	100	.	.
3.	1991	5764.9	5452.8	.	105	100	.
4.	1992	5932.4	5764.9	5452.8	108	105	100
			(output omitted)				
8.	1996	7330.1	6892.2	6519.1	122	119	112

In *operator#*, making # zero returns the variable itself. L0.gnp is gnp. Thus the listing above could have been produced by typing list year l(0/3).gnp.

The parenthetical notation may be used with any operator. Typing D(1/3).gnp would return the first through third differences.

The parenthetical notation may be used in operator lists with multiple operators, such as L(0/3)D2S12.gnp.

Operator lists may include up to one set of parentheses, and the parentheses may enclose a *numlist*; see [U] **11.1.8 numlist**.

Before you can use these time-series operators, the dataset must satisfy two requirements:

1. the dataset must be tsset, and

2. the dataset must be sorted by *timevar* or, if it is a cross-sectional time-series dataset, by *panelvar timevar*.

tsset handles both requirements. As you use Stata, however, you may later use a command that re-sorts that data, and if you do, the time-series operators will not work:

```
. tsset time
(output omitted )
. regress y x l.x
(output omitted )
. (you continue to use Stata and, sometime later:)
. regress y x l.x
not sorted
r(5);
```

In that case, typing `tsset` without arguments will re-establish the sort order:

```
. tsset
(output omitted)
. regress y x l.x
(output omitted)
```

In this case, typing `tsset` is the same as typing `sort time`. Had we previously `tsset country time`, however, typing `tsset` would be the same as typing `sort country time`. You can type the `sort` command or type `tsset` without arguments; it makes no difference.

timevar must take on integer values. If *panelvar* is also specified, the dataset is declared to be a cross-section of time series (e.g., time series of different countries). `tsset` must be used before time-series operators may be used in expressions and varlists. After `tsset`, the data is sorted on *timevar* or on *panelvar timevar*.

If you have annual data with the variable `year` representing time, using `tsset` is as simple as typing

```
. tsset year, yearly
```

although you could omit the `yearly` option because it affects only how results are displayed.

If you are using one of the `xt` commands that require the data be `tsset`, specifying `tsset` overrides any settings previously specified by `iis` and `tis`.

▷ Example 1: Numeric time variable

We have monthly data on personal income. Variable `month` records the time of an observation:

```
. list t income
```

	t	income
1.	1	1153
2.	2	1181
	(output omitted)	
9.	9	1282

```
. tsset t
        time variable:  t, 1 to 9
. regress income l.income
(output omitted)
```

◁

▷ Example 2: Adjusting the starting date

In the example above, it is not important that t start at 1. The t variable could just as well be recorded 21, 22, ..., 29, or 426, 427, ..., 434, or any other way we liked. What is important is that the difference in t between observations when there are no gaps be 1.

Although the way time is measured makes no difference, Stata has formats to display time nicely if it is recorded in certain ways. In particular, Stata likes time variables in which 1jan1960 is recorded as 0. In our example above, if our first observation is July 1995, such that t = 1 corresponds to July 1995, then we could make a time variable that fits Stata's preference by typing

```
. generate newt = m(1995m7) + t - 1
```

`m()` is the function that returns a month equivalent; `m(1995m6)` evaluates to the constant 425, meaning 425 months after January 1960. We now have variable `newt` containing

```
. list t newt income
```

	t	newt	income
1.	1	426	1153
2.	2	427	1181
3.	3	428	1208
(output omitted)			
9.	9	434	1282

If we put a %tm format on newt, it will display more cleanly:

```
. format newt %tm
. list t newt income
```

	t	newt	income
1.	1	1995m7	1153
2.	2	1995m8	1181
3.	3	1995m9	1208
(output omitted)			
9.	9	1996m3	1282

We could now tsset newt rather than t:

```
. tsset newt
        time variable:  newt, 1995m7 to 1996m3
```

◁

▷ Example 3: Time-series data but no time variable

Perhaps we have the same time-series data but no time variable:

```
. list income
```

	income
1.	1153
2.	1181
3.	1208
4.	1272
5.	1236
6.	1297
7.	1265
8.	1230
9.	1282

Say that we know that the first observation corresponds to July 1995 and continues without gaps. We can create a monthly time variable and format it by typing

```
. generate t = m(1995m7) + _n - 1
. format t %tm
```

We can now tsset our dataset and list it:

```
. tsset t
        time variable:  t, 1995m7 to 1996m3

. list t income
```

	t	income
1.	1995m7	1153
2.	1995m8	1181
3.	1995m9	1208
	(output omitted)	
9.	1996m3	1282

◁

❑ Technical Note

Your data do not have to be monthly. Stata understands daily, weekly, monthly, quarterly, half-yearly, and yearly data. Correspondingly, there are the d(), w(), m(), q(), h(), and y() functions and the %td, %tw, %tm, %tq, %th, and %ty formats. Here is what we would have typed in the above examples had our data been on a different time scale:

Daily: if your t variable had t=1 corresponding to 15mar1993
```
. gen newt = d(15mar1993) + t - 1
. format newt %td
. tsset newt
```

Weekly: if your t variable had t=1 corresponding to 1994w1:
```
. gen newt = w(1994w1) + t - 1
. format newt %tw
. tsset newt
```

Monthly: if your t variable had t=1 corresponding to 2004m7:
```
. gen newt = m(2004m7) + t - 1
. format newt %tm
. tsset newt
```

Quarterly: if your t variable had t=1 corresponding to 1994q1:
```
. gen newt = q(1994q1) + t - 1
. format newt %tq
. tsset newt
```

Halfyearly: if your t variable had t=1 corresponding to 1921h2:
```
. gen newt = h(1921h2) + t - 1
. format newt %th
. tsset newt
```

Yearly: if your t variable had t=1 corresponding to 1842:
```
. gen newt = y(1842) + t - 1
. format newt %ty
. tsset newt
```

In each of the above examples, we subtracted one from our time variable in constructing the new time variable newt because we assumed that our starting time value was 1. For the quarterly example, if our starting time value had been 5 and that corresponded to 1994q1, we would have typed

```
. generate newt = q(1994q1) + t - 5
```

Had our initial time value been $t = 742$ and this corresponded to 1994q1, we would have typed

```
. generate newt = q(1994q1) + t - 742
```

The %td, %tw, %tm, %tq, %th, and %ty formats can display the date in the form you want; when you type, for instance, %td, you specify the *default* daily format, which produces dates in the form 15apr2002. If you wanted that to be displayed as "April 15, 2002", %td can do that; see [U] **24.3 Time-series dates**. Similarly, all the other %t formats can be modified to produce the results you want.

❑

▷ Example 4: Time variable as a string

Your data might include a time variable that is encoded into a string. Below each monthly observation is identified by string variable yrmo containing the year and month of the observation, sometimes with punctuation in between:

```
. list yrmo income
```

	yrmo	income
1.	1995 7	1153
2.	1995 8	1181
3.	1995,9	1208
4.	1995 10	1272
5.	1995/11	1236
6.	1995,12	1297
7.	1996-1	1265
8.	1996.2	1230
9.	1996 Mar	1282

The first step is to convert the string to a numeric representation. That is easy using the monthly() function; see [U] **24.3 Time-series dates**.

```
. gen mdate = monthly(yrmo, "ym")
(1 missing value generated)
. list yrmo mdate income
```

	yrmo	mdate	income
1.	1995 7	426	1153
2.	1995 8	427	1181
3.	1995,9	428	1208
	(output omitted)		
9.	1996 Mar	434	1282

Our new variable, mdate, contains the number of months from January, 1960. Now that we have numeric variable mdate, we can tsset the data:

```
. format mdate %tm
. tsset mdate
        time variable:  mdate, 1995m7 to 1996m3
```

In fact, we can combine the two and type

```
. tsset mdate, format(%tm)
        time variable:  mdate, 1995m7 to 1996m3
```

or type

```
. tsset mdate, monthly
        time variable:  mdate, 1995m7 to 1996m3
```

We do not have to bother to format the time variable at all, but formatting makes it display more cleanly:

```
. list yrmo mdate income
```

	yrmo	mdate	income
1.	1995 7	1995m7	1153
2.	1995 8	1995m8	1181
3.	1995,9	1995m9	1208
4.	1995 10	1995m10	1272
5.	1995/11	1995m11	1236
6.	1995,12	1995m12	1297
7.	1996-1	1996m1	1265
8.	1996.2	1996m2	1230
9.	1996 Mar	1996m3	1282

◁

❏ Technical Note

In addition to the monthly() function for translating strings to monthly dates, Stata has daily(), weekly(), quarterly(), halfyearly(), and yearly(). Stata also has the yw(), ym(), yq(), and yh() functions to convert from two numeric time variables to a Stata time variable. For example, generate qdate = yq(year,qtr) takes the variable year, containing year values, and the variable qtr, containing quarter values (1–4), and produces the variable qdate containing the number of quarters since 1960q1. See [U] **24.3 Time-series dates**.

❏

▷ Example 5: Time-series data with gaps

Gaps in the time series cause no difficulties:

```
. list yrmo income
```

	yrmo	income
1.	1995 7	1153
2.	1995 8	1181
3.	1995/11	1236
4.	1995,12	1297
5.	1996-1	1265
6.	1996 Mar	1282

```
. gen mdate = monthly(yrmo, "ym")
(1 missing value generated)
. tsset mdate, monthly
        time variable:  mdate, 1995m7 to 1996m1, but with a gap
```

Once the dataset has been tsset, we can use the time-series operators. The D operator specifies first (or higher-order) differences:

```
. list mdate income d.income
```

	mdate	income	D.income
1.	1995m7	1153	.
2.	1995m8	1181	28
3.	1995m11	1236	.
4.	1995m12	1297	61
5.	1996m1	1265	-32
6.	1996m3	1282	.

We can use the operators in an expression or varlist context; we do not have to create a new variable to hold D.income. We can use D.income with the list command, with regress or any other Stata command that allows time-series varlists.

◁

Panel data

▷ Example 6: Time-series data for multiple groups

Now assume that we have time series on annual income and that we have the series for two groups: individuals who have not completed high school (edlevel = 1) and individuals who have (edlevel = 2).

```
. list edlevel year income, sep(0)
```

	edlevel	year	income
1.	1	1988	14500
2.	1	1989	14750
3.	1	1990	14950
4.	1	1991	15100
5.	2	1989	22100
6.	2	1990	22200
7	2	1992	22800

We declare the data to be a panel by typing

```
. tsset edlevel year, yearly
        panel variable:  edlevel, 1 to 2
         time variable:  year, 1988 to 1992, but with a gap
```

Having tsset the data, we can now use time-series operators. The difference operator, for example, can be used to list annual changes in income:

```
. list edlevel year income d.income, sep(0)
```

	edlevel	year	income	D.income
1.	1	1988	14500	.
2.	1	1989	14750	250
3.	1	1990	14950	200
4.	1	1991	15100	150
5.	2	1989	22100	.
6.	2	1990	22200	100
7.	2	1992	22800	.

We see that in addition to producing missing values due to missing times, the difference operator correctly produced a missing value at the start of each panel. Once we have `tsset` our panel data, we can use time-series operators and be assured that they will handle missing time periods and panel changes correctly.

◁

Saved Results

`tsset` saves in `r()`:

Scalars

r(tmin)	minimum elapsed time
r(tmax)	maximum elapsed time
r(imin)	minimum panel id
r(imax)	maximum panel id

Macros

r(timevar)	elapsed time variable
r(panvar)	panel variable
r(tmins)	formatted minimum elapsed time
r(tmaxs)	formatted maximum elapsed time
r(tsfmt)	format for the current time variable
r(unit)	daily, weekly, monthly, quarterly, halfyearly, yearly, or generic
r(unit1)	d, w, m, q, h, y, or ""

Methods and Formulas

`tsset` is implemented as an ado-file.

Reference

Baum, C. F. 2000. sts17: Compacting time series data. *Stata Technical Bulletin* 57: 44–45. Reprinted in *Stata Technical Bulletin Reprints*, vol 10, pp. 369–370.

Also See

Complementary:	[TS] **tsfill**
Background:	[U] **11.4.3 Time-series varlists**,
	[U] **12.5.4 Time-series formats**,
	[U] **24.3 Time-series dates**,
	[U] **26.13 Models with time-series data**

Title

> **tssmooth** — Smooth and forecast univariate time-series data

Syntax

tssmooth *smoother* [*type*] *newvar* = *exp* [*if*] [*in*] [, ...]

Smoother category	*smoother*
Moving average	
with uniform weights	ma
with specified weights	ma
Recursive	
exponential	exponential
double exponential	dexponential
nonseasonal Holt–Winters	hwinters
seasonal Holt–Winters	shwinters
Nonlinear filter	nl

See [TS] **tssmooth ma**, [TS] **tssmooth exponential**, [TS] **tssmooth dexponential**, [TS] **tssmooth hwinters**, [TS] **tssmooth shwinters**, and [TS] **tssmooth nl**.

Description

tssmooth creates new variable *newvar* and fills it in by passing the specified expression (usually a variable name) through the requested smoother.

Remarks

The recursive smoothers may also be used for forecasting univariate time series, indeed, the Holt–Winters methods are used almost exclusively for this. All can perform dynamic out-of-sample forecasts, and the smoothing parameters may be chosen to minimize the in-sample sum-of-squared prediction errors.

The moving-average and nonlinear smoothers are generally used to extract the trend—or signal—from a time series while omitting the high-frequency or noise components.

All smoothers work both with time-series data and panel data. When used with panel data, the calculation is performed separately within panel.

Several texts provide good introductions to the methods available in tssmooth. Chatfield (1996) discusses how these methods fit into time-series analysis in general. Abraham and Ledolter (1983); Montgomery, Johnson, and Gardiner (1990); Bowerman and O'Connell (1993); and Chatfield (2001) discuss using these methods for modern time-series forecasting.

References

Abraham, B. and J. Ledolter. 1983. *Statistical Methods for Forecasting*. New York: Wiley.

Bowerman, B. and R. O'Connell. 1993. *Forecasting and Time Series: An Applied Approach*. 3rd ed. Pacific Grove, CA: Duxbury.

Chatfield, C. 1996. *The Analysis of Time Series: An Introduction*. 5th ed. London: Chapman & Hall.

——. 2001. *Time-Series Forecasting*. London: Chapman & Hall.

Chatfield, C. and M. Yar. 1988. Holt–Winters forecasting: Some practical issues. *The Statistician* 37: 129–140.

Montgomery, D. C., L. A. Johnson, and J. S. Gardiner. 1990. *Forecasting and Time Series Analysis*. 2nd ed. New York: McGraw–Hill.

Also See

Complementary:	[TS] **tsset**, [TS] **tssmooth dexponential**, [TS] **tssmooth exponential**, [TS] **tssmooth hwinters**, [TS] **tssmooth ma**, [TS] **tssmooth nl**, [TS] **tssmooth shwinters**
Related:	[TS] **arima**, [D] **egen**, [D] **generate**
Background:	[U] **11.4.3 Time-series varlists**

Title

> **tssmooth dexponential** — Double-exponential smoothing

Syntax

tssmooth dexponential [*type*] *newvar* = *exp* [*if*] [*in*] [, *options*]

options	description
Main	
replace	replace *newvar* if it already exists
parms($\#_\alpha$)	use $\#_\alpha$ as smoothing parameter
samp0(#)	use # observations to obtain initial values for recursions
s0($\#_1$ $\#_2$)	use $\#_1$ and $\#_2$ as initial values for recursions
forecast(#)	use # periods for the out-of-sample forecast

You must tsset your data before using tssmooth; see [TS] **tsset**.

exp may contain time-series operators; see [U] **11.4.3 Time-series varlists**.

Description

tssmooth dexponential models the trend of a variable whose difference between changes from the previous values is serially correlated. More precisely, it models a variable whose second-difference follows a low-order, moving-average process.

Options

> Main

replace replaces *newvar* if it already exists.

parms($\#_\alpha$) specifies the parameter α for the double-exponential smoothers; $0 < \#_\alpha < 1$. If parms($\#_\alpha$) is not specified, the smoothing parameter is chosen to minimize the in-sample sum-of-squared forecast errors.

samp0(#) and s0($\#_1$ $\#_2$) are mutually exclusive ways of specifying the initial values for the recursion.

By default, initial values are obtained by fitting a linear regression with a time trend using the first half of the observations in the dataset; see *Remarks*.

samp0(#) specifies that the first # be used in that regression.

s0($\#_1$ $\#_2$) specifies that $\#_1$ $\#_2$ be used as initial values.

forecast(#) specifies the number of periods for the out-of-sample prediction. $0 \le \# \le 500$. The default is 0, which is equivalent to not performing an out-of-sample forecast.

Remarks

The double-exponential smoothing procedure is designed for series that can be locally approximated as

$$\widehat{x}_t = m_t + b_t t$$

where $\widehat{x}_t$ is the smoothed or predicted value of the series x, and the terms m_t and b_t change over time. Abraham and Ledolter (1983), Bowerman and O'Connell (1993), and Montgomery, Johnson, and Gardiner (1990) all provide good introductions to double-exponential smoothing. Chatfield (2000, 2001) provides helpful discussions of how double-exponential smoothing relates to modern time-series methods.

The double-exponential method has been used both as a smoother and as a prediction method. Recall from [TS] **tssmooth exponential** that the single-exponential smoothed series is given by

$$S_t = \alpha x_t + (1 - \alpha)S_{t-1}$$

where α is the smoothing constant and x_t is the original series. The double-exponential smoother is obtained by smoothing the smoothed series,

$$S_t^{[2]} = \alpha S_t + (1 - \alpha)S_{t-1}^{[2]}$$

Values of S_0 and $S_0^{[2]}$ are necessary to begin the process. Following Montgomery, Johnson, and Gardiner(1990), the default method is to obtain S_0 and $S_0^{[2]}$ from a regression of the first N_{pre} values of x_t on $\widetilde{t} = (1, \dots, N_{\mathrm{pre}} - t_0)'$. By default, N_{pre} is equal to one-half the number of observations in the sample. N_{pre} can be specified using the samp0() option.

Alternatively, the values of S_0 and $S_0^{[2]}$ can be specified using the option s0().

▷ Example 1

Suppose that we had some data on the monthly sales of a book and that we wanted to smooth this series. The graph below illustrates that this series is locally trending over time, so we would not want to use single-exponential smoothing.

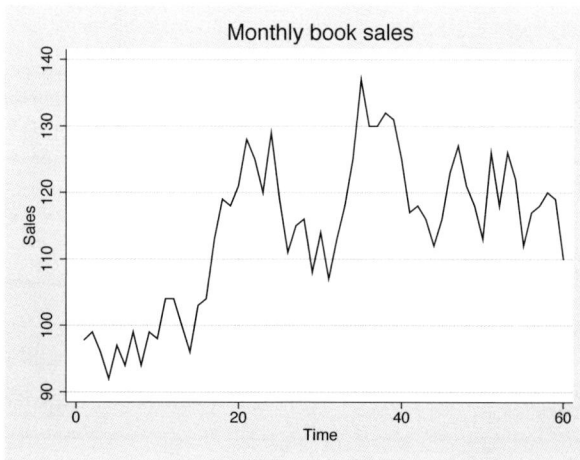

The following example illustrates that double-exponential smoothing is simply smoothing the smoothed series. Because the starting values are treated as time-zero values, we actually lose two observations when smoothing the smoothed series.

```
. use http://www.stata-press.com/data/r9/sales2

. tssmooth exponential double sm1=sales, p(.7) s0(1031)
exponential coefficient    =     0.7000
sum-of-squared residuals   =      13923
root mean squared error    =     13.192

. tssmooth exponential double sm2=sm1, p(.7) s0(1031)
exponential coefficient    =     0.7000
sum-of-squared residuals   =     7698.6
root mean squared error    =     9.8098

. tssmooth dexponential double sm2b=sales, p(.7) s0(1031 1031)
double-exponential coefficient  =     0.7000
sum-of-squared residuals        =     3724.4
root mean squared error         =     6.8231

. generate double sm2c = f2.sm2
(2 missing values generated)

. list sm2b sm2c in 1/10
```

	sm2b	sm2c
1.	1031	1031
2.	1028.3834	1028.3834
3.	1030.6306	1030.6306
4.	1017.8182	1017.8182
5.	1022.938	1022.938
6.	1026.0752	1026.0752
7.	1041.8587	1041.8587
8.	1042.8341	1042.8341
9.	1035.9571	1035.9571
10.	1030.6651	1030.6651

◁

The double-exponential method can also be viewed as a forecasting mechanism. The exponential forecast method is a constrained version of the Holt–Winters method implemented in [TS] **tssmooth hwinters** (as discussed by Gardner [1985] and Chatfield [2001]). Chatfield (2001) also notes that the double-exponential method arises when the underlying model is an ARIMA(0,2,2) with equal roots.

This method produces predictions $\widehat{x}_t$ for $t = t_1, \ldots, T + $ forecast(). These predictions are obtained as a function of the smoothed series and the smoothed-smoothed series. For $t \in [t_0, T]$,

$$\widehat{x}_t = \left(2 + \frac{\alpha}{1-\alpha}\right) S_t - \left(1 + \frac{\alpha}{1-\alpha}\right) S_t^{[2]}$$

where S_t and $S_t^{[2]}$ are as given above.

The out-of-sample predictions are obtained as a function of the constant term, the linear term of the smoothed series at the last observation in the sample, and time. The constant term is $a_T = 2S_T - S_T^{[2]}$, and the linear term is $b_T = \frac{\alpha}{1-\alpha}(S_T - S_T^{[2]})$. The τth step-ahead out-of-sample prediction is given by

$$\widehat{x}_t = a_t + \tau b_T$$

▷ Example 2

Specifying the `forecast` option puts the double-exponential forecast into the new variable instead of the double-exponential smoothed series. The code given below uses the smoothed series `sm1` and `sm2` that were generated above to illustrate how the double-exponential forecasts are computed.

```
. tssmooth dexponential double f1=sales, p(.7) s0(1031 1031) forecast(4)
double-exponential coefficient  =     0.7000
sum-of-squared residuals        =      20737
root mean squared error         =       16.1
. generate double xhat = (2 + .7/.3) * sm1 - (1 + .7/.3) * f.sm2
(5 missing values generated)
. list xhat f1 in 1/10
```

	xhat	f1
1.	1031	1031
2.	1031	1031
3.	1023.524	1023.524
4.	1034.8039	1034.8039
5.	994.0237	994.0237
6.	1032.4463	1032.4463
7.	1031.9015	1031.9015
8.	1071.1709	1071.1709
9.	1044.6454	1044.6454
10.	1023.1855	1023.1855

◁

▷ Example 3

Generally, when you are forecasting, you do not know the smoothing parameter. `tssmooth dexponential` computes the double-exponential forecasts of a series and obtains the optimal smoothing parameter by finding the smoothing parameter that minimizes the in-sample sum-of-squared forecast errors.

```
. tssmooth dexponential f2=sales, forecast(4)
computing optimal double-exponential coefficient (0,1)
optimal double-exponential coefficient =      0.3631
sum-of-squared residuals               =   16075.805
root mean squared error                =   14.175598
```

The following graph describes the fit we obtained by applying the double-exponential forecast method to our sales data. Note that the out-of-sample dynamic predictions are not constant, as in the single-exponential case.

(Continued on next page)

```
. line f2 sales t, title("Double exponential forecast with optimal alpha")
> ytitle(Sales) xtitle(time)
```

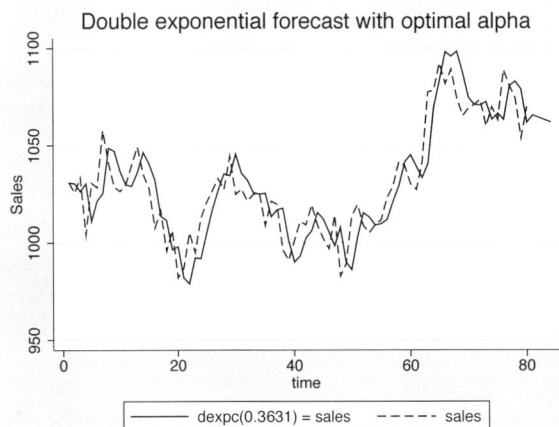

Double exponential forecast with optimal alpha

◁

tssmooth dexponential automatically detects panel data from the information provided when the dataset was tsset. The starting values are chosen separately for each series. If the smoothing parameter is chosen to minimize the sum-of-squared prediction errors, the optimization is performed separately on each panel. The saved results contain the results from the last panel. Missing values at the beginning of the sample are excluded from the sample. After at least one value has been found, missing values are filled in using the one-step predictions from the previous period.

Saved Results

tssmooth saves in r():

Scalars

r(N)	number of observations
r(alpha)	α smoothing parameter
r(rss)	sum-of-squared errors
r(rmse)	root mean squared error
r(N_pre)	number of observations used in calculating starting values, if starting values calculated
r(s2_0)	initial value for linear term, i.e., $S_0^{[2]}$
r(s1_0)	initial value for constant term, i.e., S_0
r(linear)	final value of linear term
r(constant)	final value of constant term
r(period)	period, if filter is seasonal

Macros

r(method)	smoothing method
r(exp)	expression specified
r(timevar)	time variables specified in tsset
r(panelvar)	panel variables specified in tsset

Methods and Formulas

`tssmooth dexponential` is implemented as an ado-file.

A truncated description of the specified double-exponential filter is used to label the new variable. See [D] **label** for more information on labels.

An untruncated description of the specified double-exponential filter is saved in the characteristic `tssmooth` for the new variable. See [P] **char** for more information on characteristics.

The updating equations for the smoothing and forecasting versions are as given previously.

The starting values for both the smoothing and forecasting versions of double-exponential are obtained using the same method, which begins with the model

$$x_t = \beta_0 + \beta_1 t$$

where x_t is the series to be smoothed and t is a time variable that has been normalized to equal 1 in the first period included in the sample. The regression coefficient estimates $\widehat{\beta}_0$ and $\widehat{\beta}_1$ are obtained via OLS. The sample is determined by the option `samp0()`. By default, `samp0()` includes the first half of the observations. Given the estimates $\widehat{\beta}_0$ and $\widehat{\beta}_1$, the starting values are

$$S_0 = \widehat{\beta}_0 - \{(1-\alpha)/\alpha\}\widehat{\beta}_1$$
$$S_0^{[2]} = \widehat{\beta}_0 - 2\{(1-\alpha)/\alpha\}\widehat{\beta}_1$$

References

Abraham, B. and J. Ledolter. 1983. *Statistical Methods for Forecasting.* New York: Wiley.

Bowerman, B. and R. O'Connell. 1993. *Forecasting and Time Series: An Applied Approach.* 3rd ed. Pacific Grove, CA: Duxbury.

Chatfield, C. 1996. *The Analysis of Time Series: An Introduction.* 5th ed. London: Chapman & Hall.

——. 2001. *Time-Series Forecasting.* London: Chapman & Hall.

Chatfield, C. and M. Yar. 1988. Holt–Winters Forecasting: Some practical issues. *The Statistician* 37: 129–140.

Gardner, E. S., Jr. 1985. Exponential smoothing: The state of the art. *Journal of Forecasting* 4: 1–28.

Montgomery, D. C., L. A. Johnson, and J. S. Gardiner. 1990. *Forecasting and Time Series Analysis.* 2nd ed. New York: McGraw–Hill.

Also See

Complementary:	[TS] **tsset**
Related:	[TS] **arima**, [TS] **tssmooth exponential**, [TS] **tssmooth hwinters**, [TS] **tssmooth ma**, [TS] **tssmooth nl**, [TS] **tssmooth shwinters**, [D] **egen**, [D] **generate**
Background:	[U] **11.4.3 Time-series varlists**, [TS] **tssmooth**

Title

tssmooth exponential — Single-exponential smoothing

Syntax

tssmooth exponential $[type]$ *newvar* = *exp* $[if]$ $[in]$ $[$, *options* $]$

options	description
Main	
replace	replace *newvar* if it already exists
parms($\#_\alpha$)	use $\#_\alpha$ as smoothing parameter
samp0(#)	use # observations to obtain initial value for recursion
s0(#)	use # as initial value for recursion
forecast(#)	use # periods for the out-of-sample forecast

You must tsset your data before using tssmooth; see [TS] **tsset**.

exp may contain time-series operators; see [U] **11.4.3 Time-series varlists**.

Description

tssmooth exponential models the trend of a variable whose change from the previous value is serially correlated. More precisely, it models a variable whose first-difference follows a low-order, moving-average process.

Options

> Main

replace replaces *newvar* if it already exists.

parms($\#_\alpha$) specifies the parameter α for the exponential smoother; $0 < \#_\alpha < 1$. If parms($\#_\alpha$) is not specified, the smoothing parameter is chosen to minimize the in-sample sum-of-squared forecast errors.

samp0(#) and s0(#) are mutually exclusive ways of specifying the initial value for the recursion.

samp0(#) specifies that the initial value be obtained by calculating the mean over the first # observations of the sample.

s0(#) specifies the initial value to be used.

If neither option is specified, the default is to use the mean calculated over the first half of the sample.

forecast(#) gives the number of observations for the out-of-sample prediction, where $0 \le \# \le 500$. The default value is 0 and is equivalent to not forecasting out-of-sample.

Remarks

Introduction
Examples
Treatment of missing values

Introduction

Exponential smoothing can be viewed either as an adaptive-forecasting algorithm or, equivalently, as a geometrically weighted moving-average filter. Exponential smoothing is most appropriate when used with time-series data that exhibit no linear or higher-order trends but that do exhibit low-velocity, aperiodic variation in the mean. Abraham and Ledolter (1983), Bowerman and O'Connell (1993), and Montgomery, Johnson, and Gardiner (1990) all provide good introductions to single-exponential smoothing. Chatfield (1996, 2001) discusses how single-exponential smoothing relates to modern time-series methods. For example, simple exponential smoothing produces optimal forecasts for several underlying models, including ARIMA(0,1,1) and the random-walk-plus-noise state-space model. (See Chatfield [2001, section 4.3.1].)

The exponential filter with smoothing parameter α creates the series S_t, where

$$S_t = \alpha X_t + (1 - \alpha) S_{t-1} \qquad \text{for } t = 1, \ldots, T$$

and S_0 is the initial value. This is the adaptive forecast-updating form of the exponential smoother. This implies that

$$S_t = \alpha \sum_{k=0}^{T-1} (1 - \alpha)^K X_{T-k} + (1 - \alpha)^T S_0$$

which is the weighted moving-average representation, with geometrically declining weights. The choice of the smoothing constant α determines how quickly the smoothed series or forecast will adjust to changes in the mean of the unfiltered series. For small values of α, the response will be slow because more weight is placed on the previous estimate of the mean of the unfiltered series, whereas larger values of α will put more emphasis on the most recently observed value of the unfiltered series.

Examples

▷ Example 1

Let's consider some examples using sales data. Here we forecast sales for three periods with a smoothing parameter of .4:

```
. use http://www.stata-press.com/data/r9/sales1

. tssmooth exponential sm1=sales, parms(.4) forecast(3)
exponential coefficient   =      0.4000
sum-of-squared residuals  =        8345
root mean squared error   =      12.919
```

To compare our forecast with the actual data, we graph the series and the forecasted series over time.

```
. line sm1 sales t, title("Single exponential forecast") ytitle(Sales) xtitle(time)
```

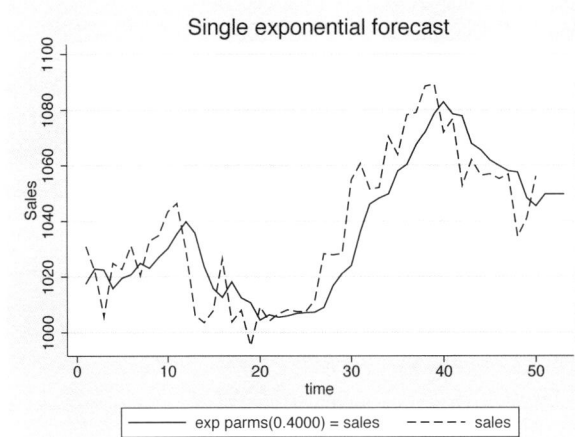

The graph indicates that our forecasted series may not be adjusting rapidly enough to the changes in the actual series. The smoothing parameter α controls the rate at which the forecast adjusts. Smaller values of α adjust the forecasts more slowly. Thus we suspect that our chosen value of .4 is too small. One way to investigate this suspicion is to ask `tssmooth exponential` to choose the smoothing parameter that minimizes the sum-of-squared forecast errors.

```
. tssmooth exponential sm2=sales, forecast(3)
computing optimal exponential  coefficient (0,1)
optimal exponential coefficient =        0.7815
sum-of-squared residuals         =     6727.7056
root mean squared error          =       11.599746
```

The output suggests that the value of $\alpha = .4$ is too small. The graph below indicates that the new forecast tracks the series much more closely than the previous forecast.

```
. line sm2 sales t, title ("Single exponential forecast with optimal alpha")
> ytitle(sales) xtitle(time)
```

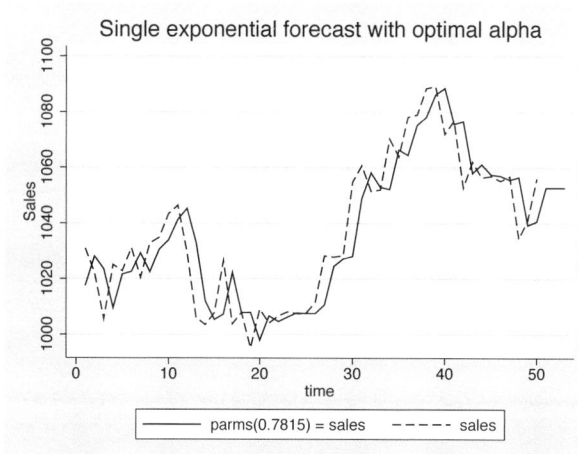

We noted above that simple exponential forecasts are optimal for an ARIMA $(0,1,1)$ model. (See [TS] **arima** for fitting ARIMA models in Stata.) Chatfield (2001, 90) gives the following useful derivation that relates the MA coefficient in an ARIMA $(0,1,1)$ model to the smoothing parameter in single-exponential smoothing. An ARIMA $(0,1,1)$ is given by

$$x_t - x_{t-1} = \epsilon_t + \theta\epsilon_{t-1}$$

where ϵ_t is an identically and independently distributed white-noise error term. Thus given $\widehat{\theta}$, an estimate of θ, an optimal one-step prediction of $\widehat{x}_{t+1}$ is $\widehat{x}_{t+1} = x_t + \widehat{\theta}\epsilon_t$. Because ϵ_t is not observable, it can be replaced by

$$\widehat{\epsilon}_t = x_t - \widehat{x}_{t-1}$$

yielding

$$\widehat{x}_{t+1} = x_t + \widehat{\theta}(x_t - \widehat{x}_{t-1})$$

Letting $\widehat{\alpha} = 1 + \widehat{\theta}$ and doing some further rearranging implies that

$$\widehat{x}_{t+1} = (1 + \widehat{\theta})x_t - \widehat{\theta}\widehat{x}_{t-1}$$
$$\widehat{x}_{t+1} = \widehat{\alpha}x_t - (1 - \widehat{\alpha})\widehat{x}_{t-1}$$

▷ Example 2

Let's compare the estimate of the optimal smoothing parameter of .7815 with the one we could obtain using [TS] **arima**. Below we fit an ARIMA$(0,1,1)$ to the sales data and then back out the estimate of α. The two estimates of α are quite close, given the large estimated standard error of $\widehat{\theta}$.

```
. arima sales, arima(0,1,1)

(setting optimization to BHHH)
Iteration 0:    log likelihood = -189.91037
Iteration 1:    log likelihood = -189.62405
Iteration 2:    log likelihood = -189.60468
Iteration 3:    log likelihood = -189.60352
Iteration 4:    log likelihood = -189.60343
(switching optimization to BFGS)
Iteration 5:    log likelihood = -189.60342

ARIMA regression

Sample:  2 to 50                          Number of obs      =         49
                                          Wald chi2(1)       =       1.41
Log likelihood = -189.6034                Prob > chi2        =     0.2347
```

D.sales	Coef.	OPG Std. Err.	z	P>|z|	[95% Conf.	Interval]
sales						
_cons	.5025469	1.382727	0.36	0.716	-2.207548	3.212641
ARMA						
ma						
L1.	-.1986561	.1671699	-1.19	0.235	-.5263031	.1289908
/sigma	11.58992	1.240607	9.34	0.000	9.158378	14.02147

```
. di 1 + _b[ARMA:L.ma]
.80134387
```

◁

▷ Example 3

tssmooth exponential automatically detects panel data. Suppose that we had sales figures for five companies in long form. Running tssmooth exponential on the variable that contains all five series puts the smoothed series and the predictions in a single variable in long form. When the smoothing parameter is chosen to minimize the squared prediction error, an optimal value for the smoothing parameter is chosen separately for each panel.

```
. use http://www.stata-press.com/data/r9/sales_cert, clear

. tsset
        panel variable:  id, 1 to 5
         time variable:  t, 1 to 100

. tssmooth exponential sm5=sales, forecast(3)


-> id = 1

computing optimal exponential  coefficient (0,1)

optimal exponential coefficient =       0.8702
sum-of-squared residuals         =     16070.567
root mean squared error          =     12.676974


-> id = 2

computing optimal exponential  coefficient (0,1)

optimal exponential coefficient =       0.7003
sum-of-squared residuals         =     20792.393
root mean squared error          =     14.419568


-> id = 3

computing optimal exponential  coefficient (0,1)

optimal exponential coefficient =       0.6927
sum-of-squared residuals         =        21629
root mean squared error          =     14.706801


-> id = 4

computing optimal exponential  coefficient (0,1)

optimal exponential coefficient =       0.3866
sum-of-squared residuals         =     22321.334
root mean squared error          =     14.940326


-> id = 5

computing optimal exponential  coefficient (0,1)

optimal exponential coefficient =       0.4540
sum-of-squared residuals         =     20714.095
root mean squared error          =     14.392392
```

tssmooth exponential computed starting values and chose an optimal α for each panel individually.

◁

Treatment of missing values

Missing values in the middle of the data are filled in with the one-step prediction using the previous values. Missing values at the beginning or end of the data are treated as if the observations were not there.

tssmooth exponential treats observations excluded from the sample by if and in just as if they were missing.

▷ Example 4

Here the 28th observation is missing. Note that the prediction for the 29th observation is repeated in the new series.

```
. use http://www.stata-press.com/data/r9/sales1, clear
. tssmooth exponential sm1=sales, parms(.7) forecast(3)
```
(output omitted)
```
. generate sales2=sales if t!=28
(4 missing values generated)
. tssmooth exponential sm3=sales2, parms(.7) forecast(3)
exponential coefficient   =      0.7000
sum-of-squared residuals  =      6842.4
root mean squared error   =      11.817
. list t sales2 sm3 if t>25 & t < 31
```

	t	sales2	sm3
26.	26	1011.5	1007.5
27.	27	1028.3	1010.3
28.	28	.	1022.9
29.	29	1028.4	1022.9
30.	30	1054.8	1026.75

Because the data for $t = 28$ are missing, the prediction for period 28 has been used in its place. This implies that the updating equation for period 29 is

$$S_{29} = \alpha S_{28} + (1 - \alpha)S_{28} = S_{28}$$

which explains why the prediction for $t = 28$ is repeated.

Because this is a single-exponential procedure, the loss of that one observation will not be noticed several periods later.

```
. generate diff = sm3-sm1 if t>28
(28 missing values generated)
. list t diff if t>28 & t < 39
```

	t	diff
29.	29	-3.5
30.	30	-1.050049
31.	31	-.3150635
32.	32	-.0946045
33.	33	-.0283203
34.	34	-.0085449
35.	35	-.0025635
36.	36	-.0008545
37.	37	-.0003662
38.	38	-.0001221

◁

▷ Example 5

Now consider an example in which there are data missing at the beginning and end of the sample.

```
. generate sales3=sales if t>2 & t<49
(7 missing values generated)
```

```
. tssmooth exponential sm4=sales3, parms(.7) forecast(3)
exponential coefficient    =        0.7000
sum-of-squared residuals   =        6215.3
root mean squared error    =        11.624

. list t sales sales3 sm4 if t<5 | t >45
```

	t	sales	sales3	sm4
1.	1	1031	.	.
2.	2	1022.1	.	.
3.	3	1005.6	1005.6	1016.787
4.	4	1025	1025	1008.956
46.	46	1055.2	1055.2	1057.2
47.	47	1056.8	1056.8	1055.8
48.	48	1034.5	1034.5	1056.5
49.	49	1041.1	.	1041.1
50.	50	1056.1	.	1041.1
51.	51	.	.	1041.1
52.	52	.	.	1041.1
53.	53	.	.	1041.1

The output above illustrates that missing values at the beginning or end of the sample cause the sample to be truncated. The new series begins with nonmissing data and begins predicting immediately after it stops.

One period after the actual data concludes, the exponential forecast becomes a constant. After the actual end of the data, the forecast at period t is substituted for the missing data. This also illustrates why the forecasted series is a constant.

◁

Saved Results

tssmooth saves in r():

Scalars

r(N)	number of observations
r(alpha)	α smoothing parameter
r(rss)	sum-of-squared prediction errors
r(rmse)	root mean squared error
r(N_pre)	number of observations used in calculating starting values
r(s1_0)	initial value for S_t

Macros

r(method)	smoothing method
r(exp)	expression specified
r(timevar)	time variables specified in tsset
r(panelvar)	panel variables specified in tsset

Methods and Formulas

`tssmooth exponential` is implemented as an ado-file.

The formulas for deriving smoothed series are as given in the text. When the value of α is not specified, an optimal value is found that minimizes the mean squared forecast error. A method of bisection is employed to find the solution to this optimization problem.

A truncated description of the specified exponential filter is used to label the new variable. See [D] **label** for more information about labels.

An untruncated description of the specified exponential filter is saved in the characteristic `tssmooth` for the new variable. See [P] **char** for more information about characteristics.

References

Abraham, B. and J. Ledolter. 1983. *Statistical Methods for Forecasting*. New York: Wiley.

Bowerman, B. and R. O'Connell. 1993. *Forecasting and Time Series: An Applied Approach*. 3rd ed. Pacific Grove, CA: Duxbury.

Chatfield, C. 1996. *The Analysis of Time Series: An Introduction*. 5th ed. London: Chapman & Hall.

——. 2001. *Time-Series Forecasting*. London: Chapman & Hall.

Chatfield, C. and M. Yar. 1988. Holt–Winters forecasting: Some practical issues. *The Statistician* 37: 129–140.

Montgomery, D. C., L. A. Johnson, and J. S. Gardiner. 1990. *Forecasting and Time Series Analysis*. 2nd ed. New York: McGraw–Hill.

Also See

Complementary:	[TS] **tsset**
Related:	[TS] **arima**, [TS] **tssmooth dexponential**, [TS] **tssmooth hwinters**, [TS] **tssmooth ma**, [TS] **tssmooth nl**, [TS] **tssmooth shwinters**, [D] **egen**, [D] **generate**
Background:	[U] **11.4.3 Time-series varlists**, [TS] **tssmooth**

Title

Syntax

tssmooth hwinters [*type*] *newvar* = *exp* [*if*] [*in*] [, *options*]

options	description
Main	
replace	replace *newvar* if it already exists
parms($\#_\alpha$ $\#_\beta$)	use $\#_\alpha$ and $\#_\beta$ as smoothing parameters
samp0(#)	use # observations to obtain initial values for recursion
s0($\#_{\text{cons}}$ $\#_{\text{lt}}$)	use $\#_{\text{cons}}$ and $\#_{\text{lt}}$ as initial values for recursion
forecast(#)	use # periods for the out-of-sample forecast
Options	
diff	alternative initial-value specification; see *Options*
Max options	
maximize_options	control the maximization process; seldom used
from($\#_\alpha$ $\#_\beta$)	use $\#_\alpha$ and $\#_\beta$ as starting values for the parameters

You must tsset your data before using tssmooth hwinters; see [TS] **tsset**.

exp may contain time-series operators; see [U] **11.4.3 Time-series varlists**.

Description

tssmooth hwinters is used in smoothing or forecasting a series that can be modeled as a linear trend in which the intercept and the coefficient on time vary over time.

Options

> Main

replace replaces *newvar* if it already exists.

parms($\#_\alpha$ $\#_\beta$), $0 \leq \#_\alpha \leq 1$ and $0 \leq \#_\beta \leq 1$, specifies the parameters. If parms() is not specified, the values are chosen by an iterative process to minimize the in-sample sum-of-squared prediction errors.

If you experience difficulty converging (many iterations and "not concave" messages), try using from() to provide better starting values.

samp0(#) and s0($\#_{\text{cons}}$ $\#_{\text{lt}}$) specify how the initial values $\#_{\text{cons}}$ and $\#_{\text{lt}}$ for the recursion are obtained.

By default, initial values are obtained by fitting a linear regression with a time trend using the first half of the observations in the dataset.

samp0(#) specifies that the first # observations be used in that regression.

s0($\#_{\text{cons}}$ $\#_{\text{lt}}$) specifies that $\#_{\text{cons}}$ and $\#_{\text{lt}}$ be used as initial values.

255

forecast(*#*) specifies the number of periods for the out-of-sample prediction. $0 \leq \# \leq 500$. The default is 0, which is equivalent to not performing an out-of-sample forecast.

⌐ Options ⌐

diff specifies the linear term is obtained by averaging the first-difference of exp_t and the intercept is obtained as the difference of *exp* in the first observation and the mean of D.exp_t.

If option diff is not specified, a linear regression of exp_t on a constant and t is fitted.

⌐ Max options ⌐

maximize_options controls the process for solving for the optimal α and β when parms() is not specified.

maximize_options: <u>nodiff</u>icult, <u>tech</u>nique(*algorithm_spec*), <u>iter</u>ate(*#*), [<u>no</u>]<u>log</u>, <u>trace</u>, gradient, showstep, <u>hess</u>ian, shownrtolerance, <u>tol</u>erance(*#*), <u>lt</u>olerance(*#*), <u>gt</u>olerance(*#*), <u>nrtol</u>erance(*#*), <u>nonrtol</u>erance, see [R] **maximize**. These options are seldom used.

from(*#$_\alpha$ #$_\beta$*), $0 < \#_\alpha < 1$ and $0 < \#_\beta < 1$, specifies starting values from which the optimal values of α and β will be obtained. If from() is not specified, from(.5 .5) is used.

Remarks

The Holt–Winters method forecasts series of the form

$$\widehat{x}_{t+1} = a_t + b_t t$$

where $\widehat{x}_t$ is the forecast of the original series x_t, a_t is a mean that drifts over time, and b_t is a coefficient on time that also drifts. In fact, as Gardner (1985) has noted, the Holt–Winters method produces optimal forecasts for an ARIMA(0,2,2) model and some local linear models. See [TS] **arima** and the references in that entry for ARIMA models, and see Harvey (1989) for a discussion of the local linear model and its relationship to the Holt–Winters method. Abraham and Ledolter (1983), Bowerman and O'Connell (1993), and Montgomery, Johnson, and Gardiner (1990) all provide good introductions to the Holt–Winters method. Chatfield (1996, 2001) provides helpful discussions of how this method relates to modern time-series analysis.

The Holt–Winters method can be viewed as an extension of double-exponential smoothing with two parameters, which may be explicitly set or chosen to minimize the in-sample sum-of-squared forecast errors. In the latter case, as discussed in *Methods and Formulas*, the smoothing parameters are chosen to minimize the in-sample sum-of-squared forecast errors plus a penalty term that helps to achieve convergence when one of the parameters is too close to the boundary.

Given the series x_t, the smoothing parameters α and β, and the starting values a_0 and b_0, the updating equations are

$$a_t = \alpha x_t + (1 - \alpha)(a_{t-1} + b_{t-1})$$

$$b_t = \beta(a_t - a_{t-1}) + (1 - \beta)b_{t-1}$$

After computing the series of constant and linear terms, a_t and b_t, respectively, the τ step-ahead prediction of x_t is given by

$$\widehat{x}_{t+\tau} = a_t + b_t \tau$$

▷ Example 1

Below we illustrate how to use `tssmooth hwinters` with specified smoothing parameters. This example also illustrates that the Holt–Winters method can closely follow a series in which both the mean and the time-coefficient drift over time.

Suppose that we have data on the monthly sales of a book and that we want to forecast this series using the Holt–Winters method.

```
. use http://www.stata-press.com/data/r9/bsales
. tssmooth hwinters hw1=sales, parms(.7 .3) forecast(3)
Specified weights:
                alpha = 0.7000
                 beta = 0.3000
sum-of-squared residuals = 2301.046
 root mean squared error = 6.192799
. line sales hw1 t, title("Holt-Winters Forecast with alpha=.7  and beta=.3")
> ytitle(Sales) xtitle(Time)
```

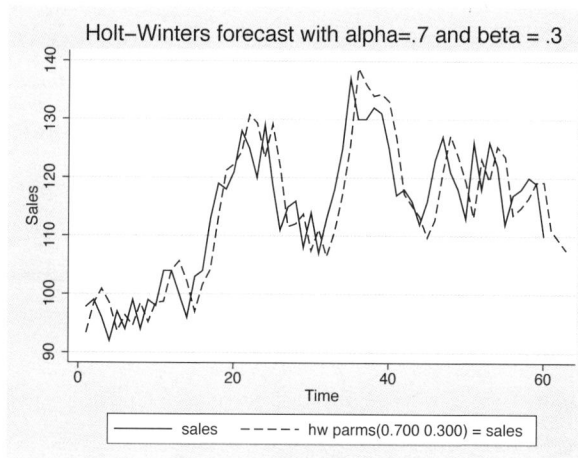

The graph indicates that the forecasts are for linearly decreasing sales. Given a_T and b_T, the out-of-sample predictions are linear functions of time. In this example, the slope appears to be too steep, probably because our choice of α and β.

◁

▷ Example 2

The graph in the previous example illustrates that the starting values for the linear and constant series can affect the in-sample fit of the predicted series for the first few observations. The previous example used the default method for obtaining the initial values for the recursion. The output below illustrates that, for some problems, the differenced-based initial values provide a better in-sample fit for the first few observations. However, the differenced-based initial values do not always outperform the regression-based initial values. Furthermore, as shown in the output below, for series of reasonable length, the predictions produced are nearly identical.

```
. tssmooth hwinters hw2=sales, parms(.7 .3) forecast(3) diff
Specified weights:
                        alpha = 0.7000
                         beta = 0.3000
sum-of-squared residuals = 2261.173
 root mean squared error = 6.13891
. list hw1 hw2 if _n<6 | _n>57
```

	hw1	hw2
1.	93.31973	97.80807
2.	98.40002	98.11447
3.	100.8845	99.2267
4.	98.50404	96.78276
5.	93.62408	92.2452
58.	116.5771	116.5771
59.	119.2146	119.2146
60.	119.2608	119.2608
61.	111.0299	111.0299
62.	109.2815	109.2815
63.	107.5331	107.5331

When the smoothing parameters are chosen to minimize the in-sample sum-of-squared forecast errors, changing the initial values can affect the choice of the optimal α and β. When changing the initial values results in different optimal values for α and β, the predictions will also differ.

◁

When the Holt–Winters model fits the data well, finding the optimal smoothing parameters generally proceeds well. When the model fits poorly, it can be difficult to find the α and β that minimize the in-sample sum-of-squared forecast errors.

▷ Example 3

In this example, we forecast the book sales data using the α and β that minimize the in-sample squared forecast errors.

```
. tssmooth hwinters hw3=sales, forecast(3)
computing optimal weights

Iteration 0:    penalized RSS = -2632.2073  (not concave)
Iteration 1:    penalized RSS = -1982.8431
Iteration 2:    penalized RSS = -1976.4236
Iteration 3:    penalized RSS = -1975.9175
Iteration 4:    penalized RSS = -1975.9036
Iteration 5:    penalized RSS = -1975.9036

Optimal weights:
                         alpha = 0.8209
                          beta = 0.0067
penalized sum of squared residuals = 1975.904
        sum of squared residuals = 1975.904
        root mean-squared error = 5.738617
```

The following graph contains the data and the forecast using the optimal α and β. Comparing this graph with the one above illustrates how different choices of α and β can lead to very different forecasts. Instead of linearly decreasing sales, the new forecast is for linearly increasing sales.

```
. line sales hw3 t, title("Holt-Winters Forecast with optimal alpha and beta")
> ytitle(Sales) xtitle(Time)
```

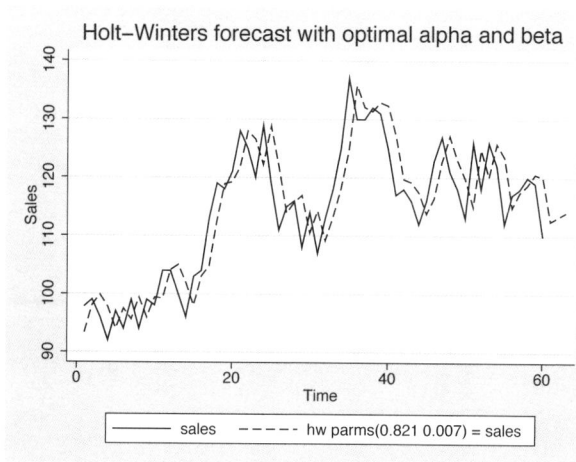

◁

Saved Results

tssmooth saves in r():

Scalars

r(N)	number of observations	r(N_pre)	number of observations used in calculating starting values
r(alpha)	α smoothing parameter		
r(beta)	β smoothing parameter	r(s2_0)	initial value for linear term
r(rss)	sum-of-squared errors	r(s1_0)	initial value for constant term
r(prss)	penalized sum-of-squared errors, if parms() not specified	r(linear)	final value of linear term
		r(constant)	final value of constant term
r(rmse)	root mean squared error		

Macros

r(method)	smoothing method	r(timevar)	time variables specified in tsset
r(exp)	expression specified	r(panelvar)	panel variables specified in tsset

Methods and Formulas

tssmooth hwinters is implemented as an ado-file.

A truncated description of the specified Holt–Winters filter is used to label the new variable. See [D] **label** for more information on labels.

An untruncated description of the specified Holt–Winters filter is saved in the characteristic named tssmooth for the new variable. See [P] **char** for more information on characteristics.

Given the series, x_t; the smoothing parameters, α and β; and the starting values, a_0 and b_0, the updating equations are

$$a_t = \alpha x_t + (1 - \alpha)(a_{t-1} + b_{t-1})$$

$$b_t = \beta(a_t - a_{t-1}) + (1 - \beta)b_{t-1}$$

By default, the initial values are found by fitting a linear regression with a time trend. The time variable in this regression is normalized to equal one in the first period included in the sample. By default, one-half of the data is used in this regression, but this sample can be changed using samp0(). a_0 is then set to the estimate of the constant, and b_0 is set to the estimate of the coefficient on the time trend. Specifying the diff option sets b_0 to the mean of D.x and a_0 to $x_1 - b_0$. Alternatively, s0() can be used to specify the initial values directly.

Sometimes, one or both of the optimal parameters may lie on the boundary of $[0, 1]$. To keep the estimates inside $[0, 1]$, tssmooth hwinters parameterizes the objective function in terms of their inverse logits, i.e., in terms of $\exp(\alpha)/\{1 + \exp(\alpha)\}$ and $\exp(\beta)/\{1 + \exp(\beta)\}$. When one of these parameters is actually on the boundary, this can complicate the optimization. For this reason, tssmooth hwinters optimizes a penalized sum-of-squared forecast errors. Let $\widehat{x}_t(\widetilde{\alpha}, \widetilde{\beta})$ be the forecast for the series x_t, given the choices of $\widetilde{\alpha}$ and $\widetilde{\beta}$. Then the in-sample penalized sum-of-squared prediction errors is

$$P = \sum_{t=1}^{T} \left[\{x_t - \widehat{x}_t(\widetilde{\alpha}, \widetilde{\beta})\}^2 + I_{|f(\widetilde{\alpha})| > 12}(|f(\widetilde{\alpha})| - 12)^2 + I_{|f(\widetilde{\beta})| > 12}(|f(\widetilde{\beta})| - 12)^2 \right]$$

where $f(x) = \ln\{x(1 - x)\}$. The penalty term is zero unless one of the parameters is very close to the boundary. When one of the parameters is very close to the boundary, the penalty term will help to obtain convergence.

Acknowledgment

We would like to thank Nick Cox of the University of Durham for his helpful comments.

References

Abraham, B. and J. Ledolter. 1983. *Statistical Methods for Forecasting.* New York: Wiley.

Bowerman, B. and R. O'Connell. 1993. *Forecasting and Time Series: An Applied Approach.* 3rd ed. Pacific Grove, CA: Duxbury.

Chatfield, C. 1996. *The Analysis of Time Series: An Introduction.* 5th ed. London: Chapman & Hall.

——. 2001. *Time-Series Forecasting.* London: Chapman & Hall.

Chatfield, C. and M. Yar. 1988. Holt–Winters forecasting: Some practical issues. *The Statistician* 37: 129–140.

Gardner, E. S., Jr. 1985. Exponential smoothing: The state of the art. *Journal of Forecasting* 4: 1–28.

Montgomery, D. C., L. A. Johnson, and J. S. Gardiner. 1990. *Forecasting and Time Series Analysis.* 2nd ed. New York: McGraw–Hill.

Also See

Complementary:	[TS] **tsset**
Related:	[TS] **arima**, [TS] **tssmooth dexponential**, [TS] **tssmooth exponential**, [TS] **tssmooth ma**, [TS] **tssmooth nl**, [TS] **tssmooth shwinters**, [D] **egen**, [D] **generate**
Background:	[U] **11.4.3 Time-series varlists**, [TS] **tssmooth**

Title

> **tssmooth ma** — Moving-average filter

Syntax

Moving average with uniform weights

> tssmooth ma $\left[\,type\,\right]$ *newvar* = *exp* $\left[\,if\,\right]$ $\left[\,in\,\right]$, <u>w</u>indow($\#_l\left[\#_c\left[\#_f\right]\right]$) $\left[\,\text{replace}\,\right]$

Moving average with specified weights

> tssmooth ma $\left[\,type\,\right]$ *newvar* = *exp* $\left[\,if\,\right]$ $\left[\,in\,\right]$, <u>we</u>ights($\left[\,numlist_l\,\right]$ <$\#_c$> $\left[\,numlist_f\,\right]$)
>
> $\left[\,\text{replace}\,\right]$

You must tsset your data before using tssmooth; see [TS] **tsset**.

exp may contain time-series operators; see [U] **11.4.3 Time-series varlists**.

Description

tssmooth ma creates a new series in which each observation is an average of nearby observations in the original series.

In the first syntax, window() is required and specifies the span the filter. tssmooth ma constructs a uniformly weighted moving average of the expression.

In the second syntax, weights() is required and specifies the weights to be used. tssmooth ma then applies the specified weights to construct a weighted moving average of the expression.

Options

window($\#_l\left[\#_c\left[\#_f\right]\right]$) describes the span of the uniformly weighted moving average.

$\#_l$ specifies the number of lagged terms to be included, $0 \leq \#_l \leq$ one-half the number of observations in the sample.

$\#_c$ is optional and specifies whether to include the current observation in the filter. 0 indicates exclusion and 1, inclusion. The current observation is excluded by default.

$\#_f$ is optional and specifies the number of forward terms to be included, $0 \leq \#_f \leq$ one-half the number of observations in the sample.

weights($\left[\,numlist_l\,\right]$ <$\#_c$> $\left[\,numlist_f\,\right]$) is required for the weighted moving average and describes the span of the moving average, as well as the weights to be applied to each term in the average. Note that the middle term literally is surrounded by < and >, so you might type weights(1/2 <3> 2/1).

numlist$_l$ is optional and specifies the weights to be applied to the lagged terms when computing the moving average.

$\#_c$ is required and specifies the weight to be applied to the current term.

numlist$_f$ is optional and specifies the weights to be applied to the forward terms when computing the moving average.

The number of elements in each *numlist* is limited to one-half the number of observations in the sample.

`replace` replaces *newvar* if it already exists.

Remarks

Moving averages are simple linear filters of the form

$$\widehat{x}_t = \frac{\sum_{i=-l}^{f} w_i x_{t+i}}{\sum_{i=-l}^{f} w_i}$$

where

     $\widehat{x}_t$ is the moving average

     x_t is the variable or expression to be smoothed

     w_i are the weights being applied to the terms in the filter

     l is the longest lag in the span of the filter

     f is the longest lead in the span of the filter

Moving averages are used primarily to reduce noise in time-series data. Using moving averages to isolate signals is problematic, however, because the moving averages themselves are serially correlated, even when the underlying data series is not. Still, Chatfield (1996) discusses moving-average filters and provides a number of specific moving-average filters for extracting certain trends.

▷ Example 1

Suppose that we have a time series of sales data, and we want to separate the data into two components: signal and noise. To eliminate the noise, we apply a moving-average filter. In this example, we use a symmetric moving average with a span of 5. This means that we will average the first two lagged values, the current value, and the first two forward terms of the series, with each term in the average receiving a weight of 1.

```
. use http://www.stata-press.com/data/r9/sales1
. tsset
        time variable:  t, 1 to 50
. tssmooth ma sm1 = sales, window(2 1 2)
The smoother applied was
        (1/5)*[x(t-2) + x(t-1) + 1*x(t) + x(t+1) + x(t+2)]; x(t)= sales
```

We would like to smooth our series so that there is no autocorrelation in the noise. Below we compute the noise as the difference between the smoothed series and the series itself. Then we use ac (see [TS] **corrgram**) to check for autocorrelation in the noise.

```
. generate noise = sales-sm1

. ac noise
```

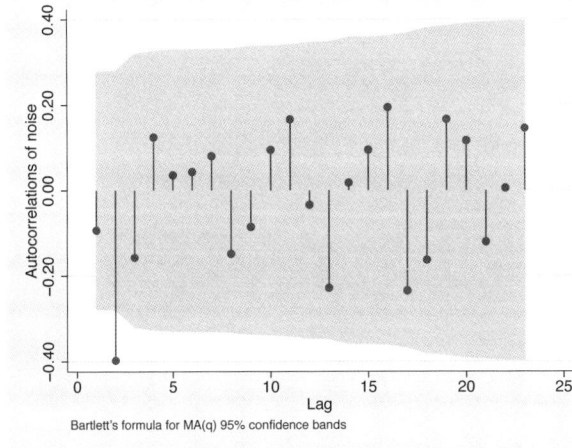

Bartlett's formula for MA(q) 95% confidence bands

◁

▷ Example 2

In the previous example, there is some evidence of negative second-order autocorrelation, possibly due to the uniform weighting or the length of the filter. We are going to specify a shorter filter in which the weights decline as the observations get farther away from the current observation.

The weighted moving-average filter requires that we supply the weights to apply to each element using the `weights()` option. In specifying the weights, we implicitly specify the span of the filter.

Below we use the filter

$$\widehat{x}_t = (1/9)(1x_{t-2} + 2x_{t-1} + 3x_t + 2x_{t+1} + 1x_{t+2})$$

In what follows, 1/2 does not mean one-half, it means the numlist 1 2:

```
. tssmooth ma sm2 = sales, weights( 1/2 <3> 2/1)
The smoother applied was
    (1/9)*[1*x(t-2) + 2*x(t-1) + 3*x(t) + 2*x(t+1) + 1*x(t+2)]; x(t)= sales

. generate noise2 = sales-sm2
```

We compute the noise and use `ac` to check for autocorrelation.

. ac noise2

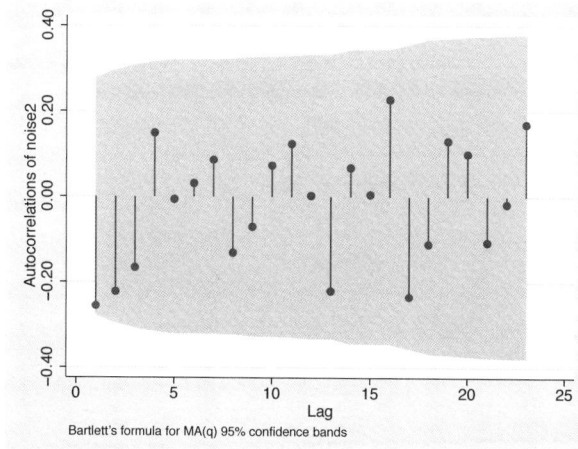

Bartlett's formula for MA(q) 95% confidence bands

The graph shows no significant evidence of autocorrelation in the noise from the second filter.

◁

❏ Technical Note

tssmooth ma gives any missing observations a coefficient of zero in both the uniformly weighted and weighted moving-average filters. This simply means that missing values or missing time periods are excluded from the moving average.

Sample restrictions, via if and in, cause the expression smoothed by tssmooth ma to be missing for the excluded observations. Thus sample restrictions have the same effect as missing values in a variable that is filtered in the expression. Also, gaps in the data that are longer than the span of the filter will generate missing values in the filtered series.

Since the first l observations and the last f observations will be outside the span of the filter, those observations will be set to missing in the moving-average series.

❏

Saved Results

tssmooth saves in r():

Scalars
    r(N)   number of observations in the sample
    r(w0)   weight on the current observation
    r(wlead*i*)   weight on lead i, if leads are specified
    r(wlag*i*)   weight on lag i, if lags are specified
Macros
    r(method)   smoothing method
    r(exp)   expression specified
    r(timevar)   time variables specified in tsset
    r(panelvar)   panel variables specified in tsset

Methods and Formulas

tssmooth ma is implemented as an ado-file. The formula for moving averages is the same as previously given.

A truncated description of the specified moving-average filter labels the new variable. See [D] **label** for more information on labels.

An untruncated description of the specified moving-average filter is saved in the characteristic tssmooth for the new variable. See [P] **char** for more information on characteristics.

Reference

Chatfield, C. 1996. *The Analysis of Time Series: An Introduction*. 5th ed. London: Chapman & Hall.

Also See

Complementary:	[TS] **tsset**
Related:	[TS] **arima**, [TS] **tssmooth dexponential**, [TS] **tssmooth exponential**, [TS] **tssmooth hwinters**, [TS] **tssmooth nl**, [TS] **tssmooth shwinters**, [D] **egen**, [D] **generate**
Background:	[U] **11.4.3 Time-series varlists**, [TS] **tssmooth**

Title

---
tssmooth nl — Nonlinear filter

Syntax

tssmooth nl $\left[\,type\,\right]$ *newvar* = *exp* $\left[\,if\,\right]$ $\left[\,in\,\right]$, smoother(*smoother*$\left[\,,\,\underline{t}wice\,\right]$)

$\left[\,replace\,\right]$

where *smoother* is specified as $Sm\left[\,Sm\left[\,\ldots\,\right]\,\right]$ and *Sm* is one of

$$\{\,1\,|\,2\,|\,3\,|\,4\,|\,5\,|\,6\,|\,7\,|\,8\,|\,9\,\}\left[\,R\,\right]$$
$$3\left[\,R\,\right]S\left[\,S\,|\,R\,\right]\left[\,S\,|\,R\,\right]\ldots$$
$$E$$
$$H$$

Letters may be specified in lowercase, if preferred. Examples of *smoother*$\left[\,,\,\text{twice}\,\right]$ include

3RSSH	3RSSH,twice	4253H	4253H,twice	43RSR2H,twice
3rssh	3rssh,twice	4253h	4253h,twice	43rsr2h,twice

You must tsset your data before using tssmooth nl; see [TS] **tsset**.

exp may contain time-series operators; see [U] **11.4.3 Time-series varlists**.

Description

tssmooth nl uses nonlinear smoothers to identify the underlying trend in a series.

Options

‾‾‾‾| Main |‾‾

smoother(*smoother*$\left[\,,\,\text{twice}\,\right]$) specifies the nonlinear smoother to be used.

replace replaces *newvar* if it already exists.

Remarks

tssmooth nl works as a front end to smooth. See [R] **smooth** for details.

Saved Results

tssmooth saves in r():

Scalars
    r(N)   number of observations

Macros
    r(method)   nl
    r(smoother)   specified smoother
    r(timevar)   time variables specified in tsset
    r(panelvar)   panel variables specified in tsset

Methods and Formulas

tssmooth nl is implemented as an ado-file. The methods are documented in [R] **smooth**.

A truncated description of the specified nonlinear filter labels the new variable. See [D] **label** for more information on labels.

An untruncated description of the specified nonlinear filter is saved in the characteristic tssmooth for the new variable. See [P] **char** for more information on characteristics.

Also See

Complementary:	[TS] **tsset**
Related:	[TS] **tssmooth dexponential**, [TS] **tssmooth exponential**, [TS] **tssmooth hwinters**, [TS] **tssmooth ma**, [TS] **tssmooth shwinters**, [D] **egen**, [D] **generate**
Background:	[U] **11.4.3 Time-series varlists**, [TS] **tssmooth**

Title

tssmooth shwinters — Holt–Winters seasonal smoothing

Syntax

tssmooth $\underline{\text{sh}}$winters $\left[type \right]$ *newvar* = *exp* $\left[if \right]$ $\left[in \right]$ $\left[\, , options \right]$

options	description
Main	
replace	replace *newvar* if it already exists
$\underline{\text{p}}$arms($\#_\alpha$ $\#_\beta$ $\#_\gamma$)	use $\#_\alpha$, $\#_\beta$, and $\#_\gamma$ as smoothing parameters
$\underline{\text{s}}$amp0(#)	use # observations to obtain initial values for recursion
s0($\#_{\text{cons}}$ $\#_{\text{lt}}$)	use $\#_{\text{cons}}$ and $\#_{\text{lt}}$ as initial values for recursion
$\underline{\text{f}}$orecast(#)	use # periods for the out-of-sample forecast
$\underline{\text{p}}$eriod(#)	use # for period of the seasonality
$\underline{\text{add}}$itive	use additive seasonal Holt–Winters method
Options	
sn0_0(*varname*)	use initial seasonal values in *varname*
sn0_v(*newvar*)	store estimated initial values for seasonal terms in *newvar*
snt_v(*newvar*)	store final year's estimated seasonal terms in *newvar*
$\underline{\text{n}}$ormalize	normalize seasonal values
$\underline{\text{alt}}$starts	use alternative method for computing the starting values
Max options	
maximize_options	control the maximization process; seldom used
$\underline{\text{fr}}$om($\#_\alpha$ $\#_\beta$ $\#_\gamma$)	use $\#_\alpha$, $\#_\beta$, and $\#_\gamma$ as starting values for the parameters

You must tsset your data before using tssmooth shwinters; see [TS] **tsset**.

exp may contain time-series operators; see [U] **11.4.3 Time-series varlists**.

Description

tssmooth shwinters performs the seasonal Holt–Winters method on a user-specified expression, which is usually just a variable name, and generates a new variable containing the forecasted series.

Options

> Main

replace replaces *newvar* if it already exists.

parms($\#_\alpha$ $\#_\beta$ $\#_\gamma$), $0 \le \#_\alpha \le 1$, $0 \le \#_\beta \le 1$, and $0 \le \#_\gamma \le 1$, specifies the parameters. If parms() is not specified, the values are chosen by an iterative process to minimize the in-sample sum-of-squared prediction errors.

If you experience difficulty converging (many iterations and "not concave" messages), try using from() to provide better starting values.

samp0(#) and s0($\#_{cons}$ $\#_{lt}$) have to do with how the initial values $\#_{cons}$ and $\#_{lt}$ for the recursion are obtained.

s0($\#_{cons}$ $\#_{lt}$) specifies the initial values to be used.

samp0(#) specifies that the initial values be obtained using the first # observations of the sample. This calculation is described under *Methods and Formulas* and depends on whether options altstart and additive are also specified.

If neither option is specified, the first half of the sample is used to obtain initial values.

forecast(#) specifies the number of periods for the out-of-sample prediction. $0 \le \# \le 500$. The default is 0, which is equivalent to not performing an out-of-sample forecast.

period(#) specifies the period of the seasonality. If period() is not specified, the seasonality is obtained from the tsset options daily, weekly, ..., yearly. If you did not specify one of those options when you tsset the data, you must specify the period() option. For instance, if your data is quarterly and you did not specify tsset's quarterly option, you must now specify period(4).

By default, seasonal values are calculated, but you may specify the initial seasonal values to be used via the sn0_0(*varname*) option. The first period() observations of *varname* are to contain the initial seasonal values.

additive uses the additive seasonal Holt–Winters method instead of the default multiplicative seasonal Holt–Winters method.

___

⌐ Options ⌐

sn0_0(*varname*) specifies the initial seasonal values to use. *varname* must contain a complete year's worth of seasonal values, beginning with the first observation in the estimation sample. For example, if you have monthly data, the first 12 observations of *varname* must contain nonmissing data. sn0_0() cannot be used with sn0_v().

sn0_v(*newvar*) stores in *newvar* the initial seasonal values after they have been estimated. sn0_v() cannot be used with sn0_0().

snt_v(*newvar*) stores in *newvar* the seasonal values for the final year's worth of data.

normalize specifies that the seasonal values be normalized. In the multiplicative model, they are normalized to sum to one. In the additive model, the seasonal values are normalized to sum to zero.

altstarts uses an alternative method to compute the starting values for the constant, the linear, and the seasonal terms. The default and the alternative methods are described in *Methods and Formulas*. altstarts may not be specified with s0().

___

⌐ Max options ⌐

maximize_options controls the process for solving for the optimal α, β, and γ when option parms() is not specified.

maximize_options: nodifficult, technique(*algorithm_spec*), iterate(#), [no]log, trace, gradient, showstep, hessian, shownrtolerance, tolerance(#), ltolerance(#), gtolerance(#), nrtolerance(#), nonrtolerance, see [R] **maximize**. These options are seldom used.

from($\#_\alpha$ $\#_\beta$ $\#_\gamma$), $0 < \#_\alpha < 1$, $0 < \#_\beta < 1$, and $0 < \#_\gamma < 1$, specifies starting values from which the optimal values of α, β, and γ will be obtained. If from() is not specified, from(.5 .5 .5) is used.

Remarks

Remarks are presented under the headings

Introduction
Holt–Winters seasonal multiplicative method
Holt–Winters seasonal additive method

Introduction

The seasonal Holt–Winters methods forecast univariate series that have a seasonal component. If the amplitude of the seasonal component grows with the series, the Holt–Winters multiplicative method should be used. If the amplitude of the seasonal component is not growing with the series, the Holt–Winters additive method should be used. Abraham and Ledolter (1983), Bowerman and O'Connell (1993), and Montgomery, Johnson, and Gardiner (1990) provide good introductions to the Holt–Winters methods in the context of recursive univariate forecasting methods. Chatfield (1996, 2001) provides introductions in the broader context of modern time-series analysis.

Like the other recursive methods in `tssmooth`, `tssmooth shwinters` uses the information stored by `tsset` to detect panel data. When applied to panel data, each series is smoothed separately, and the starting values are computed separately for each panel. If the smoothing parameters are chosen to minimize the in-sample sum-of-squared forecast errors, the optimization is performed separately on each panel.

When there are missing values at the beginning of the series, the sample begins with the first nonmissing observation. Missing values after the first nonmissing observation are filled in with forecasted values.

Holt–Winters seasonal multiplicative method

This method forecasts seasonal time series in which the amplitude of the seasonal component grows with the series. Chatfield (2001) notes that there are some nonlinear state-space models whose optimal prediction equations correspond to the multiplicative Holt–Winters method. This procedure is best applied to data that could be described by

$$x_{t+j} = (\mu_t + \beta j)S_{t+j} + \epsilon_{t+j}$$

where x_t is the series, μ_t is the time-varying mean at time t, β is a parameter, S_t is the seasonal component at time t, and ϵ_t is an idiosyncratic error. See *Methods and Formulas* for the updating equations.

▷ Example 1

We have quarterly data on turkey sales by a new producer in the 1990s. The data have a strong seasonal component and an upward trend. We use the multiplicative Holt–Winters method to forecast sales for the year 2000. As we have already `tsset` our data to the quarterly format, we do not need to specify the `period()` option.

```
. use http://www.stata-press.com/data/r9/turksales

. tssmooth shwinters shw1 = sales, forecast(4)
computing optimal weights

Iteration 0:    penalized RSS = -189.34609  (not concave)
Iteration 1:    penalized RSS = -108.68038  (not concave)
Iteration 2:    penalized RSS = -106.23342
Iteration 3:    penalized RSS = -106.14103
Iteration 4:    penalized RSS = -106.14093
Iteration 5:    penalized RSS = -106.14093

Optimal weights:
                            alpha = 0.1310
                             beta = 0.1428
                            gamma = 0.2999
penalized sum-of-squared residuals = 106.1409
         sum-of-squared residuals = 106.1409
         root mean squared error  = 1.628964
```

The graph below describes the fit and the forecast that was obtained.

```
. line sales shw1 t, title("Multiplicative Holt-Winters forecast")
> xtitle(Time) ytitle(Sales)
```

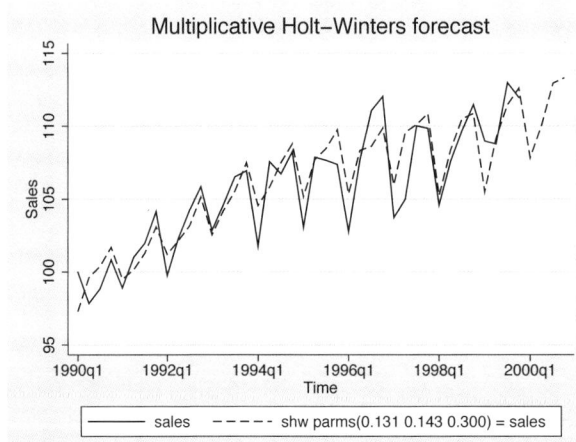

◁

Holt–Winters seasonal additive method

This method is similar to the previous one, but the seasonal effect is assumed to be additive rather than multiplicative. This method forecasts series that can be described by the equation

$$x_{t+j} = (\mu_t + \beta j) + S_{t+j} + \epsilon_{t+j}$$

See *Methods and Formulas* for the updating equations.

▷ Example 2

In this example, we fit the data from the previous example to the additive model in order to forecast sales in the coming year. We use the `snt_v()` option to save the last year's seasonal terms in the new variable `seas`.

```
. tssmooth shwinters shwa = sales, forecast(4) snt_v(seas) normalize additive
computing optimal weights

Iteration 0:    penalized RSS = -190.90242   (not concave)
Iteration 1:    penalized RSS =  -108.8357
Iteration 2:    penalized RSS = -107.77074
Iteration 3:    penalized RSS = -107.66487
Iteration 4:    penalized RSS = -107.66442
Iteration 5:    penalized RSS = -107.66442

Optimal weights:
                            alpha = 0.1219
                             beta = 0.1580
                            gamma = 0.3340
    penalized sum-of-squared residuals = 107.6644
            sum-of-squared residuals = 107.6644
            root mean squared error = 1.640613
```

The output reveals that the multiplicative model has a better in-sample fit, and the graph below shows that the forecast from the multiplicative model is higher than that of the additive model.

```
. line shw1 shwa t if t>=q(2000q1), title("Multiplicative and additive"
> "Holt-Winters forecasts") xtitle("Time") ytitle("Sales") legend(cols(1))
```

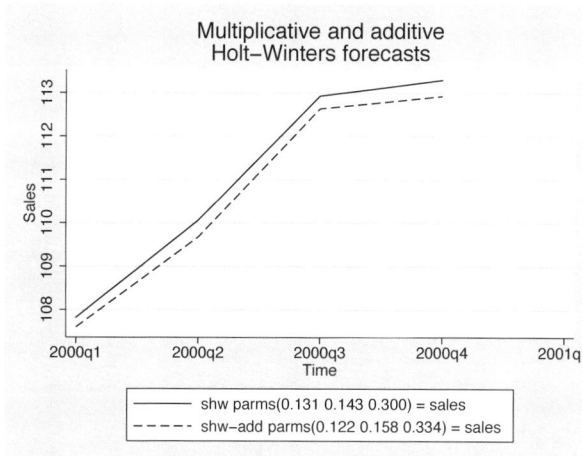

To check whether the estimated seasonal components are intuitively sound, we list the last year's seasonal components.

```
. list t seas if seas < .
```

	t	seas
37.	1999q1	-2.7533393
38.	1999q2	-.91752567
39.	1999q3	1.8082417
40.	1999q4	1.8626233

The output indicates that the signs of the estimated seasonal components agree with our intuition.

◁

Saved Results

tssmooth saves in r():

Scalars

r(N)	number of observations	r(N_pre)	number of seasons used
r(alpha)	α smoothing parameter		in calculating starting values
r(beta)	β smoothing parameter	r(s2_0)	initial value for linear term
r(gamma)	γ smoothing parameter	r(s1_0)	initial value for constant term
r(prss)	penalized sum-of-squared errors	r(linear)	final value of linear term
r(rss)	sum-of-squared errors	r(constant)	final value of constant term
r(rmse)	root mean squared error	r(period)	period, if filter is seasonal

Macros

r(method)	shwinters, additive or	r(exp)	expression specified
	shwinters, multiplicative	r(timevar)	time variables specified in tsset
r(normalize	normalize, if specified	r(panelvar)	panel variables specified in tsset

Methods and Formulas

tssmooth shwinters is implemented as an ado-file.

A truncated description of the specified seasonal Holt–Winters filter labels the new variable. See [D] **label** for more information on labels.

An untruncated description of the specified seasonal Holt–Winters filter is saved in the characteristic named tssmooth for the new variable. See [P] **char** for more information on characteristics.

When the parms() option is not specified, the smoothing parameters are chosen to minimize the in-sample sum of penalized squared-forecast errors. Sometimes, one or more of the three optimal parameters lies on the boundary $[0, 1]$. To keep the estimates inside $[0, 1]$, tssmooth shwinters parameterizes the objective function in terms of their inverse logits, i.e., $\exp(\alpha)/\{1 + \exp(\alpha)\}$, $\exp(\beta)/\{1 + \exp(\beta)\}$, and $\exp(\gamma)/\{1 + \exp(\gamma)\}$. When one of these parameters is actually on the boundary, this can complicate the optimization. For this reason, tssmooth shwinters optimizes a penalized sum-of-squared forecast errors. Let $\widehat{x}_t(\widetilde{\alpha}, \widetilde{\beta}, \widetilde{\gamma})$ be the forecast for the series x_t given the choices of $\widetilde{\alpha}$, $\widetilde{\beta}$, and $\widetilde{\gamma}$. Then the in-sample penalized sum-of-squared prediction errors is

$$P = \sum_{t=1}^{T} \left[\{x_t - \widehat{x}_t(\widetilde{\alpha}, \widetilde{\beta}, \widetilde{\gamma})\}^2 + I_{|f(\widetilde{\alpha})|>12)}(|f(\widetilde{\alpha})| - 12)^2 + I_{|f(\widetilde{\beta})|>12)}(|f(\widetilde{\beta})| - 12)^2 \right.$$

$$\left. + I_{|f(\widetilde{\gamma})|>12)}(|f(\widetilde{\gamma})| - 12)^2 \right]$$

where $f(x) = \ln\left(\frac{x}{1-x}\right)$. The penalty term is zero unless one of the parameters is very close to the boundary. When one of the parameters is very close to the boundary, the penalty term will help to obtain convergence.

Holt–Winters seasonal multiplicative procedure

As with the other recursive methods in tssmooth, there are three aspects to implementing the Holt–Winters seasonal multiplicative procedure: the forecasting equation, the initial values, and the updating equations. Unlike in the other methods, the data are now assumed to be seasonal with period L.

Given the estimates $a(t)$, $b(t)$, and $s(t + \tau - L)$, a τ step-ahead point forecast of x_t, denoted by $\widehat{y}_{t+\tau}$, is

$$\widehat{y}_{t+\tau} = \{a(t) + b(t)\tau\}\, s(t + \tau - L)$$

Given the smoothing parameters α, β, and γ, the updating equations are

$$a(t) = \alpha\frac{x_t}{s(t-L)} + (1-\alpha)\{a(t-1) + b(t-1)\}$$

$$b(t) = \beta\{a(t) - a(t-1)\} + (1-\beta)\,b(t-1)$$

and

$$s(t) = \gamma\left\{\frac{x_t}{a(t)}\right\} + (1-\gamma)s(t-L)$$

To restrict the seasonal terms to sum to 1 over each year, specify the normalize option.

The updating equations require the $L+2$ initial values $a(0)$, $b(0)$, $s(1-L)$, $s(2-L)$, ..., $s(0)$. Two methods calculate the initial values using the first m years, each of which contains L seasons. By default, m is set to the number of seasons in one-half the sample.

The initial value of the trend component, $b(0)$, can be estimated by

$$b(0) = \frac{\overline{x}_m - \overline{x}_1}{(m-1)L}$$

where $\overline{x}_m$ is the average level of x_t in year m and $\overline{x}_1$ is the average level of x_t in the first year.

The initial value for the linear term, $a(0)$, is then calculated as

$$a(0) = \overline{x}_1 - \frac{L}{2}b(0)$$

To calculate the initial values for the seasons $1, 2, \ldots, L$, we first calculate the deviation-adjusted values,

$$S(t) = \frac{x_t}{\overline{x}_i - \left\{\frac{(L+1)}{2} - j\right\}b(0)}$$

where i is the year that corresponds to time t, j is the season that corresponds to time t, and $\overline{x}_i$ is the average level of x_t in year i.

Next, for each season $l = 1, 2, \ldots, L$, we define $\bar{s}_l$ as the average S_t over the years. That is,

$$\bar{s}_l = \frac{1}{m} \sum_{k=0}^{m-1} S_{l+kL} \qquad \text{for } l = 1, 2, \ldots, L$$

Then the initial seasonal estimates are

$$s_{0l} = \bar{s}_l \left(\frac{L}{\sum_{l=1}^{L} \bar{s}_l} \right) \qquad \text{for } l = 1, 2, \ldots, L$$

and these values are used to fill in $s(1 - L), \ldots, s(0)$.

If the `altstarts` option is specified, the starting values are computed based on a regression with seasonal indicator variables. Specifically, the series x_t is regressed on a time variable normalized to equal one in the first period in the sample and on a constant. Then $b(0)$ is set to the estimated coefficient on the time variable, and $a(0)$ is set to the estimated constant term. To calculate the seasonal starting values, x_t is regressed on a set of L seasonal dummy variables. The lth seasonal starting value is set to $(\frac{1}{\mu})\hat{\beta}_l$, where μ is the mean of x_t and $\hat{\beta}_l$ is the estimated coefficient on the lth seasonal dummy variable. The sample used in both regressions and the mean computation is restricted to include the first `samp0()` years. By default, `samp0()` includes half of the data.

❏ Technical Note

If there are missing values in the first few years, a small value of m can cause the starting value methods for seasonal term to fail. In this case, you should either specify a larger value of m using `samp0()` or directly specify the seasonal starting values using the `snt0_0()` option.

❏

Holt–Winters seasonal additive procedure

This procedure is similar to the previous one, except that the data are assumed to be described by

$$x_t = (\beta_0 + \beta_1 t) + s_t + \epsilon_t$$

As in the multiplicative case, there are three smoothing parameters, α, β, and γ, which can either be set or chosen to minimize the in-sample sum-of-squared forecast errors.

The updating equations are

$$a(t) = \alpha \left\{ x_t - s(t - L) \right\} + (1 - \alpha) \left\{ a(t - 1) + b(t_1) \right\}$$

$$b(t) = \beta \left\{ a(t) - a(t - 1) \right\} + (1 - \beta) b(t - 1)$$

and

$$s(t) = \gamma \left\{ x_t - a(t) \right\} + (1 - \gamma) s(t - L)$$

To restrict the seasonal terms to sum to 0 over each year, specify the `normalize` option.

A τ step-ahead forecast, denoted by $\hat{y}_{t+\tau}$, is given by

$$\hat{x}_{t+\tau} = a(t) + b(t)\tau + s(t + \tau - L)$$

As in the multiplicative case, there are two methods for setting the initial values.

The default method is to obtain the initial values for $a(0), b(0), s(1 - L), \ldots, s(0)$ from the regression

$$x_t = a(0) + b(0)t + \beta_{s,1-L}D_1 + \beta_{s,2-L}D_2 + \cdots + \beta_{s,0}D_L + e_t$$

where the $D_1, \ldots, D_L$ are dummy variables with

$$D_i = \left\{ \begin{array}{ll} 1 & \text{if } t \text{ corresponds to season } i \\ 0 & \text{otherwise} \end{array} \right\}$$

When `altstarts` is specified, an alternative method is used that regresses the x_t series on a time variable that has been normalized to equal one in the first period in the sample and on a constant term. $b(0)$ is set to the estimated coefficient on the time variable, and $a(0)$ is set to the estimated constant term. Then the demeaned series $\widetilde{x}_t = x_t - \mu$ is created, where μ is the mean of the x_t. The $\widetilde{x}_t$ are regressed on L seasonal dummy variables. The lth seasonal starting value is then set to β_l, where β_l is the estimated coefficient on the lth seasonal dummy variable. The sample in both the regression and the mean calculation is restricted to include the first `samp0` years, where, by default, `samp0()` includes half the data.

Acknowledgment

We would like to thank Nick Cox of the University of Durham for his helpful comments.

References

Abraham, B. and J. Ledolter. 1983. *Statistical Methods for Forecasting.* New York: Wiley.

Bowerman, B. and R. O'Connell. 1993. *Forecasting and Time Series: An Applied Approach.* 3rd ed. Pacific Grove, CA: Duxbury.

Chatfield, C. 1996. *The Analysis of Time Series: An Introduction.* 5th ed. London: Chapman & Hall.

———. 2001. *Time-Series Forecasting.* London: Chapman & Hall.

Chatfield, C. and M. Yar. 1988. Holt–Winters forecasting: Some practical issues. *The Statistician* 37: 129–140.

Montgomery, D. C., L. A. Johnson, and J. S. Gardiner. 1990. *Forecasting and Time Series Analysis.* 2nd ed. New York: McGraw–Hill.

Also See

Complementary:	[TS] **tsset**
Related:	[TS] **arima**, [TS] **tssmooth dexponential**, [TS] **tssmooth exponential**, [TS] **tssmooth hwinters**, [TS] **tssmooth ma**, [TS] **tssmooth nl**, [D] **egen**, [D] **generate**
Background:	[U] **11.4.3 Time-series varlists**, [TS] **tssmooth**

Title

var intro — Introduction to vector autoregression models

Description

Stata has a suite of commands for fitting, forecasting, interpreting, and performing inference on vector autoregressions (VARs) and structural vector autoregressions (SVARs). The suite includes several commands for estimating and interpreting impulse–response functions (IRFs) and forecast-error variance decompositions (FEVDs). The table below describes the available commands.

Fitting a VAR or SVAR

var	[TS] **var**	Fit vector autoregression models
svar	[TS] **var svar**	Fit structural vector autoregression models
varbasic	[TS] **varbasic**	Fit a simple VAR and graph impulse–response functions

Model diagnostics and inference

varstable	[TS] **varstable**	Check the stability condition of VAR or SVAR estimates
varsoc	[TS] **varsoc**	Obtain lag-order selection statistics for VARs and VECMs
varwle	[TS] **varwle**	Obtain Wald lag-exclusion statistics after var or svar
vargranger	[TS] **vargranger**	Perform pairwise Granger causality tests after var or svar
varlmar	[TS] **varlmar**	Obtain LM statistics for residual autocorrelation after var or svar
varnorm	[TS] **varnorm**	Test for normally distributed disturbances after var or svar

Forecasting after fitting a VAR or SVAR

fcast compute	[TS] **fcast compute**	Compute dynamic forecasts of dependent variables after var, svar, or vec
fcast graph	[TS] **fcast graph**	Graph forecasts of variables computed by fcast compute

Working with IRFs and FEVDs

irf	[TS] **irf**	Create and analyze IRFs and FEVDs

This manual entry provides an overview of vector autoregressions and structural vector autoregressions. More rigorous treatments can be found in Hamilton (1994), Lütkepohl (1993), and Amisano and Giannini (1997). Stock and Watson (2001) provide an excellent nonmathematical treatment of vector autoregressions and their role in macroeconomics.

Remarks

Remarks are presented under the headings

> *Introduction to VARs*
> *Introduction to SVARs*
> *Short-run SVAR models*
> *Long-run restrictions*
> *IRFs and FEVDs*

Introduction to VARs

A vector autoregression (VAR) is a model in which K variables are specified as linear functions of p of their own lags, p lags of the other $K - 1$ variables, and possibly additional exogenous variables. Algebraically, a p-order vector autoregressive model, written VAR(p), with exogenous variables $\mathbf{x}_t$ is given by

$$\mathbf{y}_t = \mathbf{v} + \mathbf{A}_1\mathbf{y}_{t-1} + \cdots + \mathbf{A}_p\mathbf{y}_{t-p} + \mathbf{B}\mathbf{x}_t + \mathbf{u}_t \qquad t \in \{-\infty, \infty\} \tag{1}$$

where

> $\mathbf{y}_t = (y_{1t}, \ldots, y_{Kt})'$ is a $K \times 1$ random vector,
> $\mathbf{A}_1$ through $\mathbf{A}_p$ are $K \times K$ matrices of parameters,
> $\mathbf{x}_t$ is an $M \times 1$ vector of exogenous variables,
> $\mathbf{B}$ is a $K \times M$ matrix of coefficients,
> $\mathbf{v}$ is a $K \times 1$ vector of parameters, and
> $\mathbf{u}_t$ is assumed to be white noise; that is,
> $E(\mathbf{u}_t) = \mathbf{0}$,
> $E(\mathbf{u}_t\mathbf{u}_t') = \mathbf{\Sigma}$, and
> $E(\mathbf{u}_t\mathbf{u}_s') = \mathbf{0}$ for $t \neq s$

There are $K^2 \times p + K \times (M+1)$ parameters in the equation for $\mathbf{y}_t$, and there are $\{K \times (K + 1)\}/2$ parameters in the covariance matrix $\mathbf{\Sigma}$. One way to reduce the number of parameters is to specify an incomplete VAR, in which some of the $\mathbf{A}$ matrices are set to zero. Another way is to specify linear constraints on some of the coefficients in the VAR.

A VAR can be viewed as the reduced form of a system of dynamic simultaneous equations. Consider the system

$$\mathbf{W}_0\mathbf{y}_t = \mathbf{a} + \mathbf{W}_1\mathbf{y}_{t-1} + \cdots + \mathbf{W}_p\mathbf{y}_{t-p} + \mathbf{W}_x\mathbf{x}_t + \mathbf{e}_t \tag{2}$$

where $\mathbf{a}$ is a $K \times 1$ vector of parameters, each $\mathbf{W}_i$, $i = 0, \ldots, p$, is a $K \times K$ matrix of parameters, and $\mathbf{e}_t$ is a $K \times 1$ disturbance vector. In the traditional dynamic simultaneous equations approach, sufficient restrictions are placed on the $\mathbf{W}_i$ to obtain identification. Assuming that $\mathbf{W}_0$ is nonsingular, (2) can be rewritten as

$$\mathbf{y}_t = \mathbf{W}_0^{-1}\mathbf{a} + \mathbf{W}_0^{-1}\mathbf{W}_1\mathbf{y}_{t-1} + \cdots + \mathbf{W}_0^{-1}\mathbf{W}_p\mathbf{y}_{t-p} + \mathbf{W}_0^{-1}\mathbf{W}_x\mathbf{x}_t + \mathbf{W}_0^{-1}\mathbf{e}_t \tag{3}$$

which is a VAR with

$$\mathbf{v} = \mathbf{W}_0^{-1}\mathbf{a}$$
$$\mathbf{A}_i = \mathbf{W}_0^{-1}\mathbf{W}_i$$
$$\mathbf{B} = \mathbf{W}_0^{-1}\mathbf{W}_x$$
$$\mathbf{u}_t = \mathbf{W}_0^{-1}\mathbf{e}_t$$

The cross-equation error variance–covariance matrix Σ contains all the information about contemporaneous correlations in a VAR and may be the VAR's greatest strength and its greatest weakness. Since no questionable *a priori* assumptions are imposed, fitting a VAR allows the dataset to speak for itself. However, without imposing some restrictions on the structure of Σ, we cannot make a causal interpretation of the results.

If we make additional technical assumptions, we can derive another representation of the VAR in (1). To simplify the notation, consider the case without exogenous variables. If the VAR is stable (see [TS] **varstable**), we can rewrite $\mathbf{y}_t$ as

$$\mathbf{y}_t = \boldsymbol{\mu} + \sum_{i=0}^{\infty} \boldsymbol{\Phi}_i \mathbf{u}_{t-i} \tag{4}$$

where $\boldsymbol{\mu}$ is the $K \times 1$ time-invariant mean of the process and the $\boldsymbol{\Phi}_i$ are $K \times K$ matrices of parameters. Equation (4) states that the process by which the variables in $\mathbf{y}_t$ fluctuate about their time-invariant means, $\boldsymbol{\mu}$, is completely determined by the parameters in $\boldsymbol{\Phi}_i$ and the (infinite) past history of independent and identically distributed (i.i.d.) shocks or innovations, $\mathbf{u}_{t-1}, \mathbf{u}_{t-2}, \ldots$. Equation (4) is known as the vector moving-average representation of the VAR. The moving-average coefficients $\boldsymbol{\Phi}_i$ are also known as the simple impulse–response functions at horizon i. The precise relationship between the $\mathbf{A}_i$ and the $\boldsymbol{\Phi}_i$ is derived in *Methods and Formulas* of [TS] **irf create**.

The joint distribution of $\mathbf{y}_t$ is determined by the distributions of $\mathbf{x}_t$ and $\mathbf{u}_t$ and the parameters $\mathbf{v}$, $\mathbf{B}$, and $\mathbf{A}_i$. Estimating the parameters in a VAR requires that the variables in $\mathbf{y}_t$ and $\mathbf{x}_t$ be covariance stationary, meaning that their first two moments exist and are time invariant. If the $\mathbf{y}_t$ are not covariance stationary, but their first-differences are, a vector error-correction model (VECM) can be used. See [TS] **vec intro** and [TS] **vec** for more information about those models.

If the $\mathbf{u}_t$ form a zero mean, i.i.d. vector process, and $\mathbf{y}_t$ and $\mathbf{x}_t$ are covariance stationary and are not correlated with the $\mathbf{u}_t$, consistent and efficient estimates of $\mathbf{B}$, the $\mathbf{A}_i$, and $\mathbf{v}$ are obtained via seemingly unrelated regression, yielding estimators that are asymptotically normally distributed. When the equations for the variables $\mathbf{y}_t$ have the same set of regressors, equation-by-equation OLS estimates are the conditional maximum likelihood estimates.

Much of the interest in VAR models is focused on the forecasts, impulse–response functions, and the forecast-error variance decompositions, all of which are functions of the estimated parameters. Estimating these functions is straightforward, but their asymptotic standard errors are usually obtained by assuming that $\mathbf{u}_t$ forms a zero mean, i.i.d. Gaussian (normal) vector process. Also, some of the specification tests for VARs have been derived using the likelihood-ratio principle and the stronger Gaussian assumption.

In the absence of contemporaneous exogenous variables, the disturbance variance–covariance matrix contains all the information about contemporaneous correlations among the variables. VARs are sometimes classified into three types by how they account for this contemporaneous correlation. (See Stock and Watson [2001] for one derivation of this taxonomy.) A reduced-form VAR, aside from estimating the variance–covariance matrix of the disturbance, does not try to account for contemporaneous correlations. In a recursive VAR, the K variables are assumed to form a recursive dynamic structural equation model in which the first variable is a function of lagged variables, the second a function of contemporaneous values of the first variable and lagged values, and so on. In a structural VAR, the theory you are working with places restrictions on the contemporaneous correlations that are not necessarily recursive.

Stata has two commands for fitting reduced-form VARs: `var` and `varbasic`. `var` allows for constraints to be imposed on the coefficients. `varbasic` allows you to fit a simple VAR quickly without constraints and graph the impulse–response functions.

Because it can be important to fit a VAR of the correct order, `varsoc` offers several methods for choosing the lag order p of the VAR to fit. After fitting a VAR, and before proceeding with inference, interpretation, or forecasting, it is important to check that the VAR fits the data. `varlmar` can be used to check for autocorrelation in the disturbances. `varwle` performs Wald tests to determine if certain lags can be excluded. `varnorm` tests the null hypothesis that the disturbances are normally distributed. `varstable` checks the eigenvalue condition for stability, which is needed to interpret the impulse–response functions and forecast-error variance decompositions.

Introduction to SVARs

As discussed in [TS] **irf create**, a problem with VAR analysis is that, since Σ is not restricted to be a diagonal matrix, an increase in an innovation to one variable provides information about the innovations to other variables. This implies that no causal interpretation of the simple impulse–response functions is possible: there is no way to determine whether the shock to the first variable caused the shock in the second variable or vice versa.

However, suppose that we had a matrix $\mathbf{P}$ such that $\Sigma = \mathbf{P}\mathbf{P}'$. It can then be shown that the variables in $\mathbf{P}^{-1}\mathbf{u}_t$ have zero mean and that $E\{\mathbf{P}^{-1}\mathbf{u}_t(\mathbf{P}^{-1}\mathbf{u}_t)'\} = \mathbf{I}_K$. We could rewrite (4) as

$$
\begin{aligned}
\mathbf{y}_t &= \boldsymbol{\mu} + \sum_{s=0}^{\infty} \boldsymbol{\Phi}_s \mathbf{P}\mathbf{P}^{-1}\mathbf{u}_{t-s} \\
&= \boldsymbol{\mu} + \sum_{s=0}^{\infty} \boldsymbol{\Theta}_s \mathbf{P}^{-1}\mathbf{u}_{t-s} \\
&= \boldsymbol{\mu} + \sum_{s=0}^{\infty} \boldsymbol{\Theta}_s \mathbf{w}_{t-s}
\end{aligned}
\tag{5}
$$

where $\boldsymbol{\Theta}_s = \boldsymbol{\Phi}_s \mathbf{P}$ and $\mathbf{w}_t = \mathbf{P}^{-1}\mathbf{u}_t$. If we had such a $\mathbf{P}$, the $\mathbf{w}_k$ would be mutually orthogonal, and the $\boldsymbol{\Theta}_s$ would allow the causal interpretation that we seek.

SVAR models provide a framework for estimation of and inference about a broad class of $\mathbf{P}$ matrices. As described in [TS] **irf create**, the estimated $\mathbf{P}$ matrices can then be used to estimate structural impulse–response functions and structural forecast-error variance decompositions. There are two types of SVAR models. Short-run SVAR models identify a $\mathbf{P}$ matrix by placing restrictions on the contemporaneous correlations between the variables. Long-run SVAR models, on the other hand, do so by placing restrictions on the long-term accumulated effects of the innovations.

Short-run SVAR models

A short-run SVAR model without exogenous variables can be written as

$$
\mathbf{A}(\mathbf{I}_K - \mathbf{A}_1 L - \mathbf{A}_2 L^2 - \cdots - \mathbf{A}_p L^p)\mathbf{y}_t = \mathbf{A}\boldsymbol{\epsilon}_t = \mathbf{B}\mathbf{e}_t
\tag{6}
$$

where L is the lag operator; $\mathbf{A}$, $\mathbf{B}$, and $\mathbf{A}_1, \ldots, \mathbf{A}_p$ are $K \times K$ matrices of parameters; $\boldsymbol{\epsilon}_t$ is a $K \times 1$ vector of innovations with $\boldsymbol{\epsilon}_t \sim N(\mathbf{0}, \boldsymbol{\Sigma})$ and $E[\boldsymbol{\epsilon}_t \boldsymbol{\epsilon}_s'] = \mathbf{0}_K$ for all $s \neq t$; and $\mathbf{e}_t$ is a $K \times 1$ vector of orthogonalized disturbances; i.e., $\mathbf{e}_t \sim N(\mathbf{0}, \mathbf{I}_K)$ and $E[\mathbf{e}_t \mathbf{e}_s'] = \mathbf{0}_K$ for all $s \neq t$. These transformations of the innovations allow us to analyze the dynamics of the system in terms of a change to an element of $\mathbf{e}_t$. In a short-run SVAR model, we obtain identification by placing restrictions on $\mathbf{A}$ and $\mathbf{B}$, which are assumed to be nonsingular.

Equation (6) implies that $\mathbf{P}_{\text{sr}} = \mathbf{A}^{-1}\mathbf{B}$, where $\mathbf{P}_{\text{sr}}$ is the $\mathbf{P}$ matrix identified by a particular short-run SVAR model. Note that the latter equality in (6) implies that

$$\mathbf{A}\boldsymbol{\epsilon}_t\boldsymbol{\epsilon}_t'\mathbf{A}' = \mathbf{B}\mathbf{e}_t\mathbf{e}_t'\mathbf{B}'$$

Taking the expectation of both sides yields

$$\boldsymbol{\Sigma} = \mathbf{P}_{\text{sr}}\mathbf{P}_{\text{sr}}'$$

Assuming that the underlying VAR is stable (see [TS] **varstable** for a discussion of stability), we can invert the autoregressive representation of the model in (6) to an infinite-order, moving-average representation of the form

$$\mathbf{y}_t = \boldsymbol{\mu} + \sum_{s=0}^{\infty} \boldsymbol{\Theta}_s^{\text{sr}}\mathbf{e}_{t-s} \tag{7}$$

whereby $\mathbf{y}_t$ is expressed in terms of the mutually orthogonal, unit-variance structural innovations $\mathbf{e}_t$. The $\boldsymbol{\Theta}_s^{\text{sr}}$ contain the structural impulse–response functions at horizon s.

In a short-run SVAR model, the $\mathbf{A}$ and $\mathbf{B}$ matrices model all the information about contemporaneous correlations. The $\mathbf{B}$ matrix also scales the innovations $\mathbf{u}_t$ to have unit variance. This allows the structural impulse–response functions constructed from (7) to be interpreted as the effect on variable i of a one-time unit increase in the structural innovation to variable j after s periods.

$\mathbf{P}_{\text{sr}}$ identifies the structural impulse–response functions by defining a transformation of $\boldsymbol{\Sigma}$, and $\mathbf{P}_{\text{sr}}$ is identified by the restrictions placed on the parameters in $\mathbf{A}$ and $\mathbf{B}$. Since there are only $K(K+1)/2$ free parameters in $\boldsymbol{\Sigma}$, only $K(K+1)/2$ parameters may be estimated in an identified $\mathbf{P}_{\text{sr}}$. Since there are $2K^2$ total parameters in $\mathbf{A}$ and $\mathbf{B}$, the order condition for identification requires at least $2K^2 - K(K+1)/2$ restrictions be placed on those parameters. Just as in the simultaneous-equations framework, this order condition is necessary but not sufficient. Amisano and Giannini (1997) derive a method to check that an SVAR model is locally identified near some specified values for $\mathbf{A}$ and $\mathbf{B}$.

Before moving on to models with long-run constraints, we should note some limitations. We cannot place constraints on the elements of $\mathbf{A}$ in terms of the elements of $\mathbf{B}$, or vice versa. This limitation is imposed by the form of the check for identification derived by Amisano and Giannini (1997). As noted in *Methods and Formulas* of [TS] **var svar**, this test requires separate constraint matrices for the parameters in $\mathbf{A}$ and $\mathbf{B}$. Also, we cannot mix short-run and long-run constraints.

Long-run restrictions

Recall that a general short-run SVAR has the form

$$\mathbf{A}(\mathbf{I}_K - \mathbf{A}_1 L - \mathbf{A}_2 L^2 - \cdots - \mathbf{A}_p L^p)\mathbf{y}_t = \mathbf{B}\mathbf{e}_t$$

To simplify the notation, let $\bar{\mathbf{A}} = (\mathbf{I}_K - \mathbf{A}_1 L - \mathbf{A}_2 L^2 - \cdots - \mathbf{A}_p L^p)$. The model is assumed to be stable (see [TS] **varstable**), so $\bar{\mathbf{A}}^{-1}$, the matrix of estimated long-run effects of the reduced-form VAR shocks, is well defined. Constraining $\mathbf{A}$ to be an identity matrix allows us to rewrite this equation as

$$\mathbf{y}_t = \bar{\mathbf{A}}^{-1}\mathbf{B}\mathbf{e}_t$$

which implies that $\boldsymbol{\Sigma} = \mathbf{B}\mathbf{B}'$. Thus $\mathbf{C} = \bar{\mathbf{A}}^{-1}\mathbf{B}$ is the matrix of long-run responses to the orthogonalized shocks, and

$$\mathbf{y}_t = \mathbf{C}\mathbf{e}_t$$

In long-run models, the constraints are placed on the elements of $\mathbf{C}$, and the free parameters are estimated. These constraints are frequently exclusion restrictions. For instance, constraining $\mathbf{C}[1, 2]$ to be zero can be interpreted as setting the long-run response of variable 1 to the structural shocks driving variable 2 to be zero.

Stata's svar command estimates the parameters of structural vector autoregressions. See [TS] **var svar** for more information and examples.

IRFs and FEVDs

Impulse–response functions (IRFs) describe how the K endogenous variables react over time to a one-time shock to one of the K disturbances. Since the disturbances may be contemporaneously correlated, these functions do not explain how variable i reacts to a one-time increase in the innovation to variable j after s periods, holding everything else constant. To explain this, we must start with orthogonalized innovations so that the assumption to hold everything else constant is reasonable. Recursive VARs use a Cholesky decomposition to orthogonalize the disturbances and thereby obtain structurally interpretable impulse–response functions. Structural VARs use theory to impose sufficient restrictions, which need not be recursive, to decompose the contemporaneous correlations into orthogonal components.

Forecast-error variance decompositions (FEVDs) are another tool for interpreting how the orthogonalized innovations affect the K variables over time. The FEVD from j to i gives the fraction of the s-step forecast-error variance of variable i that can be attributed to the jth orthogonalized innovation.

irf create estimates impulse–response functions, Cholesky orthogonalized impulse–response functions, and structural impulse–response functions and their standard errors. It also estimates Cholesky and structural forecast-error variance decompositions. The irf graph, irf cgraph, irf ograph, irf table, and irf ctable commands graph and tabulate these estimates. In addition, Stata has several other commands to manage IRF and FEVD results. See [TS] **irf** for a description of these commands.

fcast compute computes dynamic forecasts and their standard errors from VARs. fcast graph graphs the forecasts that are generated using fcast compute.

VARs allow researchers to investigate whether one variable is useful in predicting another variable. A variable x is said to Granger-cause a variable y if, given the past values of y, past values of x are useful for predicting y. The Stata command vargranger performs Wald tests to investigate Granger causality between the variables in a VAR.

References

Amisano, G. and C. Giannini. 1997. *Topics in Structural VAR Econometrics*. 2nd ed. Heidelberg: Springer.

Hamilton, J. D. 1994. *Time Series Analysis*. Princeton: Princeton University Press.

Lütkepohl, H. 1993. *Introduction to Multiple Time Series Analysis*. 2nd ed. New York: Springer.

Stock, J. H. and M. W. Watson. 2001. Vector autoregressions. *Journal of Economic Perspectives*. 15(4): 101–115.

Watson, M. W. 1994. Vector autoregressions and cointegration. *Handbook of Econometrics*, Vol IV, Engle, R. F. and McFadden, D. L. eds, Amsterdam: Elsevier.

Also See

Complementary:	[TS] **fcast compute**, [TS] **fcast graph**, [TS] **irf add**, [TS] **irf cgraph**, [TS] **irf create**, [TS] **irf ctable**, [TS] **irf describe**, [TS] **irf drop**, [TS] **irf graph**, [TS] **irf ograph**, [TS] **irf rename**, [TS] **irf set**, [TS] **irf table**, [TS] **var**, [TS] **var svar**, [TS] **vargranger**, [TS] **varlmar**, [TS] **varnorm**, [TS] **varsoc**, [TS] **varstable**, [TS] **varwle**
Related:	[TS] **arima**, [TS] **vec intro**, [TS] **vec**, [R] **regress**, [R] **sureg**
Background:	[U] **11.4.3 Time-series varlists**, [TS] **irf**

Title

> **var** — Vector autoregression models

Syntax

> var *depvarlist* [*if*] [*in*] [, *options*]

options	description
Model	
<u>no</u>constant	suppress constant term
<u>lags</u>(*numlist*)	use lags *numlist* in the VAR
<u>exog</u>(*varlist*)	use exogenous variables *varlist*
Model 2	
<u>constr</u>aints(*constraints*)	apply specified linear constraints
<u>nolog</u>	suppress SURE iteration log
<u>iterate</u>(#)	set maximum number of iterations for SURE; default is iterate(1600)
<u>tolerance</u>(#)	set convergence tolerance of SURE
<u>noisure</u>	use one-step SURE
dfk	make small-sample degrees-of-freedom adjustment
<u>small</u>	calculate and report small-sample t and F statistics
<u>nobigf</u>	do not compute parameter vector for coefficients implicitly set to zero
Reporting	
<u>level</u>(#)	set confidence level; default is level(95)
<u>lutstats</u>	report Lütkepohl lag-order selection statistics

You must tsset your data before using var; see [TS] **tsset**.
depvarlist and *varlist* may contain time-series operators; see [U] **11.4.3 Time-series varlists**.
by, rolling, statsby, and xi may be used with var; see [U] **11.1.10 Prefix commands**.
See [U] **20 Estimation and postestimation commands** for additional capabilities of estimation commands.

Description

var fits vector autoregressive (VAR) models. The lag structure of the VAR need not be complete, and the model may contain exogenous variables. Linear constraints may be placed on any of the coefficients in the VAR, but var does not allow constraints on Σ, the error variance–covariance matrix; see [TS] **var svar** for an estimator that allows you to impose structure on the error variance–covariance matrix.

Options

> **Model**

noconstant; see [TS] **estimation options**.

lags(*numlist*) specifies the lags to be included in the model. The default is lags(1 2). Note that this option takes a *numlist* and not simply an integer for the maximum lag. For example, lags(2) would include only the second lag in the model, whereas lags(1/2) would include both the first and second lags in the model. See [U] **11.1.8 numlist** and [U] **13.8 Time-series operators** for further discussion of numlists and lags.

exog(*varlist*) specifies a list of exogenous variables to be included in the VAR.

     Model 2

constraints(*constraints*); see [TS] **estimation options**.

nolog suppresses the log from the iterated seemingly unrelated regression algorithm. By default, the iteration log is displayed when the coefficients are estimated through iterated seemingly unrelated regression. When the constraints() option is not specified, the estimates are obtained via OLS, and nolog has no effect. For this reason, nolog can only be specified when constraints() is specified. Similarly, nolog cannot be combined with noisure.

iterate(#) specifies an integer that sets the maximum number of iterations when the estimates are obtained through iterated seemingly unrelated regression. By default, the limit is 1,600. When constraints() is not specified, the estimates are obtained using OLS, and iterate() has no effect. For this reason, iterate() can only be specified when constraints() is specified. Similarly, iterate() cannot be combined with noisure.

tolerance(#) specifies a number greater than zero and less than 1 for the convergence tolerance of the iterated seemingly unrelated regression algorithm. By default, the tolerance is 1e-6. When the constraints() option is not specified, the estimates are obtained using OLS, and tolerance() has no effect. For this reason, tolerance() can only be specified when constraints() is specified. Similarly, tolerance() cannot be combined with noisure.

noisure specifies that the estimates in the presence of constraints be obtained through one-step seemingly unrelated regression. By default, var obtains estimates in the presence of constraints through iterated, seemingly unrelated regression. When constraints() is not specified, the estimates are obtained using OLS, and noisure has no effect. For this reason, noisure can only be specified when constraints() is specified.

dfk specifies that a small-sample degrees-of-freedom adjustment be used when estimating Σ, the error variance–covariance matrix. Specifically, $1/(T - \overline{m})$ is used instead of the large sample $1/T$, where $\overline{m}$ is the average number of parameters in the functional form for $\mathbf{y}_t$ over the K equations.

small causes var to report small-sample t and F statistics instead of the large-sample normal and chi-squared statistics.

nobigf requests that var not compute the estimated parameter vector that incorporates coefficients that have been implicitly constrained to be zero, such as when some lags have been omitted from a model. e(bf) is used for computing asymptotic standard errors in the postestimation commands irf create and fcast. Therefore, specifying nobigf implies that the asymptotic standard errors will not be available from the irf create and fcast postestimation routines. See *Fitting models with some lags excluded*.

Reporting

level(*#*); see [TS] **estimation options**.

lutstats specifies that the Lütkepohl (1993) versions of the lag-order selection statistics be reported. See *Methods and Formulas* in [TS] **varsoc** for a discussion of these statistics.

Remarks

Remarks are presented under the headings

Introduction
Fitting models with some lags excluded
Fitting models with exogenous variables
Fitting models with constraints on the coefficients

Introduction

A vector autoregression (VAR) is a model in which K variables are specified as linear functions of p of their own lags, p lags of the other $K - 1$ variables, and possibly exogenous variables. A VAR with p lags is usually denoted a VAR(p). For more information, see [TS] **var intro**.

▷ Example 1: VAR model

To illustrate the basic usage of **var**, we replicate the example in Lütkepohl (1993, 70). The data consists of three variables: the first-difference of the natural log of investment, dlinvestment; the first-difference of the natural log of income, dlincome; and the first-difference of the natural log of consumption, dlconsumption. The dataset contains data through the fourth quarter of 1982, though Lütkepohl uses only the observations through the fourth quarter of 1978.

```
. use http://www.stata-press.com/data/r9/lutkepohl
(Quarterly SA West German macro data, Bil DM, from Lutkepohl 1993 Table E.1)
. tsset
        time variable:  qtr, 1960q1 to 1982q4
```

(Continued on next page)

```
. var dlinvestment dlincome dlconsumption if qtr <= q(1978q4), lags(1/2)
> lutstats dfk

Vector autoregression
```

Sample: 1960q4 1978q4				No. of obs	=	73
Log likelihood = 606.307			(lutstats)	AIC	=	-24.63163
FPE = 2.18e-11				HQIC	=	-24.40656
Det(Sigma_ml) = 1.23e-11				SBIC	=	-24.06686

Equation	Parms	RMSE	R-sq	chi2	P>chi2
dlinvestment	7	.046148	0.1286	9.736909	0.1362
dlincome	7	.011719	0.1142	8.508289	0.2032
dlconsumption	7	.009445	0.2513	22.15096	0.0011

	Coef.	Std. Err.	z	P>|z|	[95% Conf. Interval]	
dlinvestment						
dlinvestment						
L1.	-.3196318	.1254564	-2.55	0.011	-.5655218	-.0737419
L2.	-.1605508	.1249066	-1.29	0.199	-.4053633	.0842616
dlincome						
L1.	.1459851	.5456664	0.27	0.789	-.9235013	1.215472
L2.	.1146009	.5345709	0.21	0.830	-.9331388	1.162341
dlconsumpt~n						
L1.	.9612288	.6643086	1.45	0.148	-.3407922	2.26325
L2.	.9344001	.6650949	1.40	0.160	-.369162	2.237962
_cons	-.0167221	.0172264	-0.97	0.332	-.0504852	.0170409
dlincome						
dlinvestment						
L1.	.0439309	.0318592	1.38	0.168	-.018512	.1063739
L2.	.0500302	.0317196	1.58	0.115	-.0121391	.1121995
dlincome						
L1.	-.1527311	.1385702	-1.10	0.270	-.4243237	.1188615
L2.	.0191634	.1357525	0.14	0.888	-.2469067	.2852334
dlconsumpt~n						
L1.	.2884992	.168699	1.71	0.087	-.0421448	.6191431
L2.	-.0102	.1688987	-0.06	0.952	-.3412354	.3208353
_cons	.0157672	.0043746	3.60	0.000	.0071932	.0243412
dlconsumpt~n						
dlinvestment						
L1.	-.002423	.0256763	-0.09	0.925	-.0527476	.0479016
L2.	.0338806	.0255638	1.33	0.185	-.0162235	.0839847
dlincome						
L1.	.2248134	.1116778	2.01	0.044	.005929	.4436978
L2.	.3549135	.1094069	3.24	0.001	.1404798	.5693471
dlconsumpt~n						
L1.	-.2639695	.1359595	-1.94	0.052	-.5304451	.0025062
L2.	-.0222264	.1361204	-0.16	0.870	-.2890175	.2445646
_cons	.0129258	.0035256	3.67	0.000	.0060157	.0198358

The output has two parts: a header and the standard Stata output table for the coefficients, standard errors, and confidence intervals. The header contains summary statistics for each of the equations in the VAR and statistics used in selecting the lag order of the VAR. Although there are standard formulas for all the lag-order statistics, Lütkepohl (1993) gives different versions of the three information criteria which drop the constant term from the likelihood. To obtain the Lütkepohl (1993) versions, we specified the lutstats option. The formulas for the standard and Lütkepohl versions of these statistics are given in *Methods and Formulas* of [TS] **varsoc**.

The dfk option specifies that the small-sample divisor $1/(T - \overline{m})$ be used in estimating Σ instead of the maximum likelihood (ML) divisor $1/T$, where $\overline{m}$ is the average number of parameters included in each of the K equations. All the lag-order statistics are computed using the ML estimator of Σ. Thus specifying dfk will not change the computed lag-order statistics, but it will change the estimated variance–covariance matrix. Also, when dfk is specified, a dfk-adjusted log likelihood is computed and saved in e(ll_dfk).

◁

The lag() option takes a *numlist* of lags. To specify a model that includes the first and second lags, type

        . var y1 y2 y3, lags(1/2)

not

        . var y1 y2 y3, lags(2)

because the latter specification would fit a model that included only the second lag.

Fitting models with some lags excluded

To fit a model that has only a fourth lag, that is,

$$\mathbf{y}_t = \mathbf{v} + \mathbf{A}_4\mathbf{y}_{t-4} + \mathbf{u}_t$$

you would specify the lags(4) option. Of course, this is equivalent to fitting the more general model

$$\mathbf{y}_t = \mathbf{v} + \mathbf{A}_1\mathbf{y}_{t-1} + \mathbf{A}_2\mathbf{y}_{t-2} + \mathbf{A}_3\mathbf{y}_{t-3} + \mathbf{A}_4\mathbf{y}_{t-4} + \mathbf{u}_t$$

with $\mathbf{A}_1$, $\mathbf{A}_2$, and $\mathbf{A}_3$ constrained to be $\mathbf{0}$. When you fit a model with some lags excluded, var estimates the coefficients included in the specification ($\mathbf{A}_4$ here) and saves these estimates in e(b). To obtain the asymptotic standard errors for impulse–response functions and other postestimation statistics, Stata needs the complete set of parameter estimates, including those that are constrained to be zero; var stores them in e(bf). Because you can specify models for which the full set of parameter estimates exceeds Stata's limit on the size of matrices, the nobigf option specifies that var not compute and store e(bf). This means that the asymptotic standard errors of the postestimation functions cannot be obtained, although bootstrap standard errors are still available. Building e(bf) can be time consuming, so if you do not need this full matrix, and speed is an issue, use nobigf.

Fitting models with exogenous variables

▷ Example 2: VAR model with exogenous variables

We use the exog() option to include exogenous variables in a VAR.

```
. var dlincome dlconsumption if qtr <= q(1978q4), lags(1/2) dfk exog(dlinvestment)
```

```
Vector autoregression
```

Sample: 1960q4 1978q4	No. of obs	=	73
Log likelihood = 478.5663	AIC	=	-12.78264
FPE = 9.64e-09	HQIC	=	-12.63259
Det(Sigma_ml) = 6.93e-09	SBIC	=	-12.40612

Equation	Parms	RMSE	R-sq	chi2	P>chi2
dlincome	6	.011917	0.0702	5.059587	0.4087
dlconsumption	6	.009197	0.2794	25.97262	0.0001

	Coef.	Std. Err.	z	P>|z|	[95% Conf. Interval]	
dlincome						
dlincome						
L1.	-.1343345	.1391074	-0.97	0.334	-.4069801	.1383111
L2.	.0120331	.1380346	0.09	0.931	-.2585097	.2825759
dlconsumpt~n						
L1.	.3235342	.1652769	1.96	0.050	-.0004027	.647471
L2.	.0754177	.1648624	0.46	0.647	-.2477066	.398542
dlinvestment	.0151546	.0302319	0.50	0.616	-.0440987	.074408
_cons	.0145136	.0043815	3.31	0.001	.0059259	.0231012
dlconsumpt~n						
dlincome						
L1.	.2425719	.1073561	2.26	0.024	.0321578	.452986
L2.	.3487949	.1065281	3.27	0.001	.1400036	.5575862
dlconsumpt~n						
L1.	-.3119629	.1275524	-2.45	0.014	-.5619611	-.0619648
L2.	-.0128502	.1272325	-0.10	0.920	-.2622213	.2365209
dlinvestment	.0503616	.0233314	2.16	0.031	.0046329	.0960904
_cons	.0131013	.0033814	3.87	0.000	.0064738	.0197288

All the postestimation commands for analyzing VARs work when exogenous variables are included in a model, but the asymptotic standard errors for the h-step-ahead forecasts are not available.

◁

Fitting models with constraints on the coefficients

var permits model specifications that include constraints on the coefficient, though var does not allow for constraints on Σ. See [TS] **var intro** and [TS] **var svar** for ways to constrain Σ.

▷ Example 3: VAR model with constraints

In the first example, we fitted a full VAR(2) to a three-equation model. The coefficients in the equation for dlinvestment were jointly insignificant, as were the coefficients in the equation for dlincome; and many individual coefficients were not significantly different from zero. In this example, we constrain the coefficient on L2.dlincome in the equation for dlinvestment and constrain the coefficient on L2.dlconsumption in the equation for dlincome to be zero.

```
. constraint define 1 [dlinvestment]L2.dlincome = 0
. constraint define 2 [dlincome]L2.dlconsumption = 0
```

```
. var dlinvestment dlincome dlconsumption  if qtr <= q(1978q4), lags(1/2) lutstats
> constraints(1 2) dfk
Estimating VAR coefficients

Iteration 1:    tolerance =  .00737681
Iteration 2:    tolerance =  3.998e-06
Iteration 3:    tolerance =  2.730e-09


Vector autoregression

Sample:  1960q4   1978q4                    No. of obs     =         73
Log likelihood =  606.2804                                 AIC    = -31.69254
FPE            =  1.77e-14       (lutstats) HQIC   = -31.46747
Det(Sigma_ml)  =  1.05e-14                  SBIC   = -31.12777

Equation            Parms     RMSE     R-sq      chi2     P>chi2

dlinvestment          6     .043895   0.1280   9.842338   0.0798
dlincome              6     .011143   0.1141   8.584446   0.1268
dlconsumption         7     .008981   0.2512  22.86958    0.0008


Constraints:
 ( 1)   [dlinvestment]L2.dlincome = 0
 ( 2)   [dlincome]L2.dlconsumption = 0
```

	Coef.	Std. Err.	z	P>|z|	[95% Conf. Interval]	
dlinvestment						
dlinvestment						
L1.	-.320713	.1247512	-2.57	0.010	-.5652208	-.0762051
L2.	-.1607084	.124261	-1.29	0.196	-.4042555	.0828386
dlincome						
L1.	.1195448	.5295669	0.23	0.821	-.9183873	1.157477
L2.	-2.55e-17	1.18e-16	-0.22	0.829	-2.57e-16	2.06e-16
dlconsumpt~n						
L1.	1.009281	.623501	1.62	0.106	-.2127586	2.231321
L2.	1.008079	.5713486	1.76	0.078	-.1117438	2.127902
_cons	-.0162102	.016893	-0.96	0.337	-.0493199	.0168995
dlincome						
dlinvestment						
L1.	.0435712	.0309078	1.41	0.159	-.017007	.1041495
L2.	.0496788	.0306455	1.62	0.105	-.0103852	.1097428
dlincome						
L1.	-.1555119	.1315854	-1.18	0.237	-.4134146	.1023908
L2.	.0122353	.1165811	0.10	0.916	-.2162595	.2407301
dlconsumpt~n						
L1.	.29286	.1568345	1.87	0.062	-.01453	.6002501
L2.	1.78e-19	8.28e-19	0.22	0.829	-1.45e-18	1.80e-18
_cons	.015689	.003819	4.11	0.000	.0082039	.0231741
dlconsumpt~n						
dlinvestment						
L1.	-.0026229	.0253538	-0.10	0.918	-.0523154	.0470696
L2.	.0337245	.0252113	1.34	0.181	-.0156888	.0831378
dlincome						
L1.	.2224798	.1094349	2.03	0.042	.0079912	.4369683
L2.	.3469758	.1006026	3.45	0.001	.1497984	.5441532
dlconsumpt~n						
L1.	-.2600227	.1321622	-1.97	0.049	-.519056	-.0009895
L2.	-.0146825	.1117618	-0.13	0.895	-.2337315	.2043666
_cons	.0129149	.003376	3.83	0.000	.0062981	.0195317

None of the free parameter estimates changed by much. While the coefficients in the equation dlinvestment are now significant at the 10% level, the coefficients in the equation for dlincome remain jointly insignificant.

◁

Saved Results

var saves in e():

Scalars

e(N)	number of observations
e(k_eq)	number of equations
e(k_dv)	number of dependent variables
e(df_eq)	average number of parameters in an equation
e(k)	number of parameters in all equations
e(k_#)	number of parameters in equation #
e(obs_#)	number of observations on equation #
e(df_m)	degrees of freedom in model
e(df_m#)	model degrees of freedom for equation #
e(df_r)	residual degrees of freedom (small only)
e(rmse_#)	root mean squared error for equation #
e(r2_#)	R-squared for equation #
e(chi2_#)	x^2 for equation #
e(ll)	log likelihood
e(ll_#)	log likelihood for equation #
e(ll_dfk)	dfk adjusted log likelihood (dfk only)
e(F_#)	F statistic for equation # (small only)
e(aic)	Akaike information criterion
e(sbic)	Schwarz–Bayesian information criterion
e(hqic)	Hannan–Quinn information criterion
e(fpe)	final prediction error
e(mlag)	highest lag in VAR
e(tmax)	maximum time
e(tmin)	first time period in sample
e(N_gaps)	number of gaps in sample
e(detsig)	determinant of e(Sigma)
e(detsig_ml)	determinant of $\widehat{\Sigma}_{ml}$

Macros

e(cmd)	var
e(depvar)	names of dependent variables
e(endog)	names of endogenous variables, if specified
e(exog)	names of exogenous variables, if specified
e(eqnames)	names of equations
e(lags)	lags in model
e(title)	title in estimation output
e(constraints)	constraints, if specified
e(small)	small, if specified
e(lutstats)	lutstats, if specified
e(timevar)	time variable specified in tsset
e(tsfmt)	format for the current time variable
e(dfk)	dfk, if specified
e(properties)	b V
e(predict)	program used to implement predict

Matrices

e(b)	coefficient vector
e(V)	variance–covariance matrix of the estimators
e(bf)	constrained coefficient vector
e(G)	Gamma matrix; see *Methods and Formulas*
e(Sigma)	$\widehat{\Sigma}$ matrix

Functions

e(sample)	marks estimation sample

Methods and Formulas

var is implemented as an ado-file.

When there are no constraints placed on the coefficients, the VAR(p) is a seemingly unrelated regression model with the same explanatory variables in each equation. As discussed in Lütkepohl (1993) and Greene (2003), performing linear regression on each equation produces the maximum likelihood estimates of the coefficients. The estimated coefficients can then be used to calculate the residuals, which in turn are used to estimate the cross-equation error variance–covariance matrix Σ.

Following the notation of Lütkepohl (1993), we write the VAR(p) with exogenous variables as

$$\mathbf{y}_t = \mathbf{A}\mathbf{Y}_{t-1} + \mathbf{B}_0\mathbf{x}_t + \mathbf{u}_t \tag{5}$$

where

$\mathbf{y}_t$ is the $K \times 1$ vector of endogenous variables,

$\mathbf{A}$ is a $K \times Kp$ matrix of coefficients,

$\mathbf{B}_0$ is a $K \times M$ matrix of coefficients,

$\mathbf{x}_t$ is the $M \times 1$ vector of exogenous variables,

$\mathbf{u}_t$ is the $K \times 1$ vector of white noise innovations, and

$\mathbf{Y}_t$ is the $Kp \times 1$ matrix given by $\mathbf{Y}_t = \begin{pmatrix} \mathbf{y}_t \\ \vdots \\ \mathbf{y}_{t-p+1} \end{pmatrix}$

While (5) is easier to read, the formulas are much easier to manipulate if it is instead written as

$$\mathbf{Y} = \mathbf{BZ} + \mathbf{U}$$

where

$$\mathbf{Y} = (\mathbf{y}_1, \ldots, \mathbf{y}_T) \qquad \mathbf{Y} \text{ is } K \times T$$
$$\mathbf{B} = (\mathbf{A}, \mathbf{B}_0) \qquad \mathbf{B} \text{ is } K \times (Kp + M)$$
$$\mathbf{Z} = \begin{pmatrix} \mathbf{Y}_0 \ldots, \mathbf{Y}_{T-1} \\ \mathbf{x}_1 \ldots, \mathbf{x}_T \end{pmatrix} \qquad \mathbf{Z} \text{ is } (Kp + M) \times T$$
$$\mathbf{U} = (\mathbf{u}_1, \ldots, \mathbf{u}_T) \qquad \mathbf{U} \text{ is } K \times T$$

Intercept terms in the model are included in $\mathbf{x}_t$. If there are no exogenous variables and no intercept terms in the model, $\mathbf{x}_t$ is empty.

The coefficients are estimated by iterated seemingly unrelated regression. Since the estimation is actually performed by reg3, the methods are documented in [R] reg3. See [P] makecns for more on estimation with constraints.

Let $\widehat{\mathbf{U}}$ be the matrix of residuals that are obtained via $\mathbf{Y} - \widehat{\mathbf{B}}\mathbf{Z}$, where $\widehat{\mathbf{B}}$ is the matrix of estimated coefficients. Then the estimator of $\mathbf{\Sigma}$ is

$$\widehat{\mathbf{\Sigma}} = \frac{1}{\widetilde{T}} \widehat{\mathbf{U}}' \widehat{\mathbf{U}}$$

By default, the maximum likelihood divisor of $\widetilde{T} = T$ is used. When dfk is specified, a small-sample degrees-of-freedom adjustment is used; in that case, $\widetilde{T} = T - \overline{m}$ where $\overline{m}$ is the average number of parameters per equation in the functional form for $\mathbf{y}_t$ over the K equations.

small specifies that Wald tests after var be assumed to have F or t distributions instead of chi-squared or standard normal distributions. The standard errors from each equation are computed using the degrees of freedom for the equation.

The "gamma" matrix saved in e(G) referred to in *Saved Results* is the $(Kp + 1) \times (Kp + 1)$ matrix given by

$$\frac{1}{T} \sum_{t=1}^{T} (1, \mathbf{Y}_t')(1, \mathbf{Y}_t')'$$

The formulas for the lag-order selection criteria and the log likelihood are discussed in [TS] varsoc.

Acknowledgment

We would like to thank Christopher Baum of Boston College for his helpful comments.

References

Greene, W. H. 2003. *Econometric Analysis*. 5th ed. Upper Saddle River, NJ: Prentice Hall.

Hamilton, J. D. 1994. *Time Series Analysis*. Princeton: Princeton University Press.

Lütkepohl, H. 1993. *Introduction to Multiple Time Series Analysis*. 2nd ed. New York: Springer.

Stock, J. H. and M. W. Watson. 2001. Vector autoregressions. *Journal of Economic Perspectives* 15(4): 101–115.

Watson, M. W. 1994. Vector autoregressions and cointegration. *Handbook of Econometrics*, Vol IV, Engle, R. F. and McFadden, D. L. eds, Amsterdam: Elsevier.

Also See

Complementary:	[TS] **var postestimation**; [TS] **tsset**
Related:	[TS] **arch**, [TS] **arima**, [TS] **var svar**, [TS] **varbasic**, [TS] **vec**, [R] **reg3**, [R] **regress**, [R] **sureg**
Background:	[U] **11.1.10 Prefix commands**, [U] **20 Estimation and postestimation commands**, [TS] **estimation options**, [TS] **var intro**

Title

> **var postestimation** — Postestimation tools for var

Description

The following postestimation commands are of special interest after `var`:

command	description
fcast compute	obtain dynamic forecasts
fcast graph	graph dynamic forecasts obtained from `fcast compute`
irf	create and analyze IRFs and FEVDs
vargranger	Granger causality tests
varlmar	LM test for autocorrelation in residuals
varnorm	test for normally distributed residuals
varsoc	lag-order selection criteria
varstable	check stability condition of estimates
varwle	Wald lag-exclusion statistics

For information about these commands, see the corresponding entries in this manual.

In addition, the following standard postestimation commands are available:

command	description
estat	AIC, BIC, VCE, and estimation sample summary
estimates	cataloging estimation results
lincom	point estimates, standard errors, testing, and inference for linear combinations of coefficients
lrtest	likelihood-ratio test
nlcom	point estimates, standard errors, testing, and inference for nonlinear combinations of coefficients
predict	predictions, residuals, influence statistics, and other diagnostic measures
predictnl	point estimates, standard errors, testing, and inference for generalized predictions
test	Wald tests for simple and composite linear hypotheses
testnl	Wald tests of nonlinear hypotheses

See the corresponding entries in the *Stata Base Reference Manual* for details.

Syntax for predict

predict [*type*] *newvar* [*if*] [*in*] [, <u>equation</u>(*eqno* | *eqname*) *statistic*]

statistic	description
xb	linear prediction; the default
stdp	standard error of the linear prediction
<u>residuals</u>	residuals

These statistics are available both in and out of sample; type `predict ... if e(sample) ...` if wanted only for the estimation sample.

Options for predict

equation(*eqno* | *eqname*) specifies the equation to which you are referring.

> equation() is filled in with one *eqno* or *eqname* for options xb, stdp, and residuals. For example, equation(#1) would mean the calculation is to be made for the first equation, equation(#2) would mean the second, and so on. Alternatively, you could refer to the equation by its name, thus equation(income) would refer to the equation named income and equation(hours) to the equation named hours.

> If you do not specify equation(), the results are the same as if you specified equation(#1).

xb, the default, calculates the linear prediction for the specified equation.

stdp calculates the standard error of the linear prediction for the specified equation.

residuals calculates the residuals.

For more information on using predict after multiple-equation estimation commands, see [R] **predict**.

Remarks

Remarks are presented under the headings

> Model selection and hypothesis testing
> Forecasting

Model selection and hypothesis testing

See the following sections for information on model selection and hypothesis testing after var.

[TS] **vargranger**	Perform pairwise Granger causality tests after var or svar
[TS] **varlmar**	Obtain LM statistics for residual autocorrelation after var or svar
[TS] **varnorm**	Test for normally distributed disturbances after var or svar
[TS] **varsoc**	Obtain lag-order selection statistics for VARs and VECMs
[TS] **varstable**	Check the stability condition of VAR or SVAR estimates
[TS] **varwle**	Obtain Wald lag-exclusion statistics after var or svar

Forecasting

Two types of forecasts are available after you fit a VAR(p): a one-step-ahead forecast and a dynamic h-step-ahead forecast.

The one-step-ahead forecast produces a prediction of the value of an endogenous variable in the current period using the estimated coefficients, the past values of the endogenous variables, and any exogenous variables. If you include contemporaneous values of exogenous variables in your model, you must have observations on the exogenous variables that are contemporaneous with the period in which the prediction is being made to compute the prediction. In Stata terms, these one-step-ahead predictions are just the standard linear predictions available after any estimation command. Thus predict, xb eq(*eqno* | *eqname*) produces one-step-ahead forecasts for the specified equation. predict, stdp eq(*eqno* | *eqname*) produces the standard error of the linear prediction for the specified equation. Note that the standard error of the forecast includes an estimate of the variability due to innovations, while the standard error of the linear prediction does not.

The dynamic h-step-ahead forecast begins by using the estimated coefficients, the lagged values of the endogenous variables, and any exogenous variables to predict one step ahead for each endogenous variable. Then the one-step-ahead forecast produces two-step-ahead forecasts for each endogenous variable. The process continues for h periods. Since each step uses the predictions of the previous steps, these forecasts are known as dynamic forecasts. See [TS] **fcast compute** for information on obtaining dynamic forecasts and their standard errors.

Methods and Formulas

All postestimation commands listed above are implemented as ado-files.

Formulas for predict

predict with the xb option provides the one-step forecast. If exogenous variables are specified, the forecast is conditional on the exogenous x_t variables. Specifying the residuals option causes predict to calculate the errors of the one-step forecasts. Specifying the stdp option causes predict to calculate the standard errors of the one-step forecasts.

Also See

Complementary:	[TS] **var**; [TS] **fcast compute**, [TS] **fcast graph**, [TS] **irf**, [TS] **vargranger**,
	[TS] **varlmar**, [TS] **varnorm**, [TS] **varsoc**, [TS] **varstable**, [TS] **varwle**,
	[R] **estimates**, [R] **lincom**, [R] **lrtest**, [R] **nlcom**, [R] **predictnl**,
	[R] **test**, [R] **testnl**
Background:	[U] **13.5 Accessing coefficients and standard errors**,
	[U] **20 Estimation and postestimation commands**,
	[R] **estat**, [R] **predict**

Title

> **var svar** — Structural vector autoregression models

Syntax

Short-run constraints

svar *depvarlist* $\left[\,if\,\right]$ $\left[\,in\,\right]$, { <u>acon</u>straints(*constraints$_a$*) <u>aeq</u>(*matrix*$_\text{aeq}$)

    <u>acns</u>(*matrix*$_\text{acns}$) <u>bcon</u>straints(*constraints$_b$*) <u>beq</u>(*matrix*$_\text{beq}$) <u>bcns</u>(*matrix*$_\text{bcns}$) }

    $\left[\,short\_run\_options\,\right]$

Long-run constraints

svar *depvarlist* $\left[\,if\,\right]$ $\left[\,in\,\right]$, { <u>lrcon</u>straints(*constraints$_\text{lr}$*) <u>lreq</u>(*matrix*$_\text{lreq}$)

    <u>lrcns</u>(*matrix*$_\text{lrcns}$) } $\left[\,long\_run\_options\,\right]$

short_run_options	description
Model	
<u>nocons</u>tant	suppress constant term
*<u>acon</u>straints(*constraints$_a$*)	apply previously defined *constraints$_a$* to **A**
*<u>aeq</u>(*matrix*$_\text{aeq}$)	define and apply to **A** equality constraint matrix *matrix*$_\text{aeq}$
*<u>acns</u>(*matrix*$_\text{acns}$)	define and apply to **A** cross-parameter constraint matrix *matrix*$_\text{acns}$
*<u>bcon</u>straints(*constraints$_b$*)	apply previously defined *constraints$_b$* to **B**
*<u>beq</u>(*matrix*$_\text{beq}$)	define and apply to **B** equality constraint matrix *matrix*$_\text{beq}$
*<u>bcns</u>(*matrix*$_\text{bcns}$)	define and apply to **B** cross-parameter constraint *matrix*$_\text{bcns}$
<u>lags</u>(*numlist*)	use lags *numlist* in the underlying VAR
Model 2	
<u>exog</u>(*varlist*$_\text{exog}$)	use exogenous variables *varlist*
<u>varconstraints</u>(*constraints$_v$*)	apply *constraints$_v$* to underlying VAR
noislog	suppress SURE iteration log
<u>isiterate</u>(#)	set maximum number of iterations for SURE; default is isiterate(1600)
<u>istol</u>erance(#)	set convergence tolerance of SURE
<u>noisure</u>	use one-step SURE
dfk	make small-sample degrees-of-freedom adjustment
<u>small</u>	calculate and report small-sample t and F statistics
<u>noiden</u>check	do not check for local identification
nobigf	do not compute parameter vector for coefficients implicitly set to zero

Reporting

<u>level</u>(#)	set confidence level; default is level(95)
<u>full</u>	show constrained parameters in table
var	display underlying var output
<u>lutstats</u>	report Lütkepohl lag-order selection statistics

Max options

maximize_options	control the maximization process; seldom used

---

* aconstraints(*constraints$_a$*), aeq(*matrix$_{aeq}$*), acns(*matrix$_{acns}$*), bconstraints(*constraints$_b$*), beq(*matrix$_{beq}$*), bcns(*matrix$_{bcns}$*): at least one of these options must be specified.

---

long_run_options	description

Model

<u>nocon</u>stant	suppress constant term
*<u>lrcon</u>straints(*constraints$_{lr}$*)	apply previously defined *constraints$_{lr}$* to **C**
*<u>lreq</u>(*matrix$_{lreq}$*)	define and apply to **C** equality constraint matrix *matrix$_{lreq}$*
*<u>lrcns</u>(*matrix$_{lrcns}$*)	define and apply to **C** cross-parameter constraint matrix *matrix$_{lrcns}$*
<u>lags</u>(*numlist*)	use lags *numlist* in the underlying VAR

Model 2

<u>exog</u>(*varlist$_{exog}$*)	use exogenous variables *varlist*
<u>varcon</u>straints(*constraints$_v$*)	apply *constraints$_v$* to underlying VAR
<u>noislog</u>	suppress SURE iteration log
<u>isiterate</u>(#)	set maximum number of iterations for SURE; default is isiterate(1600)
<u>istol</u>erance(#)	set convergence tolerance of SURE
<u>noisure</u>	use one-step SURE
dfk	make small-sample degrees-of-freedom adjustment
<u>small</u>	calculate and report small-sample t and F statistics
<u>noiden</u>check	do not check for local identification
<u>nobigf</u>	do not compute parameter vector for coefficients implicitly set to zero

Reporting

<u>level</u>(#)	set confidence level; default is level(95)
<u>full</u>	show constrained parameters in table
var	display underlying var output
<u>lutstats</u>	report Lütkepohl lag-order selection statistics

Max options

maximize_options	control the maximization process; seldom used

---

* lrconstraints(*constraints$_{lr}$*), lreq(*matrix$_{lreq}$*), lrcns(*matrix$_{lrcns}$*): at least one of these options must be specified.

You must tsset your data before using svar; see [TS] **tsset**.

depvarlist and *varlist$_{exog}$* may contain time-series operators; see [U] **11.4.3 Time-series varlists**.

by, rolling, statsby, or xi may be used with svar; see [U] **11.1.10 Prefix commands**.

See [U] **20 Estimation and postestimation commands** for additional capabilities of estimation commands.

Description

svar estimates the parameters of a structural vector autoregression (SVAR) and the parameters of the underlying vector autoregression (VAR).

Options

noconstant; see [TS] **estimation options**.

aconstraints(*constraints$_a$*), aeq(*matrix$_\text{aeq}$*), acns(*matrix$_\text{acns}$*)

bconstraints(*constraints$_b$*), beq(*matrix$_\text{beq}$*), bcns(*matrix$_\text{bcns}$*)

These options specify the short-run constraints in an SVAR. To specify a short-run SVAR model, you must specify at least one of these options. The first list of options specifies constraints on the parameters of the **A** matrix; the second list specifies constraints on the parameters of the **B** matrix (see *Short-run SVAR models*). If at least one option is selected from the first list and none are selected from the second list, svar sets **B** to the identity matrix. Similarly, if at least one option is selected from the second list and none are selected from the first list, svar sets **A** to the identity matrix.

None of these options may be specified with any of the options that define long-run constraints.

aconstraints(*constraints$_a$*) specifies a *numlist* of previously defined Stata constraints that are to be applied to **A** during estimation.

aeq(*matrix$_\text{aeq}$*) specifies a matrix that defines a set of equality constraints. This matrix must be square with dimension equal to the number of equations in the underlying VAR. The elements of this matrix must be *missing* or real numbers. A missing value in the (i, j) element of this matrix specifies that the (i, j) element of **A** is a free parameter. A real number in the (i, j) element of this matrix constrains the (i, j) element of **A** to this real number. For example,

$$\mathbf{A} = \begin{bmatrix} 1 & 0 \\ . & 1.5 \end{bmatrix}$$

specifies that $\mathbf{A}[1, 1] = 1$, $\mathbf{A}[1, 2] = 0$, $\mathbf{A}[2, 2] = 1.5$, and $\mathbf{A}[2, 1]$ is a free parameter.

acns(*matrix$_\text{acns}$*) specifies a matrix that defines a set of exclusion or cross-parameter equality constraints on **A**. This matrix must be square with dimension equal to the number of equations in the underlying VAR. Each element of this matrix must be *missing*, 0, or a positive integer. A missing value in the (i, j) element of this matrix specifies that no constraint be placed on this element of **A**. A zero in the (i, j) element of this matrix constrains the (i, j) element of **A** to be zero. Any strictly positive integers must be in two or more elements of this matrix. A strictly positive integer in the (i, j) element of this matrix constrains the (i, j) element of **A** to be equal to all the other elements of **A** that correspond to elements in this matrix that contain the same integer. For example, consider the matrix

$$\mathbf{A} = \begin{bmatrix} . & 1 \\ 1 & 0 \end{bmatrix}$$

Specifying, acns(A) in a two-equation SVAR constrains $\mathbf{A}[2, 1] = \mathbf{A}[1, 2]$ and $\mathbf{A}[2, 2] = 0$ while leaving $\mathbf{A}[1, 1]$ free.

bconstraints(*constraints*$_b$) specifies a *numlist* of previously defined Stata constraints to be applied to **B** during estimation.

beq(*matrix*$_{beq}$) specifies a matrix that defines a set of equality constraints. This matrix must be square with dimension equal to the number of equations in the underlying VAR. The elements of this matrix must be either *missing* or real numbers. The syntax of implied constraints is analogous to the one described in aeq(), except it applies to **B** rather than to **A**.

bcns(*matrix*$_{bcns}$) specifies a matrix that defines a set of exclusion or cross-parameter equality constraints on **B**. This matrix must be square with dimension equal to the number of equations in the underlying VAR. Each element of this matrix must be *missing*, 0, or a positive integer. The format of the implied constraints is the same as the one described in the acns() option above.

lrconstraints(*constraints*$_{lr}$), lreq(*matrix*$_{lreq}$), lrcns(*matrix*$_{lrcns}$)

These options specify the long-run constraints in an SVAR. To specify a long-run SVAR model, you must specify at least one of these options. The list of options specifies constraints on the parameters of the long-run **C** matrix (see *Long-run SVAR models* for the definition of **C**). None of these options may be specified with any of the options that define short-run constraints.

lrconstraints(*constraints*$_{lr}$) specifies a *numlist* of previously defined Stata constraints to be applied to **C** during estimation.

lreq(*matrix*$_{lreq}$) specifies a matrix that defines a set of equality constraints on the elements of **C**. This matrix must be square with dimension equal to the number of equations in the underlying VAR. The elements of this matrix must be either *missing* or real numbers. The syntax of implied constraints is analogous to the one described in option aeq() above except it applies to **C**.

lrcns(*matrix*$_{lrcns}$) specifies a matrix that defines a set of exclusion or cross-parameter equality constraints on **C**. This matrix must be square with dimension equal to the number of equations in the underlying VAR. Each element of this matrix must be *missing*, 0, or a positive integer. The syntax of the implied constraints is the same as the one described for the acns() option above.

lags(*numlist*) specifies the lags to be included in the underlying VAR model. Note that this option takes a *numlist* and not simply an integer for the maximum lag. For instance, lags(2) would include only the second lag in the model, while lags(1/2) would include both the first and second lags in the model. See [U] **11.1.8 numlist** and [U] **13.8 Time-series operators** for further discussion of *numlist*s and lags.

___

⌐ Model 2 ⌐

exog(*varlist*$_{exog}$) specifies a list of exogenous variables to be included in the underlying VAR.

varconstraints(*constraints*$_v$) specifies a list of constraints to be applied to coefficients in the underlying VAR. Since svar estimates multiple equations, the constraints must specify the equation name for all but the first equation.

noislog prevents svar from displaying the iteration log from the iterated seemingly unrelated regression algorithm. When the varconstraints() option is not specified, the VAR coefficients are estimated via OLS, a noniterative procedure. As a result, noislog may only be specified with varconstraints(). Similarly, noislog may not be combined with noisure.

isiterate(#) sets the maximum number of iterations for the iterated, seemingly unrelated regression algorithm. The default limit is 1600. When the varconstraints() option is not specified, the VAR coefficients are estimated via OLS, a noniterative procedure. As a result, isiterate() may only be specified with `varconstraints()`. Similarly, `isiterate()` may not be combined with `noisure`.

`istolerance(#)` specifies the convergence tolerance of the iterated, seemingly unrelated regression algorithm. The default tolerance is `1e-6`. When the `varconstraints()` option is not specified, the VAR coefficients are estimated via OLS, a noniterative procedure. As a result, `istolerance()` may only be specified with `varconstraints()`. Similarly, `istolerance()` may not be combined with `noisure`.

`noisure` specifies that the VAR coefficients be estimated via one-step seemingly unrelated regression when `varconstraints()` is specified. By default, `svar` estimates the coefficients in the VAR via iterated, seemingly unrelated regression when `varconstraints()` is specified. When the `varconstraints()` option is not specified, the VAR coefficient estimates are obtained via OLS, a noniterative procedure. As a result, `noisure` may only be specified with `varconstraints()`.

`dfk` specifies that a small-sample degrees-of-freedom adjustment be used when estimating Σ, the covariance matrix of the VAR disturbances. Specifically, $1/(T - \overline{m})$ is used instead of the large sample $1/T$, where $\overline{m}$ is the average number of parameters in the functional form for $\mathbf{y}_t$ over the K equations.

`small` causes `svar` to calculate and report small-sample t and F statistics instead of the large-sample normal and chi-squared statistics.

`noidencheck` requests that the Amisano and Giannini (1997) check for local identification not be performed. This check is local to the starting values used. Because of this dependence on the starting values, you may wish to suppress this check by specifying the `noidencheck` option. However, be careful in specifying this option. Models that are not structurally identified can still converge, thereby producing meaningless results that only appear to have meaning.

`nobigf` requests that `svar` not compute the estimated parameter vector that incorporates coefficients that have been implicitly constrained to be zero, such as when some lags have been omitted from a model. `e(bf_var)` is used for computing asymptotic standard errors in the postestimation commands `irf create` and `fcast`. Therefore, specifying `nobigf` implies that the asymptotic standard errors will not be available from the `irf create` and `fcast` postestimation routines. See *Fitting models with some lags excluded* in [TS] **var** for details.

___Reporting___

`level(#)`; see [TS] **estimation options**.

`full` shows constrained parameters in table.

`var` specifies that the output from `var` also be displayed. By default, the underlying VAR is fit `quietly`.

`lutstats` specifies that the Lütkepohl versions of the lag-order selection statistics be reported. See *Methods and Formulas* in [TS] **varsoc** for a discussion of these statistics.

___Max options___

maximize_options: <u>difficult</u>, <u>tech</u>nique(*algorithm_spec*), <u>iter</u>ate(#), [<u>no</u>]<u>log</u>, <u>trace</u>, gradient, showstep, <u>hess</u>ian, <u>shownr</u>tolerance, <u>tol</u>erance(#), <u>ltol</u>erance(#), <u>gtol</u>erance(#), <u>nrtol</u>erance(#), <u>nonrtol</u>erance, from(*init_specs*); see [R] **maximize**. These options are seldom used.

Remarks

Remarks are presented under the headings

Introduction
Short-run SVAR models
Long-run SVAR models

Introduction

This entry assumes that you have already read [TS] **var intro** and [TS] **var**; if not, please do. Here we illustrate how to fit SVARs in Stata subject to short-run and long-run restrictions. For more detailed information on SVARs, see Amisano and Giannini (1997) and Hamilton (1994). For good introductions to VARs, see Lütkepohl (1993), Hamilton (1994), and Stock and Watson (2001).

Short-run SVAR models

Recall that a short-run SVAR model without exogenous variables can be written as

$$\mathbf{A}(\mathbf{I}_K - \mathbf{A}_1 L - \mathbf{A}_2 L^2 - \cdots - \mathbf{A}_p L^p)\mathbf{y}_t = \mathbf{A}\boldsymbol{\epsilon}_t = \mathbf{B}\mathbf{e}_t$$

where L is the lag operator, $\mathbf{A}$, $\mathbf{B}$, and $\mathbf{A}_1, \ldots, \mathbf{A}_p$ are $K \times K$ matrices of parameters, $\boldsymbol{\epsilon}_t$ is a $K \times 1$ vector of innovations with $\boldsymbol{\epsilon}_t \sim N(\mathbf{0}, \boldsymbol{\Sigma})$ and $E[\boldsymbol{\epsilon}_t \boldsymbol{\epsilon}_s'] = \mathbf{0}_K$ for all $s \neq t$, and $\mathbf{e}_t$ is a $K \times 1$ vector of orthogonalized disturbances; i.e., $\mathbf{e}_t \sim N(\mathbf{0}, \mathbf{I}_K)$ and $E[\mathbf{e}_t \mathbf{e}_s'] = \mathbf{0}_K$ for all $s \neq t$. These transformations of the innovations allow us to analyze the dynamics of the system in terms of a change to an element of $\mathbf{e}_t$. In a short-run SVAR model, we obtain identification by placing restrictions on $\mathbf{A}$ and $\mathbf{B}$, which are assumed to be nonsingular.

▷ Example 1: Short-run just-identified SVAR model

Following Sims (1980), the Cholesky decomposition is one method of identifying the impulse–response functions in a VAR; thus, this method corresponds to an SVAR. There are several sets of constraints on $\mathbf{A}$ and $\mathbf{B}$ that are easily manipulated back to the Cholesky decomposition, and the following example illustrates this point.

One way to impose the Cholesky restrictions is to assume an SVAR model of the form

$$\widetilde{\mathbf{A}}(\mathbf{I}_K - \mathbf{A}_1 - \mathbf{A}_2 L^2 - \cdots \mathbf{A}_p L^p)\mathbf{y}_t = \widetilde{\mathbf{B}}\mathbf{e}_t$$

where $\widetilde{\mathbf{A}}$ is a lower triangular matrix with ones on the diagonal and $\widetilde{\mathbf{B}}$ is a diagonal matrix. Since the $\mathbf{P}$ matrix for this model is $\mathbf{P}_{sr} = \widetilde{\mathbf{A}}^{-1}\widetilde{\mathbf{B}}$, its estimate, $\widehat{\mathbf{P}}_{sr}$, obtained by plugging in estimates of $\widetilde{\mathbf{A}}$ and $\widetilde{\mathbf{B}}$, should equal the Cholesky decomposition of $\widehat{\boldsymbol{\Sigma}}$.

To illustrate, we use the German macroeconomic data discussed in Lütkepohl (1993) and used in [TS] **var**. In this example, $\mathbf{y}_t = (\texttt{dlinvestment}, \texttt{dlincome}, \texttt{dlconsumption})$, where `dlinvestment` is the first-difference of the log of investment, `dlincome` is the first-difference of the log of income, and `dlconsumption` is the first-difference of the log of consumption. Since the first-difference of the natural log of a variable can be treated as an approximation of the percentage change in that variable, we will refer to these variables as percentage changes in `investment`, `income`, and `consumption`, respectively.

We will impose the Cholesky restrictions on this system by applying equality constraints using the constraint matrices

$$A = \begin{bmatrix} 1 & 0 & 0 \\ . & 1 & 0 \\ . & . & 1 \end{bmatrix} \quad \text{and} \quad B = \begin{bmatrix} . & 0 & 0 \\ 0 & . & 0 \\ 0 & 0 & . \end{bmatrix}$$

With these structural restrictions, we assume that the percentage change in investment is not contemporaneously affected by the percentage changes in either income or consumption. We also assume that the percentage change of income is affected by contemporaneous changes in investment but not consumption. Finally, we assume that percentage changes in consumption are affected by contemporaneous changes in both investment and income.

The following commands fit an SVAR model with these constraints.

```
. use http://www.stata-press.com/data/r9/lutkepohl
(Quarterly SA West German macro data, Bil DM, from Lutkepohl 1993 Table E.1)
. mat A = (1,0,0\.,1,0\.,.,1)
. mat B = (.,0,0\0,.,0\0,0,.)
. svar dlinvestment dlincome dlconsumption if qtr <= q(1978q4), aeq(A) beq(B)
Estimating short-run parameters

  (output omitted )

Structural vector autoregression

Constraints:
 ( 1)   [a_1_1]_cons = 1
 ( 2)   [a_1_2]_cons = 0
 ( 3)   [a_1_3]_cons = 0
 ( 4)   [a_2_2]_cons = 1
 ( 5)   [a_2_3]_cons = 0
 ( 6)   [a_3_3]_cons = 1
 ( 7)   [b_1_2]_cons = 0
 ( 8)   [b_1_3]_cons = 0
 ( 9)   [b_2_1]_cons = 0
 (10)   [b_2_3]_cons = 0
 (11)   [b_3_1]_cons = 0
 (12)   [b_3_2]_cons = 0
Sample:  1960q4   1978q4                No. of obs      =        73
Exactly identified model               Log likelihood  =   606.307
```

	Coef.	Std. Err.	t	P>|z|	[95% Conf. Interval]	
/a_1_1	1	.	.	.	.	.
/a_2_1	-.0336288	.0294605	-1.14	0.254	-.0913702	.0241126
/a_3_1	-.0435846	.0194408	-2.24	0.025	-.0816879	-.0054812
/a_1_2	0	.	.	.	.	.
/a_2_2	1	.	.	.	.	.
/a_3_2	-.424774	.0765548	-5.55	0.000	-.5748187	-.2747293
/a_1_3	0	.	.	.	.	.
/a_2_3	0	.	.	.	.	.
/a_3_3	1	.	.	.	.	.
/b_1_1	.0438796	.0036315	12.08	0.000	.036762	.0509972
/b_2_1	0	.	.	.	.	.
/b_3_1	0	.	.	.	.	.
/b_1_2	0	.	.	.	.	.
/b_2_2	.0110449	.0009141	12.08	0.000	.0092534	.0128365
/b_3_2	0	.	.	.	.	.
/b_1_3	0	.	.	.	.	.
/b_2_3	0	.	.	.	.	.
/b_3_3	.0072243	.0005979	12.08	0.000	.0060525	.0083962

The SVAR output has four parts: an iteration log, a display of the constraints imposed, a header with sample and SVAR log-likelihood information, and a table displaying the estimates of the parameters from the **A** and **B** matrices. From the output above, we can see that the equality constraint matrices supplied to svar imposed the intended constraints, and that the SVAR header informs us that the model we fitted is just identified. The estimates of a_2_1, a_3_1, and a_3_2 are all negative. Since the off-diagonal elements of the **A** matrix contain the negative of the actual contemporaneous effects, the estimated effects are positive, as expected.

The estimates $\widehat{\mathbf{A}}$ and $\widehat{\mathbf{B}}$ are stored in e(A) and e(B), respectively, allowing us to compute the estimated Cholesky decomposition.

```
. mat Aest = e(A)

. mat Best = e(B)

. mat chol_est = inv(Aest)*Best

. mat list chol_est

chol_est[3,3]
                  dlinvestment      dlincome   dlconsumption
dlinvestment        .04387957             0               0
    dlincome        .00147562     .01104494               0
dlconsumption       .00253928      .0046916      .00722432
```

svar saves the estimated Σ from the underlying var in e(Sigma). The output below illustrates the computation of the Cholesky decomposition of e(Sigma). Note that it is the same as the output computed from the SVAR estimates.

```
. mat sig_var = e(Sigma)

. mat chol_var = cholesky(sig_var)

. mat list chol_var

chol_var[3,3]
                  dlinvestment      dlincome   dlconsumption
dlinvestment        .04387957             0               0
    dlincome        .00147562     .01104494               0
dlconsumption       .00253928      .0046916      .00722432
```

◁

At this point, we might wonder why we bother obtaining parameter estimates via nonlinear estimation if we can obtain them simply by a transform of the estimates produced by var. When the model is just identified, as in the previous example, the SVAR parameter estimates can be computed via a transform of the VAR estimates. However, when the model is overidentified, such is not the case.

▷ Example 2: Short-run over-identified SVAR model

The Cholesky decomposition example above fitted a just-identified model. This example considers an overidentified model. In the previous example, the a_2_1 parameter was not significant, which is consistent with a theory in which changes in our measure of investment only affect changes in income with a lag. We can impose the restriction that a_2_1 is zero and then test this overidentifying restriction. Our **A** and **B** matrices are now

$$\mathbf{A} = \begin{bmatrix} 1 & 0 & 0 \\ 0 & 1 & 0 \\ . & . & 1 \end{bmatrix} \quad \text{and} \quad \mathbf{B} = \begin{bmatrix} . & 0 & 0 \\ 0 & . & 0 \\ 0 & 0 & . \end{bmatrix}$$

The output below contains the commands and results we obtained by fitting this model on the Lütkepohl data.

```
. mat B = (.,0,0\0,.,0\0,0,.)
. mat A = (1,0,0\0,1,0\.,.,1)
. svar dlinvestment dlincome dlconsumption if qtr <= q(1978q4), aeq(A) beq(B)
Estimating short-run parameters
```

(*output omitted*)

```
Structural vector autoregression

Constraints:
 ( 1)  [a_1_1]_cons = 1
 ( 2)  [a_1_2]_cons = 0
 ( 3)  [a_1_3]_cons = 0
 ( 4)  [a_2_1]_cons = 0
 ( 5)  [a_2_2]_cons = 1
 ( 6)  [a_2_3]_cons = 0
 ( 7)  [a_3_3]_cons = 1
 ( 8)  [b_1_2]_cons = 0
 ( 9)  [b_1_3]_cons = 0
 (10)  [b_2_1]_cons = 0
 (11)  [b_2_3]_cons = 0
 (12)  [b_3_1]_cons = 0
 (13)  [b_3_2]_cons = 0
```

| Sample: 1960q4 1978q4 | | No. of obs | = | 73 |
| Overidentified model | | Log likelihood | = | 605.6613 |

	Coef.	Std. Err.	t	P>|z|	[95% Conf. Interval]	
/a_1_1	1	.	.	.	.	.
/a_2_1	0	.	.	.	.	.
/a_3_1	-.0435911	.0192696	-2.26	0.024	-.0813589	-.0058233
/a_1_2	0	.	.	.	.	.
/a_2_2	1	.	.	.	.	.
/a_3_2	-.4247741	.0758806	-5.60	0.000	-.5734973	-.2760508
/a_1_3	0	.	.	.	.	.
/a_2_3	0	.	.	.	.	.
/a_3_3	1	.	.	.	.	.
/b_1_1	.0438796	.0036315	12.08	0.000	.036762	.0509972
/b_2_1	0	.	.	.	.	.
/b_3_1	0	.	.	.	.	.
/b_1_2	0	.	.	.	.	.
/b_2_2	.0111431	.0009222	12.08	0.000	.0093356	.0129506
/b_3_2	0	.	.	.	.	.
/b_1_3	0	.	.	.	.	.
/b_2_3	0	.	.	.	.	.
/b_3_3	.0072243	.0005979	12.08	0.000	.0060525	.0083962

```
LR test of identifying restrictions:  chi2(  1)=    1.292  Prob > chi2 = 0.256
```

Note that the footer in this example reports a test of the overidentifying restriction. The null hypothesis of this test is that any overidentifying restrictions are valid. In the case at hand, we cannot reject this null hypothesis at any of the conventional levels.

◁

▷ Example 3: Short-run SVAR model with constraints

svar also allows us to place constraints on the parameters of the underlying VAR. We begin by looking at the underlying VAR for the SVARs that we have used in the previous examples.

```
. var dlinvestment dlincome dlconsumption if qtr <= q(1978q4)
```

Vector autoregression

```
Sample:  1960q4   1978q4                    No. of obs     =         73
Log likelihood =    606.307                 AIC            = -16.03581
FPE            =  2.18e-11                   HQIC           = -15.77323
Det(Sigma_ml)  =  1.23e-11                   SBIC           = -15.37691
```

Equation	Parms	RMSE	R-sq	chi2	P>chi2
dlinvestment	7	.046148	0.1286	10.76961	0.0958
dlincome	7	.011719	0.1142	9.410683	0.1518
dlconsumption	7	.009445	0.2513	24.50031	0.0004

	Coef.	Std. Err.	z	P>|z|	[95% Conf.	Interval]
dlinvestment						
dlinvestment						
L1.	-.3196318	.1192898	-2.68	0.007	-.5534355	-.0858282
L2.	-.1605508	.118767	-1.35	0.176	-.39333	.0722283
dlincome						
L1.	.1459851	.5188451	0.28	0.778	-.8709326	1.162903
L2.	.1146009	.508295	0.23	0.822	-.881639	1.110841
dlconsumpt~n						
L1.	.9612288	.6316557	1.52	0.128	-.2767936	2.199251
L2.	.9344001	.6324034	1.48	0.140	-.3050877	2.173888
_cons	-.0167221	.0163796	-1.02	0.307	-.0488257	.0153814
dlincome						
dlinvestment						
L1.	.0439309	.0302933	1.45	0.147	-.0154427	.1033046
L2.	.0500302	.0301605	1.66	0.097	-.0090833	.1091437
dlincome						
L1.	-.1527311	.131759	-1.16	0.246	-.4109741	.1055118
L2.	.0191634	.1290799	0.15	0.882	-.2338285	.2721552
dlconsumpt~n						
L1.	.2884992	.1604069	1.80	0.072	-.0258926	.6028909
L2.	-.0102	.1605968	-0.06	0.949	-.3249639	.3045639
_cons	.0157672	.0041596	3.79	0.000	.0076146	.0239198
dlconsumpt~n						
dlinvestment						
L1.	-.002423	.0244142	-0.10	0.921	-.050274	.045428
L2.	.0338806	.0243072	1.39	0.163	-.0137607	.0815219
dlincome						
L1.	.2248134	.1061884	2.12	0.034	.0166879	.4329389
L2.	.3549135	.1040292	3.41	0.001	.1510199	.558807
dlconsumpt~n						
L1.	-.2639695	.1292766	-2.04	0.041	-.517347	-.010592
L2.	-.0222264	.1294296	-0.17	0.864	-.2759039	.231451
_cons	.0129258	.0033523	3.86	0.000	.0063554	.0194962

The equation-level model tests reported in the header indicate that we cannot reject the null hypotheses that all the coefficients in the first equation are zero, nor can we reject the null that all the coefficients in the second equation are zero at the 5% significance level. We use a combination of theory and the p-values from the output above to place some exclusion restrictions on the underlying VAR(2). Specifically, in the equation for the percentage change of investment, we constrain the coefficients on L2.dlinvestment, L.dlincome, L2.dlincome, and L2.dlconsumption to be zero. In the equation for dlincome, we constrain the coefficients on L2.dlinvestment, L2.dlincome, and L2.dlconsumption to be zero. Finally, in the equation for dlconsumption, we constrain

L.dlinvestment and L2.dlconsumption to be zero. We then refit the SVAR from the previous example.

```
. constraint define 1 [dlinvestment]L2.dlinvestment = 0
. constraint define 2 [dlinvestment]L.dlincome = 0
. constraint define 3 [dlinvestment]L2.dlincome = 0
. constraint define 4 [dlinvestment]L2.dlconsumption = 0
. constraint define 5 [dlincome]L2.dlinvestment = 0
. constraint define 6 [dlincome]L2.dlincome = 0
. constraint define 7 [dlincome]L2.dlconsumption = 0
. constraint define 8 [dlconsumption]L.dlinvestment = 0
. constraint define 9 [dlconsumption]L2.dlconsumption = 0
. svar dlinvestment dlincome dlconsumption if qtr <= q(1978q4), aeq(A) beq(B)
> varconstraints(1/9) noislog
Estimating short-run parameters
  (output omitted)
Structural vector autoregression
Constraints:
 ( 1)   [a_1_1]_cons = 1
 ( 2)   [a_1_2]_cons = 0
 ( 3)   [a_1_3]_cons = 0
 ( 4)   [a_2_1]_cons = 0
 ( 5)   [a_2_2]_cons = 1
 ( 6)   [a_2_3]_cons = 0
 ( 7)   [a_3_3]_cons = 1
 ( 8)   [b_1_2]_cons = 0
 ( 9)   [b_1_3]_cons = 0
 (10)   [b_2_1]_cons = 0
 (11)   [b_2_3]_cons = 0
 (12)   [b_3_1]_cons = 0
 (13)   [b_3_2]_cons = 0
```

Sample: 1960q4 1978q4   No. of obs   =   73
Overidentified model   Log likelihood = 601.8591

	Coef.	Std. Err.	t	P>|z|	[95% Conf.	Interval]
/a_1_1	1	.	.	.	.	.
/a_2_1	0	.	.	.	.	.
/a_3_1	-.0418708	.0187579	-2.23	0.026	-.0786356	-.0051061
/a_1_2	0	.	.	.	.	.
/a_2_2	1	.	.	.	.	.
/a_3_2	-.4255808	.0745298	-5.71	0.000	-.5716565	-.2795051
/a_1_3	0	.	.	.	.	.
/a_2_3	0	.	.	.	.	.
/a_3_3	1	.	.	.	.	.
/b_1_1	.0451851	.0037395	12.08	0.000	.0378557	.0525145
/b_2_1	0	.	.	.	.	.
/b_3_1	0	.	.	.	.	.
/b_1_2	0	.	.	.	.	.
/b_2_2	.0113723	.0009412	12.08	0.000	.0095276	.013217
/b_3_2	0	.	.	.	.	.
/b_1_3	0	.	.	.	.	.
/b_2_3	0	.	.	.	.	.
/b_3_3	.0072417	.0005993	12.08	0.000	.006067	.0084164

LR test of identifying restrictions: chi2(1)=   .8448 Prob > chi2 = 0.358

If we displayed the underlying VAR(2) results using the `var` option, we would see that most of the unconstrained coefficients are now significant at the 10% level and that none of the equation-level model statistics fail to reject the null hypothesis at the 10% level. The `svar` output reveals that the p-value of the overidentification test rose and that the coefficient on a_3_1 is still insignificant at the 1% level but not at the 5% level.

◁

Before moving on to models with long-run constraints, we should note some limitations. We cannot place constraints on the elements of $\mathbf{A}$ in terms of the elements of $\mathbf{B}$, or vice versa. This limitation is imposed by the form of the check for identification derived by Amisano and Giannini (1997). As noted in *Methods and Formulas*, this test requires separate constraint matrices for the parameters in $\mathbf{A}$ and $\mathbf{B}$. Another limitation is that we cannot mix short-run and long-run constraints.

Long-run SVAR models

As discussed in [TS] **var intro**, a long-run SVAR has the form

$$\mathbf{y}_t = \mathbf{C} e_t$$

In long-run models, the constraints are placed on the elements of $\mathbf{C}$, and the free parameters are estimated. These constraints are frequently exclusion restrictions. For instance, constraining $\mathbf{C}[1,2]$ to be zero can be interpreted as setting the long-run response of variable 1 to the structural shocks driving variable 2 to be zero.

Similar to the short-run model, the $\mathbf{P}_{lr}$ matrix such that $\mathbf{P}_{lr}\mathbf{P}'_{lr} = \mathbf{\Sigma}$ identifies the structural impulse–response functions. $\mathbf{P}_{lr} = \mathbf{C}$ is identified by the restrictions placed on the parameters in $\mathbf{C}$. There are K^2 parameters in $\mathbf{C}$, and the order condition for identification requires that there be at least $K^2 - K(K+1)/2$ restrictions placed on those parameters. As in the short-run model, this order condition is necessary but not sufficient, so the Amisano and Giannini (1997) check for local identification is performed by default.

▷ Example 4: Long-run SVAR model

Suppose that we have a theory in which unexpected changes to the money supply have no long-run effects on changes in output and, similarly, that unexpected changes in output have no long-run effects on changes in the money supply. The $\mathbf{C}$ matrix implied by this theory is

$$\mathbf{C} = \begin{bmatrix} . & 0 \\ 0 & . \end{bmatrix}$$

```
. use http://www.stata-press.com/data/r9/m1gdp
. mat lr = (.,0\0,.)
. svar d.ln_m1 d.ln_gdp , lreq(lr)
Estimating long-run parameters
(output omitted )
Structural vector autoregression
Constraints:
 ( 1)  [c_1_2]_cons = 0
 ( 2)  [c_2_1]_cons = 0
Sample: 1959q4  2002q2                    No. of obs     =        171
Overidentified model                      Log likelihood =   1151.614
```

	Coef.	Std. Err.	t	P>|z|	[95% Conf. Interval]	
/c_1_1	.0301007	.0016277	18.49	0.000	.0269106	.0332909
/c_2_1	0	.	.	.	.	.
/c_1_2	0	.	.	.	.	.
/c_2_2	.0129691	.0007013	18.49	0.000	.0115946	.0143436

```
LR test of identifying restrictions:  chi2( 1)=    .1368  Prob > chi2 = 0.712
```

We have assumed that the underlying VAR has 2 lags; four of the five selection order criteria computed by varsoc (see [TS] **varsoc**) recommended this choice. The test of the overidentifying restrictions does not provide any indication that it is not valid.

◁

Saved Results

svar saves in e():

Scalars

e(N)	number of observations
e(k_eq)	number of equations
e(k_eq_var)	number of equations in underlying VAR
e(k_dv)	number of dependent variables
e(k_var)	number of coefficients in VAR
e(mlag_var)	highest lag in VAR
e(k_dv_var)	number of dependent variables in underlying VAR
e(tparms_var)	number of parameters in all equations
e(df_eq_var)	average number of parameters in an equation
e(df_m_var)	degrees of freedom in the model
e(df_r_var)	if small, residual degrees of freedom
e(obs_#_var)	number of observations on equation #
e(k_#_var)	number of coefficients in equation #
e(df_m#_var)	model degrees of freedom for equation #
e(r2_#_var)	R-squared for equation #
e(chi2_#_var)	χ^2 statistic for equation #
e(rmse_#_var)	root mean squared error for equation #
e(ll_#_var)	log likelihood for equation #
e(df_r#_var)	residual degrees of freedom for equation # (small only)
e(F_#_var)	F statistic for equation # (small only)
e(N_gaps_var)	number of gaps in the sample
e(detsig_ml_var)	determinant of $\widehat{\Sigma}_{ml}$
e(aic_var)	Akaike information criterion
e(sbic_var)	Schwarz–Bayesian information criterion

e(hqic_var)	Hannan–Quinn information criterion
e(fpe_var)	final prediction error
e(ll)	log likelihood from svar
e(ll_var)	log likelihood from var
e(tmin)	first time period in the sample
e(tmax)	maximum time
e(detsig_var)	determinant of e(Sigma)
e(chi2_oid)	overidentification test
e(oid_df)	number of overidentification restrictions
e(rc_ml)	return code from ml
e(N_cns)	number of constraints

Macros

e(cmd)	svar
e(lrmodel)	long-run model, if specified
e(lags_var)	lags in model
e(depvar_var)	names of dependent variables
e(endog_var)	names of endogenous variables
e(exog_var)	names of exogenous variables, if specified
e(nocons_var)	noconstant, if noconstant specified
e(dfk_var)	alternate divisor (dfk), if specified
e(eqnames_var)	names of equations
e(lutstats_var)	lutstats, if specified
e(constraints_var)	constraints_var, if there are constraints on VAR
e(small)	small, if specified
e(tsfmt)	format of timevar
e(timevar)	name of timevar
e(title)	title in estimation output
e(properties)	b V
e(predict)	program used to implement predict

Matrices

e(b)	coefficient vector
e(V)	variance–covariance matrix of the estimators
e(bf_var)	full coefficient vector with zeros in dropped lags
e(G_var)	G matrix saved by var; see [TS] **var** Methods and Formulas
e(Sigma)	$\widehat{\Sigma}$ matrix
e(aeq)	aeq(matrix), if specified
e(acns)	acns(matrix), if specified
e(beq)	beq(matrix), if specified
e(bcns)	bcns(matrix), if specified
e(lreq)	lreq(matrix), if specified
e(lrcns)	lrcns(matrix), if specified
e(Cns_var)	constraint matrix from var, if varconstraints() are specified
e(A)	estimated A matrix, if a short-run model
e(B)	estimated B matrix
e(C)	estimated C matrix, if a long-run model
e(A1)	estimated $\overline{A}$ matrix, if a long-run model

Functions

e(sample)	marks the estimation sample

Methods and Formulas

svar is implemented as an ado-file.

The log-likelihood function for models with short-run constraints is

$$L(\mathbf{A}, \mathbf{B}) = -\frac{NK}{2} \ln(2\pi) + \frac{N}{2} \ln(|\mathbf{W}|^2) - \frac{N}{2} \mathrm{tr}(\mathbf{W}'\mathbf{W}\widehat{\mathbf{\Sigma}})$$

where $\mathbf{W} = \mathbf{B}^{-1}\mathbf{A}$.

When there are long-run constraints, since $\mathbf{C} = \bar{\mathbf{A}}^{-1}\mathbf{B}$ and $\mathbf{A} = \mathbf{I}_K$, $\mathbf{W} = \mathbf{B}^{-1} = \mathbf{C}^{-1}\bar{\mathbf{A}}^{-1} = (\bar{\mathbf{A}}\mathbf{C})^{-1}$. Substituting the last term for $\mathbf{W}$ in the short-run log-likelihood produces the long-run log-likelihood

$$L(\mathbf{C}) = -\frac{NK}{2} \ln(2\pi) + \frac{N}{2} \ln(|\widetilde{\mathbf{W}}|^2) - \frac{N}{2} \mathrm{tr}(\widetilde{\mathbf{W}}'\widetilde{\mathbf{W}}\widehat{\mathbf{\Sigma}})$$

where $\widetilde{\mathbf{W}} = (\bar{\mathbf{A}}\mathbf{C})^{-1}$.

For both the short-run and the long-run models, the maximization is performed by the scoring method. See Harvey (1999) for a discussion of this method.

Using results from Amisano and Giannini (1997), the score vector for the short-run model is

$$\frac{\partial L(\mathbf{A}, \mathbf{B})}{\partial[\mathrm{vec}(\mathbf{A}), \mathrm{vec}(\mathbf{B})]} = N \left[\{\mathrm{vec}(\mathbf{W}'^{-1})\}' - \{\mathrm{vec}(\mathbf{W})\}'(\widehat{\mathbf{\Sigma}} \otimes \mathbf{I}_K) \right] \times$$
$$\left[(\mathbf{I}_K \otimes \mathbf{B}^{-1}), -(\mathbf{A}'\mathbf{B}'^{-1} \otimes \mathbf{B}^{-1}) \right]$$

and the expected information matrix is

$$I\left[\mathrm{vec}(\mathbf{A}), \mathrm{vec}(\mathbf{B})\right] = N \begin{bmatrix} (\mathbf{W}^{-1} \otimes \mathbf{B}'^{-1}) \\ -(\mathbf{I}_K \otimes \mathbf{B}'^{-1}) \end{bmatrix} (\mathbf{I}_{K^2} + \oplus) \left[(\mathbf{W}'^{-1} \otimes \mathbf{B}^{-1}), -(\mathbf{I}_K \otimes \mathbf{B}^{-1}) \right]$$

where $\oplus$ is the commutation matrix defined in Magnus and Neudecker (1999, 46–48).

Using results from Amisano and Giannini (1997), we can derive the score vector and the expected information matrix for the case with long-run restrictions. The score vector is

$$\frac{\partial L(\mathbf{C})}{\partial \mathrm{vec}(\mathbf{C})} = N \left[\{\mathrm{vec}(\mathbf{W}'^{-1})\}' - \{\mathrm{vec}(\mathbf{W})\}'(\widehat{\mathbf{\Sigma}} \otimes \mathbf{I}_K) \right] \left[-(\bar{\mathbf{A}}'^{-1}\mathbf{C}'^{-1} \otimes \mathbf{C}^{-1}) \right]$$

and the expected information matrix is

$$I\left[\mathrm{vec}(\mathbf{C})\right] = N(\mathbf{I}_K \otimes \mathbf{C}'^{-1})(\mathbf{I}_{K^2} + \oplus)(\mathbf{I}_K \otimes \mathbf{C}'^{-1})$$

Checking for identification

This section describes the methods used to check for identification of models with short-run or long-run constraints. Both methods depend on the starting values. By default, svar uses starting values constructed by taking a vector of appropriate dimension and applying the constraints. If there are m parameters in the model, the jth element of the $1 \times m$ vector is $1 + m/100$. svar also allows the user to provide starting values.

For the short-run case, the model is identified if the matrix

$$\mathbf{V}_{\text{sr}}^{*} = \begin{bmatrix} \mathbf{N}_K & \mathbf{N}_K \\ \mathbf{N}_K & \mathbf{N}_K \\ \mathbf{R}_a(\mathbf{W}' \otimes \mathbf{B}) & \mathbf{0}_{K^2} \\ \mathbf{0}_{K^2} & \mathbf{R}_a(\mathbf{I}_K \otimes \mathbf{B}) \end{bmatrix}$$

has full column rank of $2K^2$, where $\mathbf{N}_K = (1/2)(\mathbf{I}_{K^2} + \oplus)$, $\mathbf{R}_a$ is the constraint matrix for the parameters in $\mathbf{A}$ (i.e., $\mathbf{R}_a \text{vec}(\mathbf{A}) = \mathbf{r}_a$), and $\mathbf{R}_b$ is the constraint matrix for the parameters in $\mathbf{B}$ (i.e., $\mathbf{R}_b \text{vec}(\mathbf{B}) = \mathbf{r}_b$).

For the long-run case, using results from the $\mathbf{C}$ model in Amisano and Giannini (1997), the model is identified if the matrix

$$\mathbf{V}_{\text{lr}}^{*} = \begin{bmatrix} (\mathbf{I} \otimes \mathbf{C}'^{-1})(2\mathbf{N}_K)(\mathbf{I} \otimes \mathbf{C}^{-1}) \\ \mathbf{R}_c \end{bmatrix}$$

has full column rank of K^2, where $\mathbf{R}_c$ is the constraint matrix for the parameters in $\mathbf{C}$; i.e., $\mathbf{R}_c \text{vec}(\mathbf{C}) = \mathbf{r}_c$.

The test of the overidentifying restrictions is computed as

$$LR = 2(LL_{\text{var}} - LL_{\text{svar}})$$

where LR is the value of the test statistic against the null hypothesis that the overidentifying restrictions are valid, LL_{var} is the log likelihood from the underlying VAR(p) model, and LL_{svar} is the log likelihood from the SVAR model. The test statistic is asymptotically distributed as $\chi^2(q)$, where q is the number of overidentifying restrictions. Amisano and Giannini (1993, 38–39) emphasize that, since this test of the validity of the overidentifying restrictions is an omnibus test, it can be interpreted as a test of the null hypothesis that all the restrictions are valid.

Since constraints might not be independent either by construction or due to the data, the number of restrictions is not necessarily equal to the number of constraints. The rank of e(V) gives the number of parameters that were independently estimated after applying the constraints. The maximum number of parameters that can be estimated in an identified short-run or long-run SVAR is $K(K+1)/2$. This implies that the number of overidentifying restrictions, q, is equal to $K(K+1)/2$ minus the rank of e(V).

The number of overidentifying restrictions is also linked to the order condition for each model. In a short-run SVAR model, there are $2K^2$ parameters. Since no more than $K(K+1)/2$ parameters may be estimated, the order condition for a short-run SVAR model is that at least $2K^2 - K(K+1)/2$ restrictions be placed on the model. Similarly, there are K^2 parameters in long-run SVAR model. Since no more than $K(K+1)/2$ parameters may be estimated, the order condition for a long-run SVAR model is that at least $K^2 - K(K+1)/2$ restrictions be placed on the model.

Acknowledgment

We would like to thank Gianni Amisano, Università di Brescia, for his helpful comments.

References

Amisano, G. and C. Giannini. 1997. *Topics in Structural VAR Econometrics*. 2nd ed. Heidelberg: Springer.

Christiano, L. J., M. Eichenbaum, and C. L. Evans. 1999. Monetary Policy Shocks: What have we learned and to what end? In *Handbook of Macroeconomics*, vol. 1, ed. J. B. Taylor and M. Woodford. New York: Elsevier Science.

Hamilton, J. D. 1994. *Time Series Analysis*. Princeton: Princeton University Press.

Harvey, A. C. 1999. *The Econometric Analysis of Time Series*. 2nd ed. Cambridge, MA: MIT Press.

Lütkepohl, H. 1993. *Introduction to Multiple Time Series Analysis*. 2nd ed. New York: Springer.

Magnus, J. R. and H. Neudecker. 1999. *Matrix Differential Calculus with Applications in Statistics and Econometrics*. rev. ed. New York: Wiley.

Rothenberg, T. J. 1971. Identification in parametric models. *Econometrica* 39: 577–591.

Sims, C. A. 1980. Macroeconomics and reality. *Econometrica* 48: 1–48.

Stock, J. H. and M. W. Watson. 2001. Vector autoregressions. *Journal of Economic Perspectives* 15(4): 101–115.

Watson, M. W. 1994. Vector autoregressions and cointegration. *Handbook of Econometrics*, Vol IV, Engle, R. F. and McFadden, D. L. eds, Amsterdam: Elsevier.

Also See

Complementary:	[TS] **var svar postestimation**; [TS] **tsset**
Related:	[TS] **arch**, [TS] **arima**, [TS] **var**, [TS] **varbasic**, [TS] **vec**, [R] **reg3**, [R] **regress**, [R] **sureg**
Background:	[U] **11.1.10 Prefix commands**, [U] **20 Estimation and postestimation commands**, [TS] **estimation options**, [TS] **var intro**, [R] **maximize**

Title

var svar postestimation — Postestimation tools for svar

Description

The following postestimation commands are of special interest after `svar`:

command	description
fcast compute	obtain dynamic forecasts
fcast graph	graph dynamic forecasts obtained from `fcast compute`
irf	create and analyze IRFs and FEVDs
vargranger	Granger causality tests
varlmar	LM test for autocorrelation in residuals
varnorm	test for normally distributed residuals
varsoc	lag-order selection criteria
varstable	check stability condition of estimates
varwle	Wald lag-exclusion statistics

For information about these commands, see the corresponding entries in this manual.

In addition, the following standard postestimation commands are available:

command	description
estat	AIC, BIC, VCE, and estimation sample summary
estimates	cataloging estimation results
lincom	point estimates, standard errors, testing, and inference for linear combinations of coefficients
lrtest	likelihood-ratio test
nlcom	point estimates, standard errors, testing, and inference for nonlinear combinations of coefficients
predict	predictions, residuals, influence statistics, and other diagnostic measures
predictnl	point estimates, standard errors, testing, and inference for generalized predictions
test	Wald tests for simple and composite linear hypotheses
testnl	Wald tests of nonlinear hypotheses

See the corresponding entries in the *Stata Base Reference Manual* for details.

Syntax for predict

predict [*type*] *newvar* [*if*] [*in*] [, <u>equation</u>(*eqno* | *eqname*) *statistic*]

statistic	description
xb	linear prediction; the default
stdp	standard error of the linear prediction
<u>residuals</u>	residuals

These statistics are available both in and out of sample; type `predict ... if e(sample) ...` if wanted only for the estimation sample.

Options for predict

equation(*eqno* | *eqname*) specifies the equation to which you are referring.

> equation() is filled in with one *eqno* or *eqname* for options xb, stdp, and residuals. For example, equation(#1) would mean the calculation is to be made for the first equation, equation(#2) would mean the second, and so on. Alternatively, you could refer to the equation by its name, thus equation(income) would refer to the equation named income and equation(hours) to the equation named hours.

> If you do not specify equation(), the results are the same as if you specified equation(#1).

xb, the default, calculates the linear prediction for the specified equation.

stdp calculates the standard error of the linear prediction for the specified equation.

residuals calculates the residuals.

For more information on using predict after multiple-equation estimation commands, see [R] **predict**.

Methods and Formulas

All postestimation commands listed above are implemented as ado-files.

Also See

Complementary:	[TS] **var svar**; [TS] **fcast compute**, [TS] **fcast graph**, [TS] **irf**,
	[TS] **vargranger**, [TS] **varlmar**, [TS] **varnorm**, [TS] **varsoc**,
	[TS] **varstable**, [TS] **varwle**,
	[R] **estimates**, [R] **lincom**, [R] **lrtest**, [R] **nlcom**, [R] **predictnl**,
	[R] **test**, [R] **testnl**
Background:	[U] **13.5 Accessing coefficients and standard errors**,
	[U] **20 Estimation and postestimation commands**,
	[R] **estat**, [R] **predict**

Title

---
varbasic — Fit a simple VAR and graph impulse–response functions
---

Syntax

varbasic *depvarlist* [*if*] [*in*] [, *options*]

options	description
Main	
<u>lags</u>(*numlist*)	use lags *numlist* in the model; default is lags(1 2)
<u>i</u>rf	produce matrix graph of IRFs
<u>f</u>evd	produce matrix graph of FEVDs
<u>nograph</u>	do not produce a graph
<u>step</u>(#)	set forecast horizon # for estimating the IRFs, OIRFs, and FEVDs; default is step(8)

You must tsset your data before using varbasic; see [TS] **tsset**.

depvarlist may contain time-series operators; see [U] **11.4.3 Time-series varlists**.

by, rolling, statsby, and xi may be used with varbasic; see [U] **11.1.10 Prefix commands**.

See [U] **20 Estimation and postestimation commands** for additional capabilities of estimation commands.

Description

varbasic fits a basic vector autoregressive (VAR) model and graphs the impulse–response functions (IRFs), the orthogonalized impulse–response functions (OIRFs), or the forecast-error variance decompositions (FEVDs).

Options

⌐ Main ⌐

lags(*numlist*) specifies the lags to be included in the model. The default is lags(1 2). Note that this option takes a numlist and not simply an integer for the maximum lag. For instance, lags(2) would include only the second lag in the model, whereas lags(1/2) would include both the first and second lags in the model. See [U] **11.1.8 numlist** and [U] **13.8 Time-series operators** for further discussion of numlists and lags.

irf causes varbasic to produce a matrix graph of the IRFs instead of a matrix graph of the OIRFs, which is produced by default.

fevd causes varbasic to produce a matrix graph of the FEVDs instead of a matrix graph of the OIRFs, which is produced by default.

nograph specifies that no graph be produced. The IRFs, OIRFs, and FEVDs are still estimated and saved in the IRF file _varbasic.irf.

step(#) specifies the forecast horizon for estimating the IRFs, OIRFs, and FEVDs. The default is 8 periods.

Remarks

varbasic makes it easy to fit simple VARs and graph the IRFs, the OIRFs, or the FEVDs. See [TS] **var** and [TS] **var svar** for fitting more advanced VAR models and structural vector autoregressive models (SVAR). All the postestimation commands discussed in [TS] **var postestimation** work after varbasic.

This entry does not discuss the methods for fitting a VAR or the methods surrounding the IRFs, OIRFs, and FEVDs. See [TS] **var** and [TS] **irf create** for more on these methods. This entry illustrates how to use varbasic to easily obtain results. It also illustrates how varbasic serves as an entry point to further analysis.

▷ Example 1

We fit a three-variable VAR with two lags to the German macro data used by Lütkepohl (1993). The three variables are the first-difference of natural log of investment, dlinvestment; the first-difference of the natural log of income, dlincome; and the first-difference of the natural log of consumption, dlconsumption. In addition to fitting the VAR, we want to see the OIRFs. Below we use varbasic to fit a VAR(2) model on the data from the second quarter of 1961 through the fourth quarter of 1978. By default, varbasic produces graphs of the OIRFs.

(Continued on next page)

```
. use http://www.stata-press.com/data/r9/lutkepohl
(Quarterly SA West German macro data, Bil DM, from Lutkepohl 1993 Table E.1)
. varbasic dlinvestment dlincome dlconsumption if qtr<=q(1978q4)
```

Vector autoregression

Sample: 1960q4 1978q4				No. of obs		= 73
Log likelihood = 606.307				AIC		= -16.03581
FPE = 2.18e-11				HQIC		= -15.77323
Det(Sigma_ml) = 1.23e-11				SBIC		= -15.37691

Equation	Parms	RMSE	R-sq	chi2	P>chi2
dlinvestment	7	.046148	0.1286	10.76961	0.0958
dlincome	7	.011719	0.1142	9.410683	0.1518
dlconsumption	7	.009445	0.2513	24.50031	0.0004

	Coef.	Std. Err.	z	P>|z|	[95% Conf. Interval]	
dlinvestment						
dlinvestment						
L1.	-.3196318	.1192898	-2.68	0.007	-.5534355	-.0858282
L2.	-.1605508	.118767	-1.35	0.176	-.39333	.0722283
dlincome						
L1.	.1459851	.5188451	0.28	0.778	-.8709326	1.162903
L2.	.1146009	.508295	0.23	0.822	-.881639	1.110841
dlconsumpt~n						
L1.	.9612288	.6316557	1.52	0.128	-.2767936	2.199251
L2.	.9344001	.6324034	1.48	0.140	-.3050877	2.173888
_cons	-.0167221	.0163796	-1.02	0.307	-.0488257	.0153814
dlincome						
dlinvestment						
L1.	.0439309	.0302933	1.45	0.147	-.0154427	.1033046
L2.	.0500302	.0301605	1.66	0.097	-.0090833	.1091437
dlincome						
L1.	-.1527311	.131759	-1.16	0.246	-.4109741	.1055118
L2.	.0191634	.1290799	0.15	0.882	-.2338285	.2721552
dlconsumpt~n						
L1.	.2884992	.1604069	1.80	0.072	-.0258926	.6028909
L2.	-.0102	.1605968	-0.06	0.949	-.3249639	.3045639
_cons	.0157672	.0041596	3.79	0.000	.0076146	.0239198
dlconsumpt~n						
dlinvestment						
L1.	-.002423	.0244142	-0.10	0.921	-.050274	.045428
L2.	.0338806	.0243072	1.39	0.163	-.0137607	.0815219
dlincome						
L1.	.2248134	.1061884	2.12	0.034	.0166879	.4329389
L2.	.3549135	.1040292	3.41	0.001	.1510199	.558807
dlconsumpt~n						
L1.	-.2639695	.1292766	-2.04	0.041	-.517347	-.010592
L2.	-.0222264	.1294296	-0.17	0.864	-.2759039	.231451
_cons	.0129258	.0033523	3.86	0.000	.0063554	.0194962

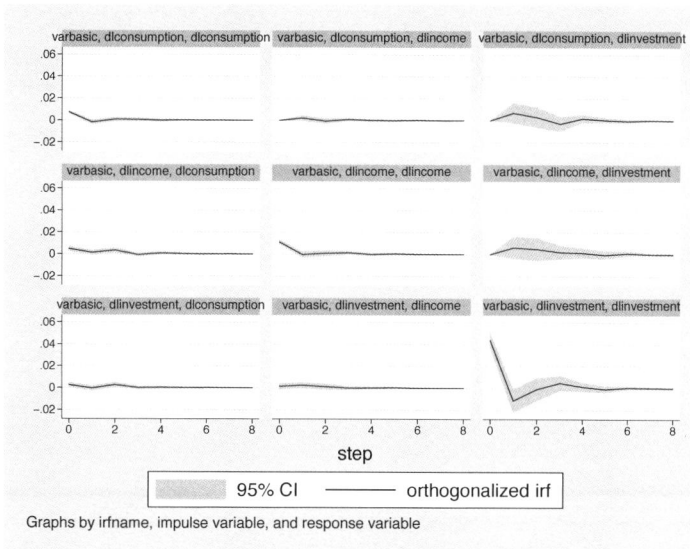

Graphs by irfname, impulse variable, and response variable

Since we are also interested in looking at the FEVDs, we can use `irf graph` to obtain the graphs. While the details are available in [TS] **irf** and [TS] **irf graph**, the command below produces what we want after the call to `varbasic`.

```
. irf graph fevd, lstep(1)
```

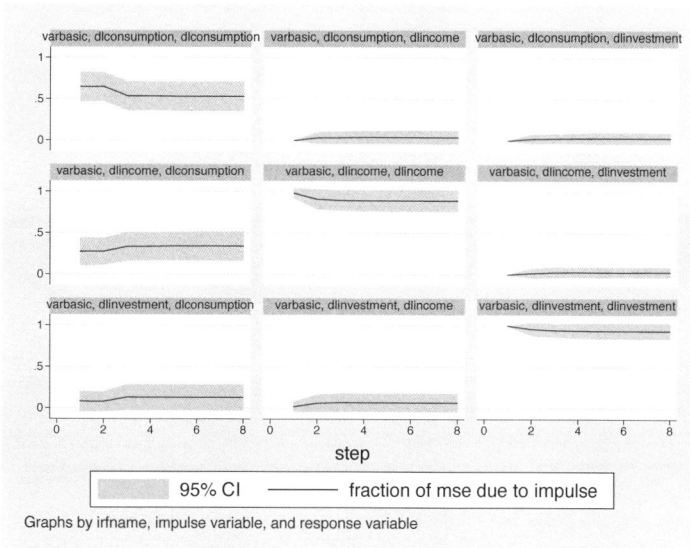

Graphs by irfname, impulse variable, and response variable

◁

❏ Technical Note

Stata stores the estimated IRFs, OIRFs, and FEVDs in a IRF file called _varbasic.irf in the current working directory. varbasic replaces any _varbasic.irf that already exists. Finally, varbasic makes _varbasic.irf the active IRF file. This means that the graph and table commands irf graph, `irf cgraph`, `irf ograph`, `irf table`, and `irf ctable` will all display results that correspond to the VAR fit by `varbasic`.

❏

Saved Results

See *Saved Results* in [TS] **var**.

Methods and Formulas

`varbasic` is implemented as an ado-file.

`varbasic` uses `var` and `irf graph` to obtain its results. See [TS] **var** and [TS] **irf graph** for a discussion of how those commands obtain their results.

Reference

Lütkepohl, H. 1993. *Introduction to Multiple Time Series Analysis.* 2nd ed. New York: Springer.

Also See

Complementary:	[TS] **varbasic postestimation**; [TS] **tsset**
Related:	[TS] **arch**, [TS] **arima**, [TS] **irf create**, [TS] **var**, [TS] **var svar**, [R] **reg3**, [R] **regress**, [R] **sureg**
Background:	[U] **11.1.10 Prefix commands**, [U] **20 Estimation and postestimation commands**, [TS] **var intro**

Title

> **varbasic postestimation** — Postestimation tools for varbasic

Description

The following postestimation commands are of special interest after `varbasic`:

command	description
fcast compute	obtain dynamic forecasts
fcast graph	graph dynamic forecasts obtained from `fcast compute`
irf	create and analyze IRFs and FEVDs
vargranger	Granger causality tests
varlmar	LM test for autocorrelation in residuals
varnorm	test for normally distributed residuals
varsoc	lag-order selection criteria
varstable	check stability condition of estimates
varwle	Wald lag-exclusion statistics

For information about these commands, see the corresponding entries in this manual.

In addition, the following standard postestimation commands are available:

command	description
estat	AIC, BIC, VCE, and estimation sample summary
estimates	cataloging estimation results
lincom	point estimates, standard errors, testing, and inference for linear combinations of coefficients
lrtest	likelihood-ratio test
nlcom	point estimates, standard errors, testing, and inference for nonlinear combinations of coefficients
predict	predictions, residuals, influence statistics, and other diagnostic measures
predictnl	point estimates, standard errors, testing, and inference for generalized predictions
test	Wald tests for simple and composite linear hypotheses
testnl	Wald tests of nonlinear hypotheses

See the corresponding entries in the *Stata Base Reference Manual* for details.

Syntax for predict

> predict [*type*] *newvar* [*if*] [*in*] [, <u>eq</u>uation(*eqno*|*eqname*) *statistic*]

statistic	description
xb	linear prediction; the default
stdp	standard error of the linear prediction
<u>residuals</u>	residuals

These statistics are available both in and out of sample; type predict ... if e(sample) ... if wanted only for the estimation sample.

Options for predict

equation(*eqno* | *eqname*) specifies the equation to which you are referring.

> equation() is filled in with one *eqno* or *eqname* for options xb, stdp, and residuals. For example, equation(#1) would mean the calculation is to be made for the first equation, equation(#2) would mean the second, and so on. Alternatively, you could refer to the equation by its name, thus equation(income) would refer to the equation named income and equation(hours) to the equation named hours.

> If you do not specify equation(), the results are the same as if you specified equation(#1).

xb, the default, calculates the linear prediction for the specified equation.

stdp calculates the standard error of the linear prediction for the specified equation.

residuals calculates the residuals.

For more information on using predict after multiple-equation estimation commands, see [R] **predict**.

Remarks

▷ Example 1

All the postestimation commands discussed in [TS] **var postestimation** work after varbasic. Suppose that we are interested in testing the hypothesis that there is no autocorrelation in the VAR disturbances. Continuing the example above, we now use varlmar to test this hypothesis.

```
. varlmar
```

Lagrange-multiplier test

lag	chi2	df	Prob > chi2
1	5.5871	9	0.78043
2	6.3189	9	0.70763

H0: no autocorrelation at lag order

Since we cannot reject the null hypothesis of no autocorrelation in the residuals, this test does not indicate any model misspecification.

◁

Methods and Formulas

All postestimation commands listed above are implemented as ado-files.

Also See

Complementary: [TS] **varbasic**; [TS] **fcast compute**, [TS] **fcast graph**, [TS] **irf**,

[TS] **vargranger**, [TS] **varlmar**, [TS] **varnorm**, [TS] **varsoc**,

[TS] **varstable**, [TS] **varwle**,

[R] **estimates**, [R] **lincom**, [R] **lrtest**, [R] **nlcom**, [R] **predictnl**,

[R] **test**, [R] **testnl**

Background: [U] **13.5 Accessing coefficients and standard errors**,

[U] **20 Estimation and postestimation commands**,

[R] **estat**, [R] **predict**

Title

> **vargranger** — Perform pairwise Granger causality tests after var or svar

Syntax

vargranger [, estimates(*estname*) separator(*#*)]

vargranger can be used only after var or svar; see [TS] **var** and [TS] **var svar**.

Description

vargranger performs a set of Granger causality tests for each equation in a VAR, providing a convenient alternative to test.

Options

estimates(*estname*) requests that vargranger use the previously obtained set of var or svar estimates saved as *estname*. By default, vargranger uses the active results. See [R] **estimates** for information about saving and restoring estimation results.

separator(*#*) specifies how often separator lines should be drawn between rows. By default, separator lines appear every K lines, where K is the number of equations in the VAR under analysis. For example, separator(1) would draw a line between each row, separator(2) between every other row, and so on. Note that separator(0) specifies that lines not appear in the table.

Remarks

After fitting a VAR, we may want to know whether one variable "Granger-causes" another (Granger 1969). A variable x is said to Granger-cause a variable y if, given the past values of y, past values of x are useful for predicting y. A common method for testing Granger causality is to regress y on its own lagged values and on lagged values of x and test the null hypothesis that the estimated coefficients on the lagged values of x are jointly zero. Failure to reject the null hypothesis is equivalent to failing to reject the hypothesis that x does not Granger-cause y.

For each equation and each endogenous variable that is not the dependent variable in that equation, vargranger computes and reports Wald tests that the coefficients on all the lags of an endogenous variable are jointly zero. In other words, for each equation in a VAR, vargranger tests the hypotheses that each of the other endogenous variables does not Granger-cause the dependent variable in that equation.

Since it may be interesting to investigate these types of hypotheses using the VAR that underlies an SVAR, vargranger can also produce these tests using the e() results from an svar. When vargranger uses svar e() results, the hypotheses concern the underlying var estimates.

See [TS] **var** and [TS] **var svar** for information about fitting VARs and SVARs in Stata. See Lütkepohl (1993), Hamilton (1994), and Amisano and Gianinni (1997) for information about Granger causality and on VARs and SVARs in general.

▷ Example 1: After var

Here we refit the model using German data described in [TS] **var** and then perform Granger causality tests using `vargranger`.

```
. use http://www.stata-press.com/data/r9/lutkepohl
(Quarterly SA West German macro data, Bil DM, from Lutkepohl 1993 Table E.1)
. var dlinvestment dlincome dlconsumption if qtr<=q(1978q4), lags(1/2) dfk small
  (output omitted)
. vargranger
  Granger causality Wald tests
```

Equation	Excluded	F	df	df_r	Prob > F
dlinvestment	dlincome	.04847	2	66	0.9527
dlinvestment	dlconsumption	1.5004	2	66	0.2306
dlinvestment	ALL	1.5917	4	66	0.1869
dlincome	dlinvestment	1.7683	2	66	0.1786
dlincome	dlconsumption	1.7184	2	66	0.1873
dlincome	ALL	1.9466	4	66	0.1130
dlconsumption	dlinvestment	.97147	2	66	0.3839
dlconsumption	dlincome	6.1465	2	66	0.0036
dlconsumption	ALL	3.7746	4	66	0.0080

Since the `estimates()` option was not specified, `vargranger` used the active `e()` results. Consider the results of the three tests for the first equation. The first is a Wald test that the coefficients on the two lags of `dlincome` that appear in the equation for `dlinvestment` are jointly zero. The null hypothesis that `dlincome` does not Granger-cause `dlinvestment` cannot be rejected. Similarly, we cannot reject the null hypothesis that the coefficients on the two lags of `dlconsumption` in the equation for `dlinvestment` are jointly zero, so we cannot reject the hypothesis that `dlconsumption` does not Granger-cause `dlinvestment`. The third test is with respect to the null hypothesis that the coefficients on the two lags of all the other endogenous variables are jointly zero. Since this cannot be rejected, we cannot reject the null hypothesis that `dlincome` and `dlconsumption`, jointly, do not Granger-cause `dlinvestment`.

Since we failed to reject most of these null hypotheses, we might be interested in imposing some constraints on the coefficients. See [TS] **var** for more on fitting VAR models with constraints on the coefficients.

◁

▷ Example 2: Using test instead of vargranger

We could have used `test` to compute these Wald tests, but `vargranger` saves a significant amount of typing. Still, it is useful to see how to use `test` to obtain the results reported by `vargranger`.

```
. test [dlinvestment]L.dlincome [dlinvestment]L2.dlincome
 ( 1)  [dlinvestment]L.dlincome = 0
 ( 2)  [dlinvestment]L2.dlincome = 0

       F(  2,    66) =    0.05
            Prob > F =    0.9527
```

```
. test [dlinvestment]L.dlconsumption [dlinvestment]L2.dlconsumption, accumulate
 ( 1)  [dlinvestment]L.dlincome = 0
 ( 2)  [dlinvestment]L2.dlincome = 0
 ( 3)  [dlinvestment]L.dlconsumption = 0
 ( 4)  [dlinvestment]L2.dlconsumption = 0
        F(  4,     66) =    1.59
            Prob > F =    0.1869
. test [dlinvestment]L.dlinvestment [dlinvestment]L2.dlinvestment, accumulate
 ( 1)  [dlinvestment]L.dlincome = 0
 ( 2)  [dlinvestment]L2.dlincome = 0
 ( 3)  [dlinvestment]L.dlconsumption = 0
 ( 4)  [dlinvestment]L2.dlconsumption = 0
 ( 5)  [dlinvestment]L.dlinvestment = 0
 ( 6)  [dlinvestment]L2.dlinvestment = 0
        F(  6,     66) =    1.62
            Prob > F =    0.1547
```

The first two calls to `test` show how `vargranger` obtains its results. The first test reproduces the first test reported for the `dlinvestment` equation. The second test reproduces the ALL entry for the first equation. The third test reproduces the standard F statistic for the `dlinvestment` equation, reported in the header of the `var` output in the previous example. Note that the standard F statistic also includes the lags of the dependent variable, as well as any exogenous variables in the equation. This illustrates that the test performed by `vargranger` of the null hypothesis that the coefficients on all the lags of all the other endogenous variables are jointly zero for a particular equation; i.e., the All test, is not the same as the standard F statistic for that equation.

◁

▷ Example 3: After svar

When `vargranger` is run on `svar` estimates, the null hypotheses are with respect to the underlying `var` estimates. We run `vargranger` after using `svar` to fit an SVAR that has the same underlying VAR as our model in the previous example.

```
. matrix A = (., 0,0 \ ., ., 0\ .,.,.)
. matrix B = I(3)
. quietly svar dlinvestment dlincome dlconsumption if qtr<=q(1978q4), lags(1/2)
> dfk small aeq(A) beq(B)
. vargranger
    Granger causality Wald tests
```

Equation	Excluded	F	df	df_r	Prob > F
dlinvestment	dlincome	.04847	2	66	0.9527
dlinvestment	dlconsumption	1.5004	2	66	0.2306
dlinvestment	ALL	1.5917	4	66	0.1869
dlincome	dlinvestment	1.7683	2	66	0.1786
dlincome	dlconsumption	1.7184	2	66	0.1873
dlincome	ALL	1.9466	4	66	0.1130
dlconsumption	dlinvestment	.97147	2	66	0.3839
dlconsumption	dlincome	6.1465	2	66	0.0036
dlconsumption	ALL	3.7746	4	66	0.0080

As we expected, the `vargranger` results are identical to those in the first example.

◁

Saved Results

vargranger saves in r():

Matrices

    r(gstats) χ^2, df, and p-values (if e(small)=="")

    r(gstats) F, df, df_r, and p-values (if e(small)!="")

Methods and Formulas

vargranger is implemented as an ado-file.

vargranger uses test to obtain Wald statistics of the hypothesis that all coefficients on the lags of variable x are jointly zero in the equation for variable y. vargranger uses the e() results saved by var or svar to determine whether to calculate and report small-sample F statistics or large-sample χ^2 statistics.

Clive William John Granger (1934–) was born in Swansea, Wales, and earned degrees at the University of Nottingham in mathematics and statistics. Joining the staff there, he also worked at Princeton on the spectral analysis of economic time series, before moving in 1973 to the University of California, San Diego. He was awarded the 2003 Nobel Prize in Economics for methods of analyzing economic time series with common trends (cointegration). He was knighted in 2005, thus becoming Sir Clive Granger.

References

Amisano, G. and C. Giannini. 1997. *Topics in Structural* VAR *Econometrics*. 2nd ed. Heidelberg: Springer.

Granger, C. W. J. 1969. Investigating causal relations by econometric models and cross-spectral methods. *Econometrica* 37: 424–438.

Hamilton, J. D. 1994. *Time Series Analysis*. Princeton: Princeton University Press.

Lütkepohl, H. 1993. *Introduction to Multiple Time Series Analysis*. 2nd ed. New York: Springer.

Phillips, P. C. B. 1997. The *ET* Interview: Professor Clive Granger. *Econometric Theory* 13: 253–303.

Also See

Complementary:	[TS] **var**, [TS] **var svar**, [TS] **varbasic**
Background:	[TS] **var intro**

Title

> **varlmar** — Obtain LM statistics for residual autocorrelation after var or svar

Syntax

varlmar [, *options*]

options	description
mlag(#)	use # for the maximum order of autocorrelation; default is mlag(2)
estimates(*estname*)	use previously saved results *estname*; default is to use active results
separator(#)	draw separator line after every # rows

varlmar can be used only after var or svar; see [TS] **var** and [TS] **var svar**.

You must tsset your data before using varlmar; see [TS] **tsset**.

Description

varlmar implements a Lagrange-multiplier (LM) test for autocorrelation in the residuals of VAR models, which was presented in Johansen (1995).

Options

mlag(#) specifies the maximum order of autocorrelation to be tested. The integer specified in mlag() must be greater than 0; the default is 2.

estimates(*estname*) requests that varlmar use the previously obtained set of var or svar estimates saved as *estname*. By default, varlmar uses the active results. See [R] **estimates** for information on saving and restoring estimation results.

separator(#) specifies how often separator lines should be drawn between rows. By default, separator lines do not appear. For example, separator(1) would draw a line between each row, separator(2) between every other row, and so on.

Remarks

Most postestimation analyses of VAR models and SVAR models assume that the disturbances are not autocorrelated. varlmar implements the LM test for autocorrelation in the residuals of a VAR model discussed in Johansen (1995, 21–22). The test is performed at lags $j = 1, \ldots, \text{mlag}()$. For each j, the null hypothesis of the test is that there is no autocorrelation at lag j.

varlmar uses the estimation results saved by var or svar. By default, varlmar uses the active estimation results. However, varlmar can use any previously saved var or svar estimation results specified in the estimates() option.

▷ Example 1: After var

Here we refit the model using German data described in [TS] **var** and then call `varlmar`.

```
. use http://www.stata-press.com/data/r9/lutkepohl
(Quarterly SA West German macro data, Bil DM, from Lutkepohl 1993 Table E.1)
. var dlinvestment dlincome dlconsumption if qtr>=q(1961q2) & qtr <= q(1978q4),
> lags(1/2) dfk
```

(*output omitted*)

```
. varlmar, mlag(5)
```

    Lagrange-multiplier test

lag	chi2	df	Prob > chi2
1	8.0181	9	0.53233
2	8.1691	9	0.51720
3	8.4216	9	0.49228
4	12.0881	9	0.20839
5	5.2337	9	0.81348

    H0: no autocorrelation at lag order

Since we cannot reject the null hypothesis that there is no autocorrelation in the residuals for any of the five orders tested, this test does not provide any hint of model misspecification. Although we fit the VAR with the `dfk` option to be consistent with the example in [TS] **var**, `varlmar` always uses the ML estimator of Σ. In other words, the results obtained from `varlmar` are the same whether or not `dfk` is specified.

◁

▷ Example 2: After svar

When `varlmar` is applied to estimation results produced by `svar`, the sequence of LM tests is applied to the underlying VAR. See [TS] **var svar** for a description of how an SVAR model builds on a VAR. In this example, we fit an SVAR that has an underlying VAR with two lags that is identical to the one fitted in the previous example.

```
. mat A = (.,.,,0\0,.,,0\.,.,,.)
. mat B = I(3)
. qui svar dlinvestment dlincome dlconsumption if qtr>=q(1961q2) &
> qtr <= q(1978q4), lags(1/2) dfk aeq(A) beq(B)
. varlmar, mlag(5)
```

    Lagrange-multiplier test

lag	chi2	df	Prob > chi2
1	8.0181	9	0.53233
2	8.1691	9	0.51720
3	8.4216	9	0.49228
4	12.0881	9	0.20839
5	5.2337	9	0.81348

    H0: no autocorrelation at lag order

Since the underlying VAR(2) is the same as the previous example (we assure you that this is true), the output from `varlmar` is also the same.

◁

Saved Results

varlmar saves in r():

Matrices

r(lm) χ^2, df, and p-values

Methods and Formulas

varlmar is implemented as an ado-file.

The formula for the LM test statistic at lag j is

$$\text{LM}_s = (T - d - .5) \ln \left(\frac{|\widehat{\boldsymbol{\Sigma}}|}{|\widetilde{\boldsymbol{\Sigma}}_s|} \right)$$

where T is the number of observations in the VAR; d is explained below; $\widehat{\boldsymbol{\Sigma}}$ is the maximum likelihood estimate of $\boldsymbol{\Sigma}$, the variance–covariance matrix of the disturbances from the VAR; and $\widetilde{\boldsymbol{\Sigma}}_s$ is the maximum likelihood estimate of $\boldsymbol{\Sigma}$ from the following augmented VAR.

If there are K equations in the VAR, we can define $\mathbf{e}_t$ to be a $K \times 1$ vector of residuals. After we create the K new variables e1, e2, ..., eK containing the residuals from the K equations, we can augment the original VAR with lags of these K new variables. For each lag s, we form an augmented regression in which the new residual variables are lagged s times. Following the method of Davidson and MacKinnon (1993, 358), the missing values from these s lags are replaced with zeros. $\widetilde{\boldsymbol{\Sigma}}_s$ is the maximum likelihood estimate of $\boldsymbol{\Sigma}$ from this augmented VAR, and d is the number of coefficients estimated in the augmented VAR. See [TS] **var** for a discussion of the maximum likelihood estimate of $\boldsymbol{\Sigma}$ in a VAR.

The asymptotic distribution of LM_s is χ^2 with K^2 degrees of freedom.

References

Davidson, R. and J. G. MacKinnon. 1993. *Estimation and Inference in Econometrics*. Oxford: Oxford University Press.

Johansen, S. 1995. *Likelihood-Based Inference in Cointegrated Vector Auto-Regressive Models*. Oxford: Oxford University Press.

Also See

Complementary: [TS] **var**, [TS] **var svar**, [TS] **varbasic**

Background: [TS] **var intro**

Title

> **varnorm** — Test for normally distributed disturbances after var or svar

Syntax

varnorm [, *options*]

options	description
jbera	report Jarque–Bera statistic; default is to report all three statistics
skewness	report skewness statistic; default is to report all three statistics
kurtosis	report kurtosis statistic; default is to report all three statistics
estimates(*estname*)	use previously saved results *estname*; default is to use active results
cholesky	use Cholesky decomposition
separator(#)	draw separator line after every # rows

varnorm can be used only after var or svar; see [TS] **var** and [TS] **var svar**.
You must tsset your data before using varnorm; see [TS] **tsset**.

Description

varnorm computes and reports a series of statistics against the null hypothesis that the disturbances in a VAR are normally distributed. For each equation, and for all equations jointly, up to three statistics may be computed: a skewness statistic, a kurtosis statistic, and the Jarque–Bera statistic. By default, all three statistics are reported.

Options

jbera requests that the Jarque–Bera statistic and any other explicitly requested statistics be reported. By default, the Jarque–Bera, skewness, and kurtosis statistics are reported.

skewness requests that the skewness statistic and any other explicitly requested statistic be reported. By default, the Jarque–Bera, skewness, and kurtosis statistics are reported.

kurtosis requests that the kurtosis statistic and any other explicitly requested statistic be reported. By default, the Jarque–Bera, skewness, and kurtosis statistics are reported.

estimates(*estname*) specifies that varnorm use the previously obtained set of var or svar estimates saved as *estname*. By default, varnorm uses the active results. See [R] **estimates** for information on saving and restoring estimation results.

cholesky specifies that varnorm use the Cholesky decomposition of the estimated variance–covariance matrix of the disturbances, $\widehat{\Sigma}$, to orthogonalize the residuals when varnorm is applied to svar results. By default, when varnorm is applied to svar results, it uses the estimated structural decomposition $\widehat{A}^{-1}\widehat{B}$ on $\widehat{C}$ to orthogonalize the residuals. When applied to var e() results, varnorm always uses the Cholesky decomposition of $\widehat{\Sigma}$. For this reason, the cholesky option may not be specified when using var results.

separator(#) specifies how often separator lines should be drawn between rows. By default, separator lines do not appear. For example, separator(1) would draw a line between each row, separator(2) between every other row, and so on.

Remarks

Some of the postestimation statistics for VAR and SVAR assume that the K disturbances have a K-dimensional multivariate normal distribution. varnorm uses the estimation results produced by var or svar to produce a series of statistics against the null hypothesis that the K disturbances in the VAR are normally distributed.

Following the notation in Lütkepohl (1993), call the skewness statistic $\widehat{\lambda}_1$, the kurtosis statistic $\widehat{\lambda}_2$, and the Jarque–Bera statistic $\widehat{\lambda}_3$. The Jarque–Bera statistic is a combination of the other two statistics. The single-equation results are from tests against the null hypothesis that the disturbance for that particular equation is normally distributed. The results for all the equations are from tests against the null hypothesis that the K disturbances follow a K-dimensional multivariate normal distribution. Failure to reject the null hypothesis indicates a lack of model misspecification.

▷ Example 1: After var

We refit the model using German data described in [TS] **var** and then call varnorm.

```
. use http://www.stata-press.com/data/r9/lutkepohl
(Quarterly SA West German macro data, Bil DM, from Lutkepohl 1993 Table E.1)
. var dlinvestment dlincome dlconsumption if qtr <= q(1978q4), lags(1/2) dfk
```

(output omitted)

```
. varnorm
```

Jarque-Bera test

Equation	chi2	df	Prob > chi2
dlinvestment	2.821	2	0.24397
dlincome	3.450	2	0.17817
dlconsumption	1.566	2	0.45702
ALL	7.838	6	0.25025

Skewness test

Equation	Skewness	chi2	df	Prob > chi2
dlinvestment	.11935	0.173	1	0.67718
dlincome	-.38316	1.786	1	0.18139
dlconsumption	-.31275	1.190	1	0.27532
ALL		3.150	3	0.36913

Kurtosis test

Equation	Kurtosis	chi2	df	Prob > chi2
dlinvestment	3.9331	2.648	1	0.10367
dlincome	3.7396	1.664	1	0.19710
dlconsumption	2.6484	0.376	1	0.53973
ALL		4.688	3	0.19613

dfk estimator used in computations

In this example, neither the single-equation Jarque–Bera statistics nor the joint Jarque–Bera statistic come close to rejecting the null hypothesis.

The skewness and kurtosis results have similar structures.

The Jarque–Bera results use the sum of the skewness and kurtosis statistics. The skewness and kurtosis results are based on the skewness and kurtosis coefficients, respectively. See *Methods and Formulas*.

◁

▷ Example 2: After svar

The test statistics are computed on the orthogonalized VAR residuals; see *Methods and Formulas*. When `varnorm` is applied to `var` results, `varnorm` uses a Cholesky decomposition of the estimated variance–covariance matrix of the disturbances, $\widehat{\boldsymbol{\Sigma}}$, to orthogonalize the residuals.

By default, when `varnorm` is applied to `svar` estimation results, it uses the estimated structural decomposition $\widehat{\mathbf{A}}^{-1}\widehat{\mathbf{B}}$ on $\widehat{\mathbf{C}}$ to orthogonalize the residuals of the underlying VAR. Alternatively, when `varnorm` is applied to `svar` results and the `cholesky` option is specified, `varnorm` uses the Cholesky decomposition of $\widehat{\boldsymbol{\Sigma}}$ to orthogonalize the residuals of the underlying VAR.

We fit an SVAR that is based on an underlying VAR with two lags that is the same as the one fitted in the previous example. We impose a structural decomposition that is the same as the Cholesky decomposition, as illustrated in [TS] **var svar**.

```
. matrix a = (.,0,0\.,.,0\.,.,.)
. matrix b = I(3)
. svar dlinvestment dlincome dlconsumption if qtr <= q(1978q4),
> lags(1/2) dfk aeq(a) beq(b)
```
(output omitted)
```
. varnorm
```
Jarque-Bera test

Equation	chi2	df	Prob > chi2
dlinvestment	2.821	2	0.24397
dlincome	3.450	2	0.17817
dlconsumption	1.566	2	0.45702
ALL	7.838	6	0.25025

Skewness test

Equation	Skewness	chi2	df	Prob > chi2
dlinvestment	.11935	0.173	1	0.67718
dlincome	-.38316	1.786	1	0.18139
dlconsumption	-.31275	1.190	1	0.27532
ALL		3.150	3	0.36913

Kurtosis test

Equation	Kurtosis	chi2	df	Prob > chi2
dlinvestment	3.9331	2.648	1	0.10367
dlincome	3.7396	1.664	1	0.19710
dlconsumption	2.6484	0.376	1	0.53973
ALL		4.688	3	0.19613

dfk estimator used in computations

Since the estimated structural decomposition is the same as the Cholesky decomposition, the varnorm results are the same as those from the previous example.

◁

❏ Technical Note

The statistics computed by varnorm depend on $\widehat{\Sigma}$, the estimated variance–covariance matrix of the disturbances. var uses the maximum likelihood estimator of this matrix by default, but the dfk option produces an estimator that uses a small-sample correction. Thus specifying dfk in the call to var or svar will affect the test results produced by varnorm.

❏

Saved Results

varnorm saves in r():

Macros
    r(dfk)   dfk, if specified
Matrices
    r(kurtosis) kurtosis test, df, and p-values
    r(skewness) skewness test, df, and p-values
    r(jb)   Jarque–Bera test, df, and p-values

Methods and Formulas

varnorm is implemented as an ado-file.

varnorm is based on the derivations found in Lütkepohl (1993, 152–158). Let $\widehat{\mathbf{u}}_t$ be the $K \times 1$ vector of residuals from the K equations in a previously fitted VAR or the residuals from the K equations of the VAR underlying a previously fitted SVAR. Similarly, let $\widehat{\Sigma}$ be the estimated covariance matrix of the disturbances. (Note that $\widehat{\Sigma}$ depends on whether the dfk option was specified.) The skewness, kurtosis, and Jarque–Bera statistics must be computed using the orthogonalized residuals.

Since

$$\widehat{\Sigma} = \widehat{\mathbf{P}}\widehat{\mathbf{P}}'$$

implies that

$$\widehat{\mathbf{P}}^{-1}\widehat{\Sigma}\widehat{\mathbf{P}}^{-1\prime} = \mathbf{I}_K$$

Premultiplying $\widehat{\mathbf{u}}_t$ by $\widehat{\mathbf{P}}$ is one way of performing the orthogonalization. When varnorm is applied to var results, $\widehat{\mathbf{P}}$ is defined to be the Cholesky decomposition of $\widehat{\Sigma}$. When varnorm is applied to svar results, $\widehat{\mathbf{P}}$ is set, by default, to the estimated structural decomposition; i.e., $\widehat{\mathbf{P}} = \widehat{\mathbf{A}}^{-1}\widehat{\mathbf{B}}$, where $\widehat{\mathbf{A}}$ and $\widehat{\mathbf{B}}$ are the svar estimates of the $\mathbf{A}$ and $\mathbf{B}$ matrices, or $\widehat{\mathbf{C}}$, where $\widehat{\mathbf{C}}$ is the long-run SVAR estimation of $\mathbf{C}$. (See [TS] **var svar** for more on the origin and estimation of the $\mathbf{A}$ and $\mathbf{B}$ matrices.) When varnorm is applied to svar results and the cholesky option is specified, $\widehat{\mathbf{P}}$ is set to the Cholesky decomposition of $\widehat{\Sigma}$.

Define $\widehat{\mathbf{w}}_t$ to be the orthogonalized VAR residuals given by

$$\widehat{\mathbf{w}}_t = (\widehat{w}_{1t}, \ldots, \widehat{w}_{Kt})' = \widehat{\mathbf{P}}^{-1}\widehat{\mathbf{u}}_t$$

The $K \times 1$ vectors of skewness and kurtosis coefficients are then computed using the orthogonalized residuals by

$$\widehat{\mathbf{b}}_1 = (\widehat{b}_{11}, \ldots, \widehat{b}_{K1})'; \qquad \widehat{b}_{k1} = \frac{1}{T}\sum_{i=1}^{T} \widehat{w}_{kt}^3$$

$$\widehat{\mathbf{b}}_2 = (\widehat{b}_{12}, \ldots, \widehat{b}_{K2})'; \qquad \widehat{b}_{k2} = \frac{1}{T}\sum_{i=1}^{T} \widehat{w}_{kt}^4$$

Under the null hypothesis of multivariate Gaussian disturbances,

$$\widehat{\lambda}_1 = \frac{T\widehat{\mathbf{b}}_1'\widehat{\mathbf{b}}_1}{6} \quad \overset{d}{\to} \quad \chi^2(K)$$

$$\widehat{\lambda}_2 = \frac{T(\widehat{\mathbf{b}}_2 - 3)'(\widehat{\mathbf{b}}_2 - 3)}{24} \quad \overset{d}{\to} \quad \chi^2(K)$$

and

$$\widehat{\lambda}_3 = \widehat{\lambda}_1 + \widehat{\lambda}_2 \quad \overset{d}{\to} \quad \chi^2(2K)$$

$\widehat{\lambda}_1$ is the skewness statistic, $\widehat{\lambda}_2$ is the kurtosis statistic, and $\widehat{\lambda}_3$ is the Jarque–Bera statistic.

$\widehat{\lambda}_1$, $\widehat{\lambda}_2$, and $\widehat{\lambda}_3$ are for tests of the null hypothesis that the $K \times 1$ vector of disturbances follows a multivariate normal distribution. The corresponding statistics against the null hypothesis that the disturbances from the kth equation come from a univariate normal distribution are

$$\widehat{\lambda}_{1k} = \frac{T\widehat{b}_{k1}^2}{6} \quad \overset{d}{\to} \quad \chi^2(1)$$

$$\widehat{\lambda}_{2k} = \frac{T(\widehat{b}_{k2}^2 - 3)^2}{24} \quad \overset{d}{\to} \quad \chi^2(1)$$

and

$$\widehat{\lambda}_{3k} = \widehat{\lambda}_1 + \widehat{\lambda}_2 \quad \overset{d}{\to} \quad \chi^2(2)$$

References

Hamilton, J. D. 1994. *Time Series Analysis.* Princeton: Princeton University Press.

Jarque, C. M. and A. K. Bera. 1987. A test for normality of observations and regression residuals. *International Statistical Review* 55: 163–172.

Lütkepohl, H. 1993. *Introduction to Multiple Time Series Analysis.* 2nd ed. New York: Springer.

Also See

Complementary:	[TS] **var**, [TS] **var svar**, [TS] **varbasic**
Background:	[TS] **var intro**

Title

> **varsoc** — Obtain lag-order selection statistics for VARs and VECMs

Syntax

Pre-estimation syntax

varsoc *depvarlist* $\big[$ *if* $\big]$ $\big[$ *in* $\big]$ $\big[$, *pre-estimation_options* $\big]$

Postestimation syntax

varsoc $\big[$, <u>est</u>imates(*estname*) $\big]$

pre-estimation_options	description
Main	
<u>maxlag</u>(#)	set maximum lag order to #; default is maxlag(4)
<u>exog</u>(*varlist*)	use exogenous variables *varlist*
<u>constraints</u>(*constraints*)	apply constraints to exogenous variables
<u>noconst</u>ant	suppress constant term
<u>lutstats</u>	use Lütkepohl's version of information criteria
<u>level</u>(#)	set confidence level; default is level(95)
<u>separator</u>(#)	draw separator line after every # rows

You must tsset your data before using varsoc; see [TS] **tsset**.

Description

varsoc reports the final prediction error (FPE), Akaike's information criterion (AIC), Schwarz's Bayesian information criterion (SBIC), and the Hannan and Quinn information criterion (HQIC) lag-order selection statistics for a series of vector autoregressions of order 1, ..., maxlag(). A sequence of likelihood-ratio test statistics for all of the full VARs of order less than or equal to the highest lag order is also reported. In the postestimation version, the maximum lag and estimation options are based on the model just fit or the model specified in estimates(*estname*).

The pre-estimation version of varsoc can also be used to select the lag order for a vector error-correction model (VECM). As shown by Nielsen (2001), the lag-order selection statistics discussed here can be used in the presence of I(1) variables.

Pre-estimation options

⌐ Main ⌐

maxlag(#) specifies the maximum lag order for which the statistics are to be obtained.

exog(*varlist*) specifies exogenous variables to include in the VARs fit by varsoc.

constraints(*constraints*) specifies a list of constraints on the exogenous variables to be applied. Do not specify constraints on the lags of the endogenous variables since specifying one would mean that at least one of the VAR models considered by varsoc will not contain the lag specified in the constraint. Use [TS] **var** directly to obtain selection-order criteria with constraints on lags of the endogenous variables.

noconstant suppresses the constant terms from the model. By default, constant terms are included.

lutstats specifies that the Lütkepohl (1993) versions of the lag-order selection statistics be reported. See *Methods and Formulas* for a discussion of these statistics.

level(#) specifies the confidence level, as a percentage, that is used to identify the first likelihood-ratio test that rejects the null hypothesis that the additional parameters from adding a lag are jointly zero. The default is level(95) or as set by set level; see [U] **20.6 Specifying the width of confidence intervals**.

separator(#) specifies how often separator lines should be drawn between rows. By default, separator lines do not appear. For example, separator(1) would draw a line between each row, separator(2) between every other row, and so on.

Postestimation option

estimates(*estname*) specifies the name of a previously stored set of var or svar estimates. When no *depvarlist* is specified, varsoc uses the *postestimation syntax* and uses the currently active estimation results or the results specified in estimates(*estname*). See [R] **estimates** for information about saving and restoring estimation results.

Remarks

Many selection-order statistics have been developed to assist researchers in fitting a VAR of the correct order. Several of these selection-order statistics appear in the [TS] **var** output. The varsoc command computes these statistics over a range of lags p while maintaining a common sample and option specification.

varsoc can be used as a pre-estimation or a postestimation command. When it is used as a pre-estimation command, a *depvarlist* is required, and the default maximum lag is 4. When it is used as a postestimation command, varsoc uses the model specification stored in *estname* or the previously fit model.

varsoc computes four information criteria as well as a sequence of likelihood ratio (LR) tests. The information criteria include the final prediction error (FPE), Akaike's information criterion (AIC), the Hannan Quinn information criterion (HQIC), and Schwarz's Bayesian information criterion (SBIC).

For a given lag p, the LR test compares a VAR with p lags to one with $p - 1$ lags. The null hypothesis is that all of the coefficients on the pth lags of the endogenous variables are zero. To use this sequence of LR tests to select a lag order, we start by looking at the results of the test for the model with the most lags, which is at the bottom of the table. Proceeding up the table, the first test that rejects the null hypothesis is the lag order selected by this process. See Lütkepohl (1993, 125–126) for more information on this procedure. An '*' appears next to the LR statistic indicating the optimal lag.

For the remaining statistics, the lag with the smallest value is the order selected by that criterion. An '*' indicates the optimal lag. Strictly speaking, the FPE is not an information criterion, though we include it in this discussion because, as with an information criterion, we select the lag length corresponding to the lowest value; and, naturally, we want to minimize the prediction error. The AIC measures the discrepancy between the given model and the true model, which, of course, we want to minimize. Amemiya (1985) provides an intuitive discussion of the arguments in Akaike (1973). The SBIC and the HQIC can be interpreted similarly to the AIC, though the SBIC and the HQIC have a theoretical advantage over the AIC and the FPE. As Lütkepohl (1993, 130–133) demonstrates, choosing p to minimize the SBIC or the HQIC provides consistent estimates of the true lag order, p. In contrast, minimizing the AIC or the FPE will overestimate the true lag order with positive probability, even with an infinite sample size.

▷ Example 1: Pre-estimation

Here we use `varsoc` as a pre-estimation command.

```
. use http://www.stata-press.com/data/r9/lutkepohl
(Quarterly SA West German macro data, Bil DM, from Lutkepohl 1993 Table E.1)
. varsoc dlinvestment dlincome dlconsumption if qtr<=q(1978q4), lutstats
```

Selection order criteria (lutstats)
Sample: 1961q2   1978q4   Number of obs   =   71

lag	LL	LR	df	p	FPE	AIC	HQIC	SBIC
0	564.784				2.7e-11	-24.423	-24.423*	-24.423*
1	576.409	23.249	9	0.006	2.5e-11	-24.497	-24.3829	-24.2102
2	588.859	24.901*	9	0.003	2.3e-11*	-24.5942*	-24.3661	-24.0205
3	591.237	4.7566	9	0.855	2.7e-11	-24.4076	-24.0655	-23.5472
4	598.457	14.438	9	0.108	2.9e-11	-24.3575	-23.9012	-23.2102

Endogenous: dlinvestment dlincome dlconsumption
Exogenous: _cons

Notice that the sample used begins in 1961q2 because all the VARs are fitted to the sample defined by any `if` or `in` conditions and the available data for the maximum lag specified. The default maximum number of lags is 4. Since we specified the `lutstats` option, the table contains the Lütkepohl (1993) versions of the information criteria, which differ from the standard definitions in that they drop the constant term from the log likelihood. In this example, the likelihood-ratio tests selected a model with 2 lags. AIC and FPE have also both chosen a model with 2 lags, whereas SBIC and HQIC have both selected a model with 0 lags.

◁

▷ Example 2: Postestimation

`varsoc` works as a postestimation command when no dependent variables are specified.

```
. var dlincome dlconsumption if qtr<=q(1978q4), lutstats exog(l.dlinvestment)
(output omitted)
. varsoc
```

Selection order criteria (lutstats)
Sample: 1960q4   1978q4   Number of obs   =   73

lag	LL	LR	df	p	FPE	AIC	HQIC	SBIC
0	460.646				1.3e-08	-18.2962	-18.2962	-18.2962*
1	467.606	13.919	4	0.008	1.2e-08	-18.3773	-18.3273	-18.2518
2	477.087	18.962*	4	0.001	1.0e-08*	-18.5275*	-18.4274*	-18.2764

Endogenous: dlincome dlconsumption
Exogenous: L.dlinvestment _cons

Because we included one lag of `dlinvestment` in our original model, `varsoc` did likewise with each model it fit.

◁

Based on the work of Tsay (1984), Paulsen (1984), and Nielsen (2001), these lag-order selection criteria can be used to determine the lag length of the VAR underlying a VECM. See [TS] **vec intro** for an example in which we use `varsoc` to choose the lag order for a VECM.

Saved Results

varsoc saves in r():

Scalars

r(N)	number of observations	r(mlag)	maximum lag order
r(tmax)	last time period in sample	r(N_gaps)	the number of gaps in
r(tmin)	first time period in sample		the sample

Macros

r(endog)	names of endogenous variables	r(exog)	names of exogenous variables
r(lutstats)	lutstats, if specified	r(rmlutstats)	rmlutstats, if specified
r(cns#)	the #th constraint		

Matrices

r(stats)	LL, LR, FPE, AIC, HQIC, SBIC, and p-values

Methods and Formulas

varsoc is implemented as an ado-file.

As shown by Hamilton (1994, 295–296), the log likelihood for a VAR(p) is

$$LL = \left(\frac{T}{2}\right)\left\{\ln\left(|\widehat{\boldsymbol{\Sigma}}^{-1}|\right) - K\ln(2\pi) - K\right\}$$

where T is the number of observations, K is the number of equations, and $\widehat{\boldsymbol{\Sigma}}$ is the maximum likelihood estimate of $E[\mathbf{u}_t\mathbf{u}_t']$, where $\mathbf{u}_t$ is the $K \times 1$ vector of disturbances. Because

$$\ln\left(|\widehat{\boldsymbol{\Sigma}}^{-1}|\right) = -\ln\left(|\widehat{\boldsymbol{\Sigma}}|\right)$$

the log likelihood can be rewritten as

$$LL = -\left(\frac{T}{2}\right)\left\{\ln\left(|\widehat{\boldsymbol{\Sigma}}|\right) + K\ln(2\pi) + K\right\}$$

Letting LL(j) be the value of the log likelihood with j lags yields the LR statistic for lag order j as

$$LR(j) = 2\left\{LL(j) - LL(j-1)\right\}$$

Model-order statistics

The formula for the FPE given in Lütkepohl (1993, 128) is

$$FPE = |\boldsymbol{\Sigma}_u|\left(\frac{T + Kp + 1}{T - Kp - 1}\right)^K$$

This formula, however, assumes that there is a constant in the model and that none of the variables are dropped due to collinearity. To deal with these problems, the FPE is implemented as

$$FPE = |\boldsymbol{\Sigma}_u|\left(\frac{T + \overline{m}}{T - \overline{m}}\right)^K$$

where $\overline{m}$ is the average number of parameters over the K equations. This implementation accounts for variables dropped due to collinearity.

By default, the AIC, SBIC, and HQIC are computed according to their standard definitions, which include the constant term from the log likelihood. That is,

$$\text{AIC} = -2\left(\frac{\text{LL}}{T}\right) + \frac{2t_p}{T}$$

$$\text{SBIC} = -2\left(\frac{\text{LL}}{T}\right) + \frac{\ln(T)}{T}t_p$$

$$\text{HQIC} = -2\left(\frac{\text{LL}}{T}\right) + \frac{2\ln\{\ln(T)\}}{T}t_p$$

where t_p is the total number of parameters in the model and LL is the log likelihood.

Lutstats

Lütkepohl (1993) advocates dropping the constant term from the log likelihood since it does not affect inference. The Lütkepohl versions of the information criteria are

$$\text{AIC} = \ln\left(|\mathbf{\Sigma}_u|\right) + \frac{2pK^2}{T}$$

$$\text{SBIC} = \ln\left(|\mathbf{\Sigma}_u|\right) + \frac{\ln(T)}{T}pK^2$$

$$\text{HQIC} = \ln\left(|\mathbf{\Sigma}_u|\right) + \frac{2\ln\{\ln(T)\}}{T}pK^2$$

References

Akaike, H. 1973. Information theory and an extension of the maximum likelihood principle. In *Second International Symposium on Information Theory*, ed. B. Petrov and F. Csaki, 267–281. Budapest: Académiai Kiadó.

Amemiya, T. 1985. *Advanced Econometrics*. Cambridge, MA: Harvard University Press.

Hamilton, J. D. 1994. *Time Series Analysis*. Princeton: Princeton University Press.

Lütkepohl, H. 1993. *Introduction to Multiple Time Series Analysis*. 2nd ed. New York: Springer.

Nielsen, B. 2001. Order determination in general vector autoregressions. Working Paper, Department of Economics, University of Oxford and Nuffield College.

Paulsen, J. 1984. Order determination of multivariate autoregressive time series with unit roots. *Journal of Time Series Analysis* 5(2): 115–127.

Tsay, R. S. 1984. Order selection in nonstationary autoregressive models. *The Annals of Statistics* 12(4): 1425–1433.

Also See

Complementary:	[TS] **var**, [TS] **var svar**, [TS] **varbasic**, [TS] **vec**
Background:	[TS] **var intro**, [TS] **vec intro**

Title

> **varstable** — Check the stability condition of VAR or SVAR estimates

Syntax

varstable $\left[\,,\ options\,\right]$

options	description
Main	
est̲imates(*estname*)	use previously saved results *estname*; default is to use active results
amat(*matrix_name*)	save the companion matrix as *matrix_name*
graph	graph eigenvalues of the companion matrix
dlabel	label eigenvalues with the distance from the unit circle
modlabel	label eigenvalues with the modulus
rl̲opts(*cline_options*)	affect rendition of reference unit circle
nog̲rid	suppress polar grid circles
pgrid($\left[\ldots\right]$) $\left[\ldots\right]$	specify radii and appearance of polar grid circles; see *Options* for details
Add plot	
addplot(*plot*)	add other plots to the generated graph
Y-Axis, X-Axis, Title, Caption, Legend, Overall	
twoway_options	any options other than by() documented in [G] ***twoway_options***

varstable can be used only after var or svar; see [TS] **var** and [TS] **var svar**.

Description

varstable checks the eigenvalue stability condition after estimating the parameters of a vector autoregression using var or svar.

Options

> Main

estimates(*estname*) requests that varstable use the previously obtained set of var estimates saved as *estname*. By default, varstable uses the active estimation results. See [R] **estimates** for information about saving and restoring estimation results.

amat(*matrix_name*) specifies a valid Stata matrix name by which the companion matrix $\mathbf{A}$ can be saved (see *Methods and Formulas* for the definition of the matrix $\mathbf{A}$). The default is not to save the $\mathbf{A}$ matrix.

graph causes varstable to draw a graph of the eigenvalues of the companion matrix.

dlabel labels each eigenvalue with its distance from the unit circle. dlabel can only be specified with graph and cannot be specified with modlabel.

modlabel labels the eigenvalues with their moduli. modlabel can only be specified with graph and cannot be specified with dlabel.

rlopts(*cline_options*) affect the rendition of the reference unit circle; see [G] ***cline_options***. rlopts() can only be specified with graph.

nogrid suppresses the polar grid circles. nogrid can only be specified with graph.

pgrid([*numlist*] [,*line_options*]) determines the radii and appearance of the polar grid circles. By default, the graph includes nine polar grid circles with radii .1, .2, ..., .9 that have the grid linestyle. The *numlist* specifies the radii for the polar grid circles. The *line_options* determine the appearance of the polar grid circles; see [G] ***line_options***. Since the pgrid() option can be repeated, circles with different radii can have distinct appearances. pgrid() can only be specified with graph.

⌐ Add plot ⌐

addplot(*plot*) adds specified plots to the generated graph. See [G] ***addplot_option***. addplot() can only be specified with graph.

⌐ Y-Axis, X-Axis, Title, Caption, Legend, Overall ⌐

twoway_options are any of the options documented in [G] ***twoway_options***, except by(). These include options for titling the graph (see [G] ***title_options***) and options for saving the graph to disk (see [G] ***saving_option***). *twoway_options* can only be specified with graph.

Remarks

Inference after var and svar requires that variables be covariance stationary. The variables in $\mathbf{y}_t$ are covariance stationary if their first two moments exist and are independent of time. More explicitly, a variable y_t is covariance stationary if

1. $E[y_t]$ is finite and independent of t.

2. $\text{Var}[y_t]$ is finite and independent of t

3. $\text{Cov}[y_t, y_s]$ is a finite function of $|t - s|$, but not of t or s alone.

Interpretation of VAR models, however, requires that an even stricter stability condition be met. If a VAR is stable, it is invertible and has an infinite-order vector moving-average representation. If the VAR is stable, impulse–response functions and forecast-error variance decompositions have known interpretations.

Lütkepohl (1993) and Hamilton (1994) both show that if the modulus of each eigenvalue of the matrix $\mathbf{A}$ is strictly less than one, the estimated VAR is stable (see *Methods and Formulas* for the definition of the matrix $\mathbf{A}$).

▷ Example 1

After fitting a VAR with var, we can use varstable to check the stability condition. Using the same VAR model that was used in [TS] **var**, we demonstrate the use of varstable.

```
. use http://www.stata-press.com/data/r9/lutkepohl
(Quarterly SA West German macro data, Bil DM, from Lutkepohl 1993 Table E.1)

. var dlinvestment dlincome dlconsumption if qtr>=q(1961q2) &
> qtr <= q(1978q4), lags(1/2)
  (output omitted )
```

```
. varstable, graph
Eigenvalue stability condition
```

Eigenvalue	Modulus
.5456253	.545625
-.3785754 + .3853982i	.540232
-.3785754 - .3853982i	.540232
-.0643276 + .4595944i	.464074
-.0643276 - .4595944i	.464074
-.3698058	.369806

```
All the eigenvalues lie inside the unit circle
VAR satisfies stability condition
```

Since the modulus of each eigenvalue is strictly less than 1, the estimates satisfy the eigenvalue stability condition.

Specifying the `graph` option produced a graph of the eigenvalues with the real components on the x-axis and the complex components on the y-axis. The graph below indicates visually that these eigenvalues are well inside the unit circle.

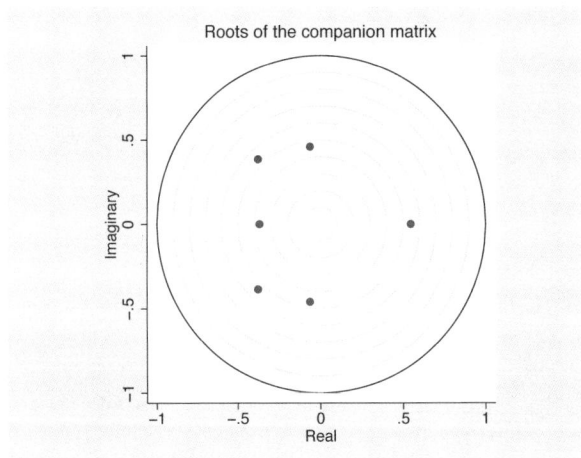

▷ Example 2

This example illustrates two other features of the `varstable` command. First, `varstable` can check the stability of the estimates of the VAR underlying an SVAR fitted by `var svar`. Second, `varstable` can check the stability of any previously stored `var` or `var svar` estimates.

We begin by `quietly` refitting the previous VAR and storing the results as `var1`. Since this is the same VAR that was fitted in the previous example, the stability results should be identical.

```
. quietly var dlinvestment dlincome dlconsumption if qtr>=q(1961q2) &
> qtr <= q(1978q4), lags(1/2)
. estimates store var1
```

Now we use `svar` to fit an SVAR with a different underlying VAR and check the estimates of that underlying VAR for stability.

```
. matrix A = (.,0\.,.)
. matrix B = I(2)
. quietly svar d.lincome d.lconsumption, aeq(A) beq(B)
. varstable
```

```
Eigenvalue stability condition
```

Eigenvalue		Modulus
.548711		.548711
-.2979493 +	.4328013i	.525443
-.2979493 -	.4328013i	.525443
-.3570825		.357082

```
All the eigenvalues lie inside the unit circle
VAR satisfies stability condition
```

The `estimates()` option allows us to check the stability of the `var` results stored as `var1`.

```
. varstable, est(var1)
```

```
Eigenvalue stability condition
```

Eigenvalue		Modulus
.5456253		.545625
-.3785754 +	.3853982i	.540232
-.3785754 -	.3853982i	.540232
-.0643276 +	.4595944i	.464074
-.0643276 -	.4595944i	.464074
-.3698058		.369806

```
All the eigenvalues lie inside the unit circle
VAR satisfies stability condition
```

The results are identical to those obtained in the previous example, confirming that we were checking the results in `var1`.

◁

Saved Results

`varstable` saves in `r()`:

Matrices

`r(Re)`	real part of the eigenvalues of A
`r(Im)`	imaginary part of the eigenvalues of A
`r(Modulus)`	modulus of the eigenvalues of A

Methods and Formulas

`varstable` is implemented as an ado-file.

`varstable` forms the companion matrix

$$
\mathbf{A} = \begin{pmatrix}
\mathbf{A}_1 & \mathbf{A}_2 & \ldots & \mathbf{A}_{p-1} & \mathbf{A}_p \\
\mathbf{I} & \mathbf{0} & \ldots & \mathbf{0} & \mathbf{0} \\
\mathbf{0} & \mathbf{I} & \ldots & \mathbf{0} & \mathbf{0} \\
\vdots & \vdots & \ddots & \vdots & \vdots \\
\mathbf{0} & \mathbf{0} & \ldots & \mathbf{I} & \mathbf{0}
\end{pmatrix}
$$

and obtains its eigenvalues using `matrix eigenvalues`. The modulus of the complex eigenvalue $r + ci$ is $\sqrt{r^2 + c^2}$. As shown by Lütkepohl (1993) and Hamilton (1994), the VAR is stable if the modulus of each eigenvalue of $\mathbf{A}$ is strictly less than 1.

References

Hamilton, J. D. 1994. *Time Series Analysis*. Princeton: Princeton University Press.

Lütkepohl, H. 1993. *Introduction to Multiple Time Series Analysis*. 2nd ed. New York: Springer.

Also See

Complementary:	[TS] **var**, [TS] **var svar**, [TS] **varbasic**
Background:	[TS] **var intro**

Title

> **varwle** — Obtain Wald lag-exclusion statistics after var or svar

Syntax

varwle [, est̲imates(*estname*) sep̲arator(#)]

varwle can be used only after var or svar; see [TS] **var** and [TS] **var svar**.

Description

varwle reports Wald tests the hypothesis that the endogenous variables at a given lag are jointly zero for each equation and for all equations jointly.

Options

est̲imates(*estname*) requests that varwle use the previously obtained set of var or svar estimates saved as *estname*. By default, varwle uses the active estimation results. See [R] **estimates** for information about saving and restoring estimation results.

sep̲arator(#) specifies how often separator lines should be drawn between rows. By default, separator lines do not appear. For example, separator(1) would draw a line between each row, separator(2) between every other row, and so on.

Remarks

After fitting a VAR, one hypothesis of interest is that all the endogenous variables at a given lag are jointly zero. varwle reports Wald tests of this hypothesis for each equation and for all equations jointly. varwle uses the estimation results from a previously fitted var or svar. By default, varwle uses the active estimation results, but you may also use a stored set of estimates by specifying the estimates() option.

If the VAR was fitted with the small option, varwle also presents small-sample F statistics; otherwise, varwle presents large-sample chi-squared statistics.

▷ Example 1: After var

We analyze the model using the German data described in [TS] **var** using varwle.

```
. use http://www.stata-press.com/data/r9/lutkepohl
(Quarterly SA West German macro data, Bil DM, from Lutkepohl 1993 Table E.1)
. var dlinvestment dlincome dlconsumption if qtr <= q(1978q4), lags(1/2) dfk small
```
(output omitted)
```
. varwle
```

Equation: dlinvestment

lag	F	df	df_r	Prob > F
1	2.64902	3	66	0.0560
2	1.25799	3	66	0.2960

Equation: dlincome

lag	F	df	df_r	Prob > F
1	2.19276	3	66	0.0971
2	.907499	3	66	0.4423

Equation: dlconsumption

lag	F	df	df_r	Prob > F
1	1.80804	3	66	0.1543
2	5.57645	3	66	0.0018

Equation: All

lag	F	df	df_r	Prob > F
1	3.78884	9	66	0.0007
2	2.96811	9	66	0.0050

Since the VAR was fitted with the `dfk` and `small` options, `varwle` used the small-sample estimator of $\widehat{\Sigma}$ in constructing the VCE, producing an F statistic. It appears that the first two equations have a different lag structure than the third. In the first two equations, we cannot reject the null hypothesis that all three endogenous variables have zero coefficients at the second lag. The hypothesis that all three endogenous variables have zero coefficients at the first lag can be rejected at the 10% level for each of the first two equations. In contrast, in the third equation, the coefficients on the second lag of the endogenous variables are jointly significant, but not those on the first lag. However, we strongly reject the hypothesis that the coefficients on the first lag of the endogenous variables are zero in all three equations jointly. Similarly, we can also strongly reject the hypothesis that the coefficients on the second lag of the endogenous variables are zero in all three equations jointly.

If we believe these results strongly enough, we might want to refit the original VAR, placing some constraints on the coefficients. See [TS] **var** for details on how to fit VAR models with constraints.

◁

▷ Example 2: After svar

Here we fit a simple SVAR and then run `varwle`:

```
. matrix a = (.,0\.,.)
. matrix b = I(2)
```

```
. svar dlincome dlconsumption, aeq(a) beq(b)
Estimating short-run parameters

Iteration 0:    log likelihood = -159.21683
Iteration 1:    log likelihood =  490.92264
Iteration 2:    log likelihood =  528.66126
Iteration 3:    log likelihood =  573.96363
Iteration 4:    log likelihood =  578.05136
Iteration 5:    log likelihood =  578.27633
Iteration 6:    log likelihood =  578.27699
Iteration 7:    log likelihood =  578.27699

Structural vector autoregression

Constraints:
 ( 1)   [a_1_2]_cons = 0
 ( 2)   [b_1_1]_cons = 1
 ( 3)   [b_1_2]_cons = 0
 ( 4)   [b_2_1]_cons = 0
 ( 5)   [b_2_2]_cons = 1

Sample:   1960q4   1982q4                  No. of obs        =         89
Exactly identified model                   Log likelihood  =    578.277
```

	Coef.	Std. Err.	t	P>|z|	[95% Conf.	Interval]
/a_1_1	89.72411	6.725107	13.34	0.000	76.54315	102.9051
/a_2_1	-64.73622	10.67698	-6.06	0.000	-85.66271	-43.80973
/a_1_2	0	.	.	.	.	.
/a_2_2	126.2964	9.466318	13.34	0.000	107.7428	144.8501
/b_1_1	1	.	.	.	.	.
/b_2_1	0	.	.	.	.	.
/b_1_2	0	.	.	.	.	.
/b_2_2	1	.	.	.	.	.

The output table from var svar provides information about the estimates of the parameters in the **A** and **B** matrices in the structural VAR. But, as discussed in [TS] **var svar**, an SVAR model builds on an underlying VAR. When varwle uses the estimation results produced by svar, it performs Wald lag-exclusion tests on the underlying VAR model. Next we run varwle on these svar results.

```
. varwle
```

Equation: dlincome

lag	chi2	df	Prob > chi2
1	6.88775	2	0.032
2	1.873546	2	0.392

Equation: dlconsumption

lag	chi2	df	Prob > chi2
1	9.938547	2	0.007
2	13.89996	2	0.001

Equation: All

lag	chi2	df	Prob > chi2
1	34.54276	4	0.000
2	19.44093	4	0.001

Now we `quietly` fit the underlying VAR with two lags and apply `varwle` to these results.

```
. qui var dlincome dlconsumption
. varwle
```

Equation: dlincome

lag	chi2	df	Prob > chi2
1	6.88775	2	0.032
2	1.873546	2	0.392

Equation: dlconsumption

lag	chi2	df	Prob > chi2
1	9.938547	2	0.007
2	13.89996	2	0.001

Equation: All

lag	chi2	df	Prob > chi2
1	34.54276	4	0.000
2	19.44093	4	0.001

Since `varwle` produces the same results in these two cases, we can conclude that when `varwle` is applied to `svar` results, it performs Wald lag-exclusion tests on the underlying VAR.

◁

Saved Results

`varwle` saves in `r()`:

Matrices
if e(small)==""
    r(chi2) χ^2 test statistics
    r(df)   degrees of freedom
    r(p) p-values
if e(small)!=""
    r(F) F test statistics
    r(df_r)   numerator degrees of freedom
    r(df)   denominator degree of freedom
    r(p) p-values

Methods and Formulas

`varwle` is implemented as an ado-file.

`varwle` uses `test` to obtain Wald statistics of the hypotheses that all the endogenous variables at a given lag are jointly zero for each equation and for all equations jointly. Like the `test` command, `varwle` uses estimation results saved by `var` or `var svar` to determine whether to calculate and report small-sample F statistics or large-sample chi-squared statistics.

Abraham Wald (1902–1950) was born in Cluj, in what is now Romania. He studied mathematics at the University of Vienna, publishing at first on geometry, but then became interested in economics and econometrics. He moved to the United States in 1938 and later joined the faculty at Columbia. His major contributions to statistics include work in decision theory, optimal sequential sampling, large-sample distributions of likelihood-ratio tests, and nonparametric inference. Wald was killed in a plane crash in India.

References

Amisano, G. and C. Giannini. 1997. *Topics in Structural VAR Econometrics.* 2nd ed. Heidelberg: Springer.

Hamilton, J. D. 1994. *Time Series Analysis.* Princeton: Princeton University Press.

Lütkepohl, H. 1993. *Introduction to Multiple Time Series Analysis.* 2nd ed. New York: Springer.

Mangel, M. and F. J. Samaniego. 1984. Abraham Wald's work on aircraft survivability. *Journal of the American Statistical Association* 79: 259–271.

Wolfowitz, J. 1952. Abraham Wald, 1902–1950. *Annals of Mathematical Statistics* 23: 1–13 (and other papers in same issue).

Also See

Complementary:	[TS] **var**, [TS] **var svar**, [TS] **varbasic**
Background:	[TS] **var intro**

Title

> **vec intro** — Introduction to vector error-correction models

Description

Stata has a suite of commands for fitting, forecasting, interpreting, and performing inference on vector error-correction models with cointegrating variables (VECMs). After fitting a VECM, the `irf` commands can be used to obtain impulse–response functions (IRFs) and forecast-error variance decompositions (FEVDs). The table below describes the available commands.

Fitting a VECM

vec	[TS] **vec**	Fit vector error-correction models

Model diagnostics and inference

vecrank	[TS] **vecrank**	Estimate the cointegrating rank using Johansen's framework
veclmar	[TS] **veclmar**	Obtain LM statistics for residual autocorrelation after vec
vecnorm	[TS] **vecnorm**	Test for normally distributed disturbances after vec
vecstable	[TS] **vecstable**	Check the stability condition of VECM estimates
varsoc	[TS] **varsoc**	Obtain lag-order selection statistics for VARs and VECMs

Forecasting from a VECM

fcast compute	[TS] **fcast compute**	Compute dynamic forecasts of dependent variables after var, svar, or vec
fcast graph	[TS] **fcast graph**	Graph forecasts of variables computed by fcast compute

Working with IRFs and FEVDs

irf	[TS] **irf**	Create and analyze IRFs and FEVDs

This manual entry provides an overview of the commands for VECMs; provides an introduction to integration, cointegration, and estimation, inference, and interpretation of VECM models; and gives an example of how to use Stata's vec commands.

Remarks

`vec` estimates the parameters of cointegrating VECMs. You may specify any of the five trend specifications in Johansen (1995, chapter 5.7). By default, identification is obtained via the Johansen normalization, but `vec` allows you to obtain identification by placing your own constraints on the parameters of the cointegrating vectors. You may also put additional restrictions on the adjustment coefficients.

`vecrank` is the command for determining the number of cointegrating equations. `vecrank` implements Johansen's multiple trace test procedure, the maximum eigenvalue test, and a method based on minimizing either of two different information criteria.

Since Nielsen (2001) has shown that the methods implemented in `varsoc` can be used to choose the order of the autoregressive process, no separate `vec` command is needed; you can simply use `varsoc`. `veclmar` tests that the residuals do not have any serial correlation, and `vecnorm` tests that they are normally distributed.

All the `irf` routines described in [TS] **irf** are available for estimating, interpreting, and managing estimated IRFs and FEVDs for VECMs.

Remarks are presented under the headings

> An introduction to cointegrating VECMs
> > What is cointegration?
> > The multivariate VECM specification
> > Trends in the Johansen VECM framework
> VECM estimation in Stata
> > Selecting the number of lags
> > Testing for cointegration
> > Fitting a VECM
> > Fitting VECMs with Johansen's normalization
> > Postestimation specification testing
> > Impulse–response functions for VECMs
> > Forecasting with VECMs

An introduction to cointegrating VECMs

This section provides a brief introduction to integration, cointegration, and cointegrated vector error-correction models. For more details about these topics, see Hamilton (1994), Johansen (1995), Lütkepohl (1993), and Watson (1994).

What is cointegration?

Standard regression techniques, such as ordinary least squares (OLS), require that the variables be covariance stationary. A variable is covariance stationary if its mean and all its autocovariances are finite and do not change over time. Cointegration analysis provides a framework for estimation, inference, and interpretation when the variables are not covariance stationary.

Instead of being covariance stationary, many economic time series appear to be "first-difference stationary". This means that the level of a time series is not stationary but its first-difference is. First-difference stationary processes are also known as integrated processes of order 1, or I(1) processes. Covariance-stationary processes are I(0). In general, a process whose dth difference is stationary is an integrated process of order d, or I(d).

The canonical example of a first-difference stationary process is the random walk. This is a variable x_t that can be written as

$$x_t = x_{t-1} + \epsilon_t \tag{1}$$

where the ϵ_t are independently and identically distributed (i.i.d.) with mean zero and a finite variance σ^2. Although $E[x_t] = 0$ for all t, $\mathrm{Var}[x_t] = T\sigma^2$ is not time invariant, so x_t is not covariance stationary. Because $\Delta x_t = x_t - x_{t-1} = \epsilon_t$ and ϵ_t is covariance stationary, x_t is first-difference stationary.

These concepts are important because, while conventional estimators are well behaved when applied to covariance-stationary data, they have nonstandard asymptotic distributions and different rates of convergence when applied to I(1) processes. To illustrate, consider several variants of the model

$$y_t = a + bx_t + e_t \tag{2}$$

Throughout the discussion, we maintain the assumption that $E[e_t] = 0$.

If both y_t and x_t are covariance-stationary processes, e_t must also be covariance stationary. As long as $E[x_t e_t] = 0$, we can consistently estimate the parameters a and b using OLS. Furthermore, the distribution of the OLS estimator converges to a normal distribution centered at the true value as the sample size grows.

If y_t and x_t are independent random walks and $b = 0$, there is no relationship between y_t and x_t, and (2) is called a spurious regression. Granger and Newbold (1974) performed Monte Carlo experiments and showed that the usual t-statistics from OLS regression provide spurious results: given a large enough dataset, we can almost always reject the null hypothesis of the test that $b = 0$ even though b is in fact zero. In this case, the OLS estimate does not converge to any well-defined population parameter.

Phillips (1986) later provided the asymptotic theory that explained the Granger and Newbold (1974) results. He showed that the random walks y_t and x_t are first-difference stationary processes and that the OLS estimator does not have its usual asymptotic properties when the variables are first-difference stationary.

Since Δy_t and Δx_t are covariance stationary, a simple regression of Δy_t on Δx_t appears to be a viable alternative. However, if y_t and x_t cointegrate, as defined below, the simple regression of Δy_t on Δx_t is misspecified.

If y_t and x_t are I(1) and $b \neq 0$, e_t could be either I(0) or I(1). Phillips and Durlauf (1986) have derived the asymptotic theory for the OLS estimator when e_t is I(1), though it has not been widely used in applied work. More interesting is the case in which $e_t = y_t - a - bx_t$ is I(0). In that case, y_t and x_t are said to be cointegrated. Two variables are cointegrated if each is an I(1) process but a linear combination of them is an I(0) process.

It is not possible for y_t to be a random walk and x_t and e_t to be covariance stationary. As Granger (1981) pointed out, because a random walk cannot be equal to a covariance-stationary process, the equation does not "balance". An equation balances when the processes on each side of the equals sign are of the same order of integration. Before attacking any applied problem with integrated variables, make sure that the equation balances before proceeding.

An example from Engle and Granger (1987) provides more intuition. Redefine y_t and x_t to be

$$y_t + \beta x_t = c_t, \qquad c_t = c_{t-1} + \xi_t \tag{3}$$
$$y_t + \alpha x_t = \nu_t, \qquad \nu_t = \rho \nu_{t-1} + \zeta_t, \quad |\rho| < 1 \tag{4}$$

where ξ_t and ζ_t are i.i.d. disturbances over time that are correlated with each other. Since ϵ_t is I(1), (3) and (4) imply that both x_t and y_t are I(1). The condition that $|\rho| < 1$ implies that ν_t and $y_t + \alpha x_t$ are I(0). Thus y_t and x_t cointegrate, and $(1, \alpha)$ is the cointegrating vector.

Using a bit of algebra, we can rewrite (3) and (4) as

$$\Delta y_t = \beta \delta z_{t-1} + \eta_{1t} \tag{5}$$
$$\Delta x_t = -\delta z_{t-1} + \eta_{2t} \tag{6}$$

where $\delta = (1 - \rho)/(\alpha - \beta)$, $z_t = y_t + \alpha x_t$, and η_{1t} and η_{2t} are distinct, stationary, linear combinations of ξ_t and ζ_t. This representation is known as the vector error-correction model (VECM). One can think of $z_t = 0$ as being the point at which y_t and x_t are in equilibrium. The coefficients on z_{t-1} describe how y_t and x_t adjust to z_{t-1} being nonzero, or out of equilibrium. z_t is the "error" in the system, and (5) and (6) describe the way in which the system adjusts or corrects back to the equilibrium. As ρ goes to 1, the system degenerates into a pair of correlated random walks. The VECM parameterization highlights this point, since $\delta \to 0$ as $\rho \to 1$.

If we knew α, we would know z_t, and we could work with the stationary system of (5) and (6). While knowing α seems silly, we can conduct much of the analysis as if we knew α because there is an estimator for the cointegrating parameter α that converges to its true value at a faster rate than the estimator for the adjustment parameters β and δ.

The definition of a bivariate cointegrating relation requires simply that there exist a linear combination of the I(1) variables that is I(0). In other words, if y_t and x_t are I(1) and there are two finite real numbers $a \neq 0$ and $b \neq 0$, such that $ay_t + bx_t$ is I(0), then y_t and x_t are cointegrated. Although there are two parameters, a and b, only one will be identifiable because if $ay_t + bx_t$ is I(0), so is $cay_t + cbx_t$ for any finite, nonzero, real number c. It is relatively simple to obtain identification in the bivariate case. Note that the coefficient on y_t in (4) is unity. This natural construction of the model placed the necessary identification restriction on the cointegrating vector. As we discuss below, identification in the multivariate case is more involved.

If $\mathbf{y}_t$ is a $K \times 1$ vector of I(1) variables and there exists a vector $\boldsymbol{\beta}$, such that $\boldsymbol{\beta}\mathbf{y}_t$ is a vector of I(0) variables, then $\mathbf{y}_t$ is said to be cointegrating of order (1,0) with cointegrating vector $\boldsymbol{\beta}$. We say that the parameters in $\boldsymbol{\beta}$ are the parameters in the cointegrating equation. For a vector of length K, there may be at most $K - 1$ distinct cointegrating vectors. Engle and Granger (1987) provide a more general definition of cointegration, but this one is sufficient for our purposes.

The multivariate VECM specification

In practice, most empirical applications analyze multivariate systems, so the rest of our discussion focuses on that case. Consider a VAR with p lags

$$\mathbf{y}_t = \mathbf{v} + \mathbf{A}_1 \mathbf{y}_{t-1} + \mathbf{A}_2 \mathbf{y}_{t-2} + \cdots + \mathbf{A}_p \mathbf{y}_{t-p} + \boldsymbol{\epsilon}_t \tag{7}$$

where $\mathbf{y}_t$ is a $K \times 1$ vector of variables, $\mathbf{v}$ is a $K \times 1$ vector of parameters, $\mathbf{A}_1$ through $\mathbf{A}_p$ are $K \times K$ matrices of parameters, and $\boldsymbol{\epsilon}_t$ is a $K \times 1$ vector of disturbances. $\boldsymbol{\epsilon}_t$ has mean $\mathbf{0}$, has covariance matrix $\boldsymbol{\Sigma}$, and is i.i.d. normal over time. Any VAR(p) can be rewritten as a VECM. Using some algebra, (7) can be rewritten in VECM form as

$$\Delta \mathbf{y}_t = \mathbf{v} + \boldsymbol{\Pi} \mathbf{y}_{t-1} + \sum_{i=1}^{p-1} \boldsymbol{\Gamma}_i \Delta \mathbf{y}_{t-i} + \boldsymbol{\epsilon}_t \tag{8}$$

where $\boldsymbol{\Pi} = \sum_{j=1}^{j=p} \mathbf{A}_j - \mathbf{I}_k$ and $\boldsymbol{\Gamma}_i = -\sum_{j=i+1}^{j=p} \mathbf{A}_j$. The $\mathbf{v}$ and $\boldsymbol{\epsilon}_t$ in (7) and (8) are identical.

Engle and Granger (1987) show that if the variables $\mathbf{y}_t$ are I(1) the matrix $\boldsymbol{\Pi}$ in (8) has rank $0 \leq r < K$, where r is the number of linearly independent cointegrating vectors. If the variables cointegrate, $0 < r < K$ and (8) shows that a VAR in first-differences is misspecified because it omits the lagged level term $\boldsymbol{\Pi}\mathbf{y}_{t-1}$.

Assume that $\boldsymbol{\Pi}$ has reduced rank $0 < r < K$ so that it can be expressed as $\boldsymbol{\Pi} = \boldsymbol{\alpha}\boldsymbol{\beta}'$, where $\boldsymbol{\alpha}$ and $\boldsymbol{\beta}$ are both $K \times r$ matrices of rank r. Without further restrictions, the cointegrating vectors are not identified: the parameters $(\boldsymbol{\alpha}, \boldsymbol{\beta})$ are indistinguishable from the parameters $(\boldsymbol{\alpha}\mathbf{Q}, \boldsymbol{\beta}\mathbf{Q}^{-1\prime})$ for any $r \times r$ nonsingular matrix $\mathbf{Q}$. Since only the rank of $\boldsymbol{\Pi}$ is identified, the VECM is said to identify the rank of the cointegrating space, or equivalently, the number of cointegrating vectors. In practice, the estimation of the parameters of a VECM requires at least r^2 identification restrictions. Stata's vec command can apply the conventional Johansen restrictions discussed below or use constraints that the user supplies.

The VECM in (8) also nests two important special cases. If the variables in $\mathbf{y}_t$ are I(1) but not cointegrated, $\boldsymbol{\Pi}$ is a matrix of zeros, and thus has rank 0. If all the variables are I(0), $\boldsymbol{\Pi}$ has full rank K.

There are several different frameworks for estimation and inference in cointegrating systems. While the methods in Stata are based on the maximum likelihood (ML) methods developed by Johansen (1988, 1991, 1995), other useful frameworks have been developed by Park and Phillips (1988a, 1988b), Sims, Stock, and Watson (1990), Stock (1987), and Stock and Watson (1998), among others. Note that the ML framework developed by Johansen was independently developed by Ahn and Reinsel (1990). Maddala and Kim (1998) and Watson (1994) survey all these methods. The cointegration methods in Stata are based on Johansen's maximum likelihood framework because it has been found to be particularly useful in several comparative studies, including Gonzalo (1994) and Hubrich, Lütkepohl, and Saikkonen (2001).

Trends in the Johansen VECM framework

Deterministic trends in a cointegrating VECM can stem from two distinct sources; the mean of the cointegrating relationship and the mean of the differenced series. Allowing for a constant and a linear trend and assuming that there are r cointegrating relations, the VECM in (8) can be rewritten as

$$\Delta \mathbf{y}_t = \alpha \beta \mathbf{y}_{t-1} + \sum_{i=1}^{p-1} \mathbf{\Gamma}_i \Delta \mathbf{y}_{t-i} + \mathbf{v} + \delta t + \epsilon_t \tag{9}$$

where δ is a $K \times 1$ vector of parameters. Because (9) models the differences of the data, the constant implies a linear time trend in the levels, and the time trend δt implies a quadratic time trend in the levels of the data. In many cases, we may want to include a constant or a linear time trend for the differences without allowing for the higher order trend that is implied for the levels of the data. VECMs exploit the properties of the matrix α to achieve this flexibility.

Since α is a $K \times r$ rank matrix, the deterministic components in (9) can be rewritten as

$$\mathbf{v} = \alpha \mu + \gamma \tag{10a}$$
$$\delta t = \alpha \rho t + \tau t \tag{10b}$$

where μ and ρ are $r \times 1$ vectors of parameters and γ and τ are $K \times 1$ vectors of parameters. γ is orthogonal to $\alpha \mu$, and τ is orthogonal to $\alpha \rho$; i.e., $\gamma' \alpha \mu = \mathbf{0}$ and $\tau' \alpha \rho = \mathbf{0}$, allowing us to rewrite (9) as

$$\Delta \mathbf{y}_t = \alpha(\beta \mathbf{y}_{t-1} + \mu + \rho t) + \sum_{i=1}^{p-1} \mathbf{\Gamma}_i \Delta \mathbf{y}_{t-i} + \gamma + \tau t + \epsilon_t \tag{11}$$

Placing restrictions on the trend terms in (11) yields five cases.

CASE 1: Unrestricted trend

> If no restrictions are placed on the trend parameters, (11) implies that there are quadratic trends in the levels of the variables and that the cointegrating equations are stationary around time trends (trend stationary).

CASE 2: Restricted trend, $\tau = \mathbf{0}$

> By setting $\tau = \mathbf{0}$, we assume that the trends in the levels of the data are linear but not quadratic. This specification allows the cointegrating equations to be trend stationary.

CASE 3: Unrestricted constant, $\tau = \mathbf{0}$ and $\rho = \mathbf{0}$

> By setting $\tau = \mathbf{0}$ and $\rho = \mathbf{0}$, we exclude the possibility that the levels of the data have quadratic trends, and we restrict the cointegrating equations to be stationary around constant means. Since γ is not restricted to zero, this specification still puts a linear time trend in the levels of the data.

CASE 4: Restricted constant, $\tau = 0$, $\rho = 0$, and $\gamma = 0$

By adding the restriction that $\gamma = 0$, we assume there are no linear time trends in the levels of the data. This specification allows the cointegrating equations to be stationary around a constant mean, but it does not allow any other trends or constant terms.

CASE 5: No trend, $\tau = 0$, $\rho = 0$, $\gamma = 0$, and $\mu = 0$

This specification assumes that there are no nonzero means or trends. It also assumes that the cointegrating equations are stationary with means of zero and that the differences and the levels of the data have means of zero.

This flexibility does come at a price. Below we discuss testing procedures for determining the number of cointegrating equations. The asymptotic distribution of the LR for hypotheses about r changes with the trend specification, so we must first specify a trend specification. A combination of theory and graphical analysis will aid in specifying the trend before proceeding with the analysis.

VECM estimation in Stata

We provide an overview of the `vec` commands in Stata through the use of an extended example. We have monthly data on the average selling prices of houses in four cities in Texas: Austin, Dallas, Houston, and San Antonio. In the dataset, these average housing prices are contained in the variables `austin`, `dallas`, `houston`, and `sa`. The series begin in January of 1990 and go through December 2003, for a total of 168 observations. The following graph depicts our data.

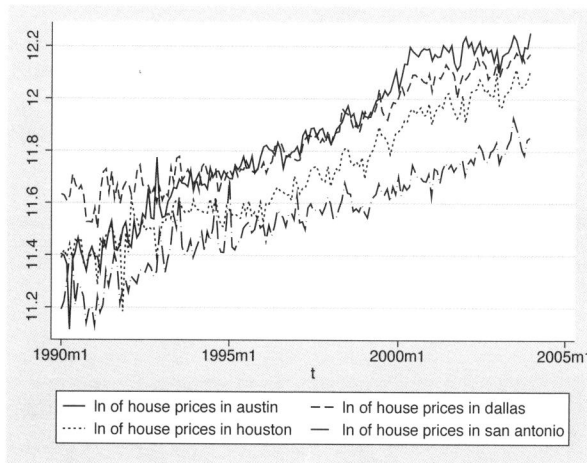

The plots on the graph indicate that all the series are trending and potential I(1) processes. In a competitive market, the current and past prices contain all the information available, so tomorrow's price will be a random walk from today's price. Some researchers may opt to use [TS] **dfgls** to investigate the presence of a unit root in each series, but the test for cointegration we use includes the case in which all the variables are stationary, so we defer formal testing until we test for cointegration. The time trends in the data appear to be approximately linear, so we will specify `trend(constant)` when modeling these series, which is the default with `vec`.

The next graph shows just Dallas' and Houston's data, so we can more carefully examine their relationship.

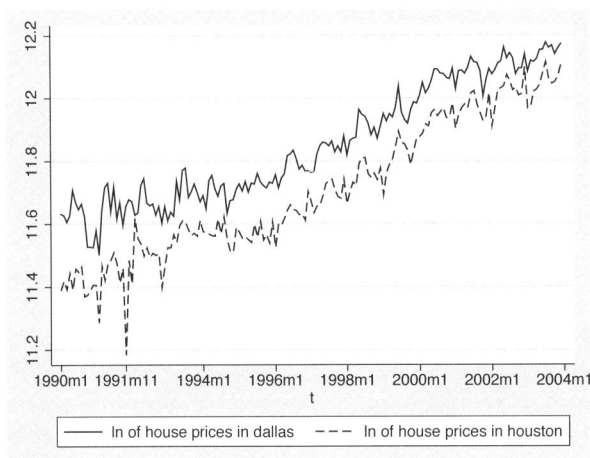

Except for the crash at the end of 1991, housing prices in Dallas and Houston appear closely related. Although average prices in the two cities will differ due to resource variations and other factors, if the housing markets become too dissimilar, people and businesses will migrate, bringing the average housing prices back toward each other. We therefore expect the series of average housing prices in Houston to be cointegrated with the series of average housing prices in Dallas.

Selecting the number of lags

To test for cointegration or fit cointegrating VECMs, we must specify how many lags to include. Building on the work of Tsay (1984) and Paulsen (1984), Nielsen (2001) has shown that the methods implemented in varsoc can be used to determine the lag order for a VAR model with I(1) variables. As can be seen from (9), the order of the corresponding VECM is always one less than the VAR. vec makes this adjustment automatically, so we will always refer to the order of the underlying VAR. The output below uses varsoc to determine the lag order of the VAR of the average housing prices in Dallas and Houston.

```
. use http://www.stata-press.com/data/r9/txhprice, clear
. varsoc dallas houston
   Selection order criteria
   Sample:  1990m5   2003m12                      Number of obs      =       164
```

lag	LL	LR	df	p	FPE	AIC	HQIC	SBIC
0	299.525				.000091	-3.62835	-3.61301	-3.59055
1	577.483	555.92	4	0.000	3.2e-06	-6.9693	-6.92326	-6.85589
2	590.978	26.991*	4	0.000	2.9e-06*	-7.0851*	-7.00837*	-6.89608*
3	593.437	4.918	4	0.296	2.9e-06	-7.06631	-6.95888	-6.80168
4	596.364	5.8532	4	0.210	3.0e-06	-7.05322	-6.9151	-6.71299

```
   Endogenous:   dallas houston
   Exogenous:    _cons
```

We will use two lags for this bivariate model because the HQIC method, SBIC method, and sequential likelihood-ratio (LR) test all chose two lags, as indicated by the "*" in the output.

The reader can verify that when all four cities' data are used, the LR test selects three lags, the HQIC method selects two lags, and the SBIC method selects one lag. We will use three lags in our four-variable model.

Testing for cointegration

The tests for cointegration implemented in `vecrank` are based on Johansen's method. If the log likelihood of the unconstrained model that includes the cointegrating equations is significantly different from the log likelihood of the constrained model that does not include the cointegrating equations, we reject the null hypothesis of no cointegration.

Here we use `vecrank` to determine the number of cointegrating equations:

```
. vecrank dallas houston
                      Johansen tests for cointegration
Trend: constant                                      Number of obs =     166
Sample:    1990m3  2003m12                                    Lags =       2
                                                         5%
maximum                                      trace   critical
   rank     parms       LL      eigenvalue  statistic   value
      0         6    576.26444        .      46.8252    15.41
      1         9    599.58781    0.24498    0.1785*     3.76
      2        10    599.67706    0.00107
```

Besides presenting information about the sample size and time span, the header indicates that test statistics are based on a model with two lags and a constant trend. The body of the table presents test statistics and their critical values for the null hypotheses of no cointegration (line 1) and one or fewer cointegrating equations (line 2). The eigenvalue shown on the last line is used to compute the trace statistic in the line above it. Johansen's testing procedure starts with the test for zero cointegrating equations (a maximum rank of zero) and then accepts the first null hypothesis that is not rejected.

In the output above, we strongly reject the null hypothesis of no cointegration and fail to reject the null hypothesis of at most one cointegrating equation. Thus we accept the null hypothesis that there is a single cointegrating equation in the bivariate model.

Using all four series and a model with three lags, we find there are two cointegrating relationships.

```
. vecrank austin dallas houston sa, lag(3)
                      Johansen tests for cointegration
Trend: constant                                      Number of obs =     165
Sample:    1990m4  2003m12                                    Lags =       3
                                                         5%
maximum                                      trace   critical
   rank     parms       LL      eigenvalue  statistic   value
      0        36    1107.7833       .      101.6070   47.21
      1        43    1137.7484   0.30456     41.6768   29.68
      2        48    1153.6435   0.17524      9.8865*  15.41
      3        51    1158.4191   0.05624      0.3354    3.76
      4        52    1158.5868   0.00203
```

Fitting a VECM

`vec` estimates the parameters of cointegrating VECMs. There are four types of parameters of interest:

(1) the parameters in the cointegrating equations β

(2) the adjustment coefficients α

(3) the short-run coefficients

(4) some standard functions of β and α that have useful interpretations

While all four types are discussed in [TS] **vec**, here we only discuss types 1–3 and how they appear in the output of vec.

Having determined that there is a cointegrating equation between the Dallas and Houston series, we now want to estimate the parameters of a bivariate cointegrating VECM for these two series using vec.

```
. vec dallas houston

Vector error-correction model

Sample:  1990m3  2003m12                        No. of obs    =        166
                                                AIC           = -7.115516
Log likelihood =  599.5878                      HQIC          =  -7.04703
Det(Sigma_ml)  =  2.50e-06                       SBIC          = -6.946794
```

Equation	Parms	RMSE	R-sq	chi2	P>chi2
D_dallas	4	.038546	0.1692	32.98959	0.0000
D_houston	4	.045348	0.3737	96.66399	0.0000

	Coef.	Std. Err.	z	P>|z|	[95% Conf. Interval]
D_dallas					
_ce1					
L1.	-.3038799	.0908504	-3.34	0.001	-.4819434 -.1258165
dallas					
LD.	-.1647304	.0879356	-1.87	0.061	-.337081 .0076202
houston					
LD.	-.0998368	.0650838	-1.53	0.125	-.2273988 .0277251
_cons	.0056128	.0030341	1.85	0.064	-.0003339 .0115595
D_houston					
_ce1					
L1.	.5027143	.1068838	4.70	0.000	.2932258 .7122028
dallas					
LD.	-.0619653	.1034547	-0.60	0.549	-.2647327 .1408022
houston					
LD.	-.3328437	.07657	-4.35	0.000	-.4829181 -.1827693
_cons	.0033928	.0035695	0.95	0.342	-.0036034 .010389

```
Cointegrating equations
```

Equation	Parms	chi2	P>chi2
_ce1	1	1640.088	0.0000

```
Identification:  beta is exactly identified
                 Johansen normalization restriction imposed
```

beta	Coef.	Std. Err.	z	P>|z|	[95% Conf. Interval]
_ce1					
dallas	1	.	.	.	. .
houston	-.8675936	.0214231	-40.50	0.000	-.9095821 -.825605
_cons	-1.688897	.	.	.	. .

The header contains information about the sample, the fit of each equation, and overall model fit statistics. The first estimation table contains the estimates of the short-run parameters, along with their standard errors, z statistics, and confidence intervals. The two coefficients on L._ce1 are the parameters in the adjustment matrix α for this model. The second estimation table contains the estimated parameters of the cointegrating vector for this model, along with their standard errors, z statistics, and confidence intervals.

Using our previous notation, we have estimated

$$\widehat{\alpha} = (-0.304, 0.503) \qquad \widehat{\beta} = (1, -0.868) \qquad \widehat{\mathbf{v}} = (0.0056, 0.0034)$$

and

$$\widehat{\Gamma} = \begin{pmatrix} -0.165 & -0.0998 \\ -0.062 & -0.333 \end{pmatrix}$$

Overall, the output indicates that the model fits quite well. The coefficient on `houston` in the cointegrating equation is statistically significant, as are the adjustment parameters. The adjustment parameters in this bivariate example are easy to interpret, and we can see that the estimates have the correct signs and imply rapid adjustment toward equilibrium. When the predictions from the cointegrating equation are positive, `dallas` is above its equilibrium value because the coefficient on `dallas` in the cointegrating equation is positive. The estimate of the coefficient [D_dallas]L._ce1 is −.3. Thus when the average housing price in Dallas is too high, it quickly falls back toward the Houston level. The estimated coefficient [D_houston]L._ce1 of .5 implies that when the average housing price in Dallas is too high, the average price in Houston quickly adjusts toward the Dallas level at the same time the Dallas prices are adjusting.

Fitting VECMs with Johansen's normalization

As discussed by Johansen (1995), if there are r cointegrating equations, then at least r^2 restrictions are required to identify the free parameters in β. Johansen proposed a default identification scheme that has become the conventional method of identifying models in the absence of theoretically justified restrictions. Johansen's identification scheme is

$$\beta' = (\mathbf{I}_r, \widetilde{\beta}')$$

where $\mathbf{I}_r$ is the $r \times r$ identity matrix and $\widetilde{\beta}$ is an $(K - r) \times r$ matrix of identified parameters. `vec` applies Johansen's normalization by default.

To illustrate, we fit a VECM with two cointegrating equations and three lags on all four series. We are only interested in the estimates of the parameters in the cointegrating equations, so we can specify the `noetable` option to suppress the estimation table for the adjustment and short-run parameters.

```
. vec austin dallas houston sa, lags(3) rank(2) noetable
Vector error-correction model
Sample:  1990m4  2003m12                      No. of obs    =        165
                                              AIC           =  -13.40174
Log likelihood =  1153.644                    HQIC          =  -13.03496
Det(Sigma_ml)  =  9.93e-12                    SBIC          =  -12.49819
Cointegrating equations

Equation          Parms       chi2     P>chi2

_ce1                  2    586.3044     0.0000
_ce2                  2   2169.826      0.0000


Identification:  beta is exactly identified
                 Johansen normalization restrictions imposed
```

beta	Coef.	Std. Err.	z	P>|z|	[95% Conf. Interval]	
_ce1						
austin	1	.	.	.	.	.
dallas	-3.48e-17	.	.	.	.	.
houston	-.2623782	.1893625	-1.39	0.166	-.6335219	.1087655
sa	-1.241805	.229643	-5.41	0.000	-1.691897	-.7917128
_cons	5.577099	.	.	.	.	.
_ce2						
austin	-1.26e-17	.	.	.	.	.
dallas	1	.	.	.	.	.
houston	-1.095652	.0669898	-16.36	0.000	-1.22695	-.9643545
sa	.2883986	.0812396	3.55	0.000	.1291718	.4476253
_cons	-2.351372	.	.	.	.	.

We can see that the Johansen identification scheme has placed four constraints on the parameters in β: [_ce1]austin=1, [_ce1]dallas=0, [_ce2]austin=0, and [_ce2]dallas=1. (Note that the computational method used imposes zero restrictions that are numerical rather than exact. The values $-3.48e-17$ and $-1.26e-17$ are indistinguishable from zero.) We interpret the results of the first equation as indicating the existence of an equilibrium relationship between the average housing price in Austin and the average prices of houses in Houston and San Antonio.

The Johansen normalization restricted the coefficient on dallas to be unity in the second cointegrating equation, but we could instead constrain the coefficient on houston. Both sets of restrictions define just-identified models, so fitting the model with the latter set of restrictions will yield the same maximized log likelihood. To impose the alternative set of constraints, we use the constraint command.

```
. constraint define 1 [_ce1]austin = 1
. constraint define 2 [_ce1]dallas = 0
. constraint define 3 [_ce2]austin = 0
. constraint define 4 [_ce2]houston = 1
```

```
. vec austin dallas houston sa, lags(3) rank(2) noetable bconstraints(1/4)

Iteration 1:     log likelihood = 1148.8745
  (output omitted )
Iteration 25:    log likelihood = 1153.6435

Vector error-correction model

Sample:  1990m4  2003m12                       No. of obs    =         165
                                               AIC           = -13.40174
Log likelihood =  1153.644                     HQIC          = -13.03496
Det(Sigma_ml)  =  9.93e-12                      SBIC          = -12.49819

Cointegrating equations
```

Equation	Parms	chi2	P>chi2
_ce1	2	586.3392	0.0000
_ce2	2	3455.469	0.0000

```
Identification:  beta is exactly identified
Identifying constraints:
 ( 1)   [_ce1]austin = 1
 ( 2)   [_ce1]dallas = 0
 ( 3)   [_ce2]austin = 0
 ( 4)   [_ce2]houston = 1
```

beta	Coef.	Std. Err.	z	P>|z|	[95% Conf. Interval]	
_ce1						
austin	1	.	.	.	.	.
dallas	(dropped)					
houston	-.2623784	.1876727	-1.40	0.162	-.6302102	.1054534
sa	-1.241805	.2277537	-5.45	0.000	-1.688194	-.7954157
_cons	5.577099	.	.	.	.	.
_ce2						
austin	(dropped)					
dallas	-.9126985	.0595804	-15.32	0.000	-1.029474	-.7959231
houston	1	.	.	.	.	.
sa	-.2632209	.0628791	-4.19	0.000	-.3864617	-.1399802
_cons	2.146094	.	.	.	.	.

Only the estimates of the parameters in the second cointegrating equation have changed, and the new estimates are simply the old estimates divided by -1.095652 because the new constraints are just an alternative normalization of the same just-identified model. With the new normalization, we can interpret the estimates of the parameters in the second cointegrating equation as indicating an equilibrium relationship between the average house price in Houston and the average prices of houses in Dallas and San Antonio.

Postestimation specification testing

Inference on the parameters in α depends crucially on the stationarity of the cointegrating equations, so we should check the specification of the model. As a first check, we can predict the cointegrating equations and graph them over time.

```
. predict ce1, ce equ(#1)
. predict ce2, ce equ(#2)
```

. twoway line ce1 t

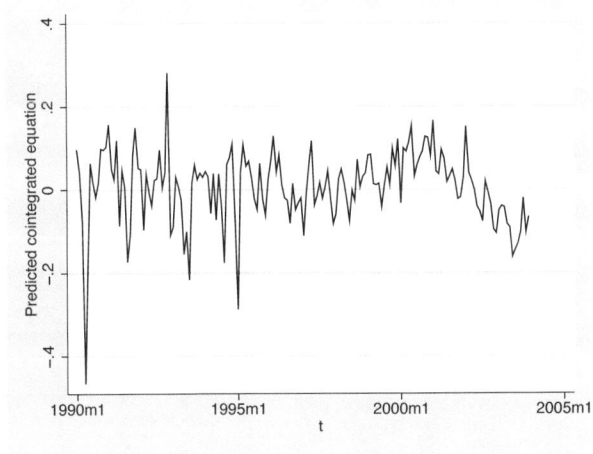

. twoway line ce2 t

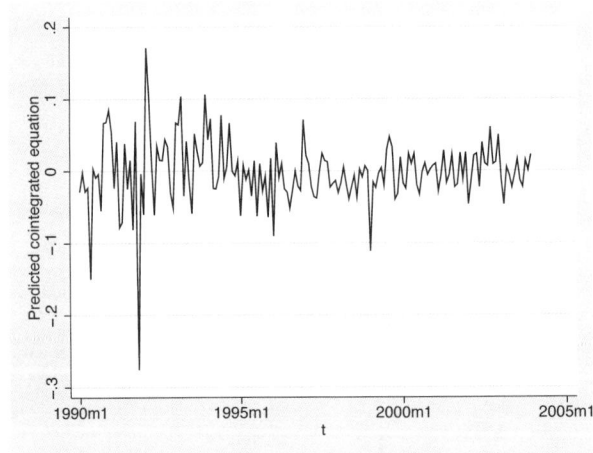

Although the large shocks apparent in the graph of the levels have clear effects on the predictions from the cointegrating equations, our only concern is the negative trend in the first cointegrating equation since the end of 2000. The graph of the levels shows that something put a significant brake on the growth of housing prices after 2000 and that the growth of housing prices in San Antonio slowed during 2000 but then recuperated while Austin maintained slower growth. We suspect that this indicates that the end of the high-tech boom affected Austin more severely than San Antonio. This difference is what causes the trend in the first cointegrating equation. While we could try to account for this effect with a more formal analysis, we will proceed as if the cointegrating equations are stationary.

We can use `vecstable` to check whether we have correctly specified the number of cointegrating equations. As discussed in [TS] **vecstable**, the companion matrix of a VECM with K endogenous variables and r cointegrating equations has $K - r$ unit eigenvalues. If the process is stable, the moduli of the remaining r eigenvalues are strictly less than one. Since there is no general distribution theory for the moduli of the eigenvalues, it can be difficult to ascertain whether the moduli are too close to one.

```
. vecstable, graph
  Eigenvalue stability condition
```

Eigenvalue		Modulus
1		1
1		1
-.6698661		.669866
.3740191 +	.4475996i	.583297
.3740191 -	.4475996i	.583297
-.386377 +	.395972i	.553246
-.386377 -	.395972i	.553246
.540117		.540117
-.0749239 +	.5274203i	.532715
-.0749239 -	.5274203i	.532715
-.2023955		.202395
.09923966		.09924

```
  The VECM specification imposes 2 unit moduli
```

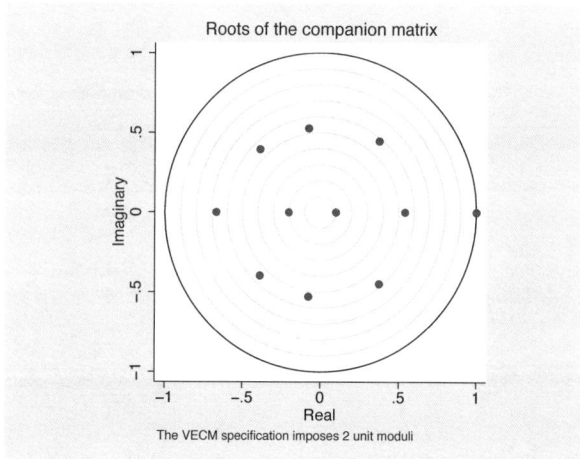

Roots of the companion matrix

The VECM specification imposes 2 unit moduli

Since we specified the `graph` option, `vecstable` plotted the eigenvalues of the companion matrix. The graph of the eigenvalues shows that none of the remaining eigenvalues appears close to the unit circle. The stability check does not indicate that our model is misspecified.

Here we use `veclmar` to test for serial correlation in the residuals.

```
. veclmar, mlag(4)
  Lagrange-multiplier test
```

lag	chi2	df	Prob > chi2
1	56.8757	16	0.00000
2	31.1970	16	0.01270
3	30.6818	16	0.01477
4	14.6493	16	0.55046

```
  H0: no autocorrelation at lag order
```

The results clearly indicate serial correlation in the residuals. The results in Gonzalo (1994) indicate that underspecifying the number of lags in a VECM can significantly increase the finite-sample bias in the parameter estimates and lead to serial correlation. For this reason, we refit the model with five lags instead of three.

```
. vec austin dallas houston sa, lags(5) rank(2) noetable bconstraints(1/4)

Iteration 1:      log likelihood = 1200.5402
  (output omitted )
Iteration 20:     log likelihood = 1203.9465

Vector error-correction model

Sample:  1990m6   2003m12                      No. of obs     =        163
                                               AIC            = -13.79075
                                               HQIC           =  -13.1743
Log likelihood =   1203.946                    SBIC           = -12.27235
Det(Sigma_ml)  =   4.51e-12

Cointegrating equations

Equation           Parms      chi2     P>chi2

_ce1                   2    498.4682   0.0000
_ce2                   2   4125.926    0.0000
```

Identification: beta is exactly identified
Identifying constraints:
(1)   [_ce1]austin = 1
(2)   [_ce1]dallas = 0
(3)   [_ce2]austin = 0
(4)   [_ce2]houston = 1

beta	Coef.	Std. Err.	z	P>|z|	[95% Conf. Interval]	
_ce1						
austin	1	.	.	.	.	.
dallas	(dropped)					
houston	-.6525574	.2047061	-3.19	0.001	-1.053774	-.2513407
sa	-.6960166	.2494167	-2.79	0.005	-1.184864	-.2071688
_cons	3.846275	.	.	.	.	.
_ce2						
austin	(dropped)					
dallas	-.932048	.0564332	-16.52	0.000	-1.042655	-.8214409
houston	1	.	.	.	.	.
sa	-.2363915	.0599348	-3.94	0.000	-.3538615	-.1189215
_cons	2.065719	.	.	.	.	.

Comparing these results with those from the previous model reveals that

(1) there is now evidence that the coefficient [_ce1]houston is not equal to zero,

(2) the two sets of estimated coefficients for the first cointegrating equation are quite different, and

(3) the two sets of estimated coefficients for the second cointegrating equation are quite similar.

The assumption that the errors are independently, identically, and normally distributed with zero mean and finite variance allows us to derive the likelihood function. If the errors do not come from a normal distribution but are just independently and identically distributed with zero mean and finite variance, the parameter estimates are still consistent, but they are not efficient.

We use `vecnorm` to test the null hypothesis that the errors are normally distributed.

```
. qui vec austin dallas houston sa, lags(5) rank(2)  bconstraints(1/4)
. vecnorm
```
Jarque-Bera test

Equation	chi2	df	Prob > chi2
D_austin	74.324	2	0.00000
D_dallas	3.501	2	0.17370
D_houston	245.032	2	0.00000
D_sa	8.426	2	0.01481
ALL	331.283	8	0.00000

Skewness test

Equation	Skewness	chi2	df	Prob > chi2
D_austin	.60265	9.867	1	0.00168
D_dallas	.09996	0.271	1	0.60236
D_houston	-1.0444	29.635	1	0.00000
D_sa	.38019	3.927	1	0.04752
ALL		43.699	4	0.00000

Kurtosis test

Equation	Kurtosis	chi2	df	Prob > chi2
D_austin	6.0807	64.458	1	0.00000
D_dallas	3.6896	3.229	1	0.07232
D_houston	8.6316	215.397	1	0.00000
D_sa	3.8139	4.499	1	0.03392
ALL		287.583	4	0.00000

The results indicate that we can strongly reject the null hypothesis of normally distributed errors. Most of the errors are both skewed and kurtotic.

Impulse–response functions for VECMs

With a model that we now consider acceptably well specified, we can use the `irf` commands to estimate and interpret the impulse–response functions (IRFs). While IRFs from a stationary VAR die out over time, IRFs from a cointegrating VECM do not always die out. Since each variable in a stationary VAR has a time-invariant mean and finite, time-invariant variance, the effect of a shock to any one of these variables must die out so that the variable can revert to its mean. In contrast, the I(1) variables modeled in a cointegrating VECM are not mean reverting, and the unit moduli in the companion matrix imply that the effects of some shocks will not die out over time.

These two possibilities gave rise to new terms. When the effect of a shock dies out over time, the shock is said to be transitory. When the effect of a shock does not die out over time, the shock is said to be permanent.

Below we use `irf create` to estimate the impulse–response functions and `irf graph` to graph two of the orthogonalized impulse–response functions.

```
. irf create vec1 , set(vecintro, replace) step(24)
(file vecintro.irf created)
(file vecintro.irf now active)
(file vecintro.irf updated)

. irf graph oirf, impulse(austin dallas) response(sa) yline(0)
```

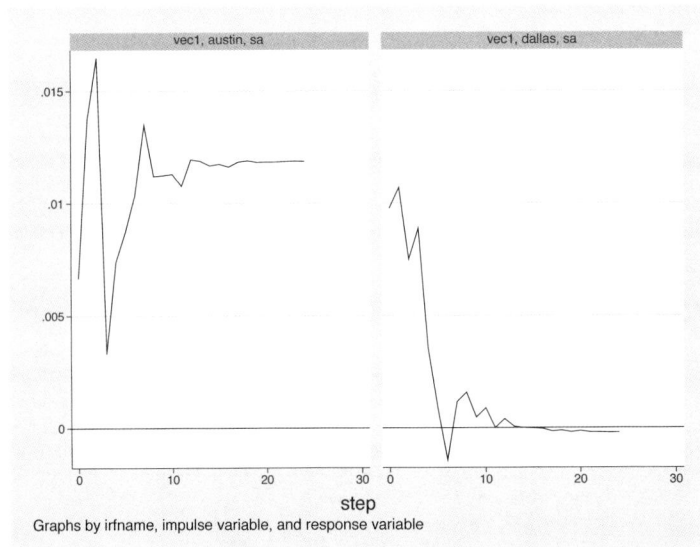

Graphs by irfname, impulse variable, and response variable

The graphs indicate that an orthogonalized shock to the average housing price in Austin has a permanent effect on the average housing price in San Antonio but that an orthogonalized shock to the average price of housing in Dallas has a transitory effect. According to this model, unexpected shocks that are local to the Austin housing market will have a permanent effect on the housing market in San Antonio, but unexpected shocks that are local to the Dallas housing market will have only a transitory effect on the housing market in San Antonio.

Forecasting with VECMs

Cointegrating VECMs are also used to produce forecasts of both the first-differenced variables and the levels of the variables. Comparing the variances of the forecast errors of stationary VARs with those from a cointegrating VECM reveals a fundamental difference between the two models. Whereas the variances of the forecast errors for a stationary VAR converge to a constant as the prediction horizon grows, the variances of the forecast errors for the levels of a cointegrating VECM diverge with the forecast horizon. (See chapter 11.3 of Lütkepohl [1993] for more about this result.) Because all the variables in the model for the first differences are stationary, the forecast errors for the dynamic forecasts of the first-differences remain finite. In contrast, the forecast errors for the dynamic forecasts of the levels diverge to infinity.

We use `fcast compute` to obtain dynamic forecasts of the levels and `fcast graph` to graph these dynamic forecasts, along with their asymptotic confidence intervals.

```
. fcast compute m1_, step(24)

. fcast graph m1_austin m1_dallas m1_houston m1_sa
```

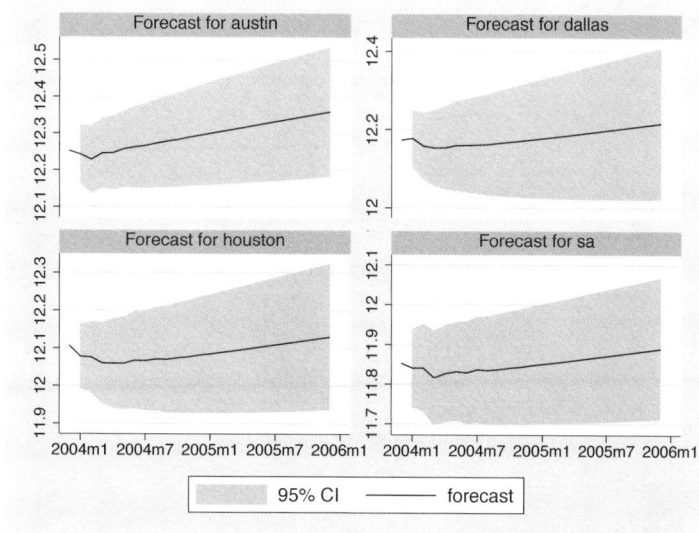

As expected, the widths of the confidence intervals grow with the forecast horizon.

References

Engle, R. F. and C. W. J. Granger. 1987. Cointegration and error correction: Representation, estimation and testing. *Econometrica* 55(2): 251–276.

Gonzalo, J. 1994. Five alternative methods of estimating long-run equilibrium relationships. *Journal of Econometrics* 60: 203–233.

Granger, C. W. J. 1981. Some properties of time series data and their use in econometric model specification. *Journal of Econometrics* 16: 121–130.

Granger, C. W. J. and P. Newbold. 1974. Spurious regressions in econometrics. *Journal of Econometrics* 26: 1045–1066.

Hamilton, J. D. 1994. *Time Series Analysis*. Princeton: Princeton University Press.

Hubrich, K., H. Lütkepohl, and P. Saikkonen. 2001. A review of systems cointegration tests. *Econometric Reviews* 20(3): 247–318.

Johansen, S. 1988. Statistical analysis of cointegration vectors. *Journal of Economic Dynamics and Control* 12: 231–254.

——. 1991. Estimation and hypothesis testing of cointegration vectors in Gaussian vector autoregressive models. *Econometrica* 59: 1551–1580.

——. 1995. *Likelihood-Based Inference in Cointegrated Vector Auto-Regressive Models*. Oxford: Oxford University Press.

Lütkepohl, H. 1993. *Introduction to Multiple Time Series Analysis*. 2nd ed. New York: Springer.

Maddala, G. S. and I. M. Kim. 1998. *Unit Roots, Cointegration, and Structural Change*. Cambridge, UK: Cambridge University Press.

Nielsen, B. 2001. Order determination in general vector autoregressions. Working Paper, Department of Economics, University of Oxford and Nuffield College.

Park, J. Y. and P. C. B. Phillips. 1988a. Statistical inference in regressions with integrated regressors: Part I. *Econometric Theory* 4: 468–497.

——. 1988b. Statistical inference in regressions with integrated regressors: Part II. *Econometric Theory* 5: 95–131.

Paulsen, J. 1984. Order determination of multivariate autoregressive time series with unit roots. *Journal of Time Series Analysis* 5(2): 115–127.

Phillips, P. C. B. 1986. Understanding spurious regressions in econometrics. *Journal of Econometrics* 33: 311–340.

Phillips, P. C. B. and S. N. Durlauf. 1986. Multiple time series regression with integrated processes. *The Review of Economic Studies* 53(4): 473–495.

Sims, C. A., J. H. Stock, and M. W. Watson. 1990. Inference in linear time series models with some unit roots. *Econometrica* 58(1): 113–144.

Stock, J. H. 1987. Asymptotic properties of least squares estimators of cointegrating vectors. *Econometrica* 55(5): 1035–1056.

Stock, J. H. and M. W. Watson. 1988. Testing for common trends. *Journal of the American Statistical Association* 404: 1099–1107.

Tsay, R. S. 1984. Order selection in nonstationary autoregressive models. *The Annals of Statistics* 12(4): 1425–1433.

Watson, M. W. 1994. Vector autoregressions and cointegration. *Handbook of Econometrics*, Vol IV, Engle, R. F. and McFadden, D. L. eds, Amsterdam: Elsevier.

Also See

Complementary:	[TS] **fcast compute**, [TS] **fcast graph**, [TS] **vec**, [TS] **veclmar**, [TS] **vecnorm**, [TS] **vecrank**, [TS] **vecstable**
Related:	[TS] **arima**, [R] **regress**, [R] **sureg**
Background:	[U] **11.4.3 Time-series varlists**, [TS] **irf**

Title

Syntax

vec *varlist* $[if]$ $[in]$ $[$, *options* $]$

options	description
Model	
<u>r</u>ank(*#*)	use *#* cointegrating equations; default is rank(1)
<u>l</u>ags(*#*)	use *#* for the maximum lag in underlying VAR model
trend(<u>c</u>onstant)	include an unrestricted constant in model; the default
trend(<u>rc</u>onstant)	include a restricted constant in model
trend(<u>t</u>rend)	include a linear trend in the cointegrating equations and a quadratic trend in the undifferenced data
trend(<u>rt</u>rend)	include a restricted trend in model
trend(<u>n</u>one)	do not include a trend or a constant
<u>b</u>constraints(*constraints*$_{bc}$)	place *constraints*$_{bc}$ on cointegrating vectors
<u>a</u>constraints(*constraints*$_{ac}$)	place *constraints*$_{ac}$ on adjustment parameters
Adv. model	
<u>si</u>ndicators(*varlist*$_{si}$)	include normalized seasonal indicator variables *varlist*$_{si}$
noreduce	do not perform checks and corrections for collinearity among lags of dependent variables
Reporting	
<u>l</u>evel(*#*)	set confidence level; default is level(95)
<u>nob</u>table	do not report parameters in the cointegrating equations
<u>noid</u>test	do not report the likelihood-ratio test of overidentifying restrictions
<u>al</u>pha	report adjustment parameters in separate table
pi	report parameters in $\Pi = \alpha\beta'$
<u>nop</u>table	do not report elements of Π matrix
<u>m</u>ai	report parameters in the moving-average impact matrix
<u>noe</u>table	do not report adjustment and short-run parameters
dforce	force reporting of short-run, beta, and alpha parameters when the parameters in beta are not identified; advanced option
Max options	
maximize_options	control the maximization process; seldom used

You must tsset your data before using vec; see [TS] **tsset**.

varlist must contain at least two variables and may contain time-series operators; see [U] **11.4.3 Time-series varlists**.

by, rolling, statsby, and xi may be used with vec; see [U] **11.1.10 Prefix commands**.

See [U] **20 Estimation and postestimation commands** for additional capabilities of estimation commands.

vec does not allow gaps in the data.

Description

vec estimates the parameters in vector error-correction models (VECM) using Johansen's (1995) maximum likelihood method. Constraints may be placed on the parameters in the cointegrating equations or on the adjustment terms.

Options

<u>Model</u>

rank(#) specifies the number of cointegrating equations; rank(1) is the default.

lags(#) specifies the maximum lag to be included in the underlying VAR model. The maximum lag in a VECM is one smaller than the maximum lag in the corresponding VAR in levels; the number of lags must be greater than zero but small enough so that the degrees of freedom used up by the model are fewer than the number of observations. The default is lags(2).

trend(*trend_spec*) specifies which of Johansen's five trend specifications to include in the model. These specifications are discussed in *Specification of constants and trends* below. The default is trend(constant).

bconstraints(*constraints*$_{bc}$) specifies the constraints to be placed on the parameters of the cointegrating equations. When no constraints are placed on the adjustment parameters—that is, when the aconstraints() option is not specified—the default is to place the constraints defined by Johansen's normalization on the parameters of the cointegrating equations. When constraints are placed on the adjustment parameters, the default is not to place constraints on the parameters in the cointegrating equations.

aconstraints(*constraints*$_{ac}$) specifies the constraints to be placed on the adjustment parameters. By default, no constraints are placed on the adjustment parameters.

<u>Adv. model</u>

sindicators(*varlist*$_{si}$) specifies the normalized seasonal indicator variables to include in the model. The indicator variables specified in this option must be normalized as discussed in Johansen (1995). If the indicators are not properly normalized, the estimator of the cointegrating vector does not converge to the asymptotic distribution derived by Johansen (1995). More details about how these variables are handled are provided in *Methods and Formulas*. sindicators() cannot be specified with trend(none) or with trend(rconstant).

noreduce causes vec to skip the checks and corrections for collinearity among the lags of the dependent variables. By default, vec checks to see if the current lag specification causes some of the regressions performed by vec to contain perfectly collinear variables; if so, it reduces the maximum lag until the perfect collinearity is removed.

<u>Reporting</u>

level(#); see [TS] **estimation options**.

nobtable suppresses the estimation table for the parameters in the cointegrating equations. By default, vec displays the estimation table for the parameters in the cointegrating equations.

noidtest suppresses the likelihood-ratio test of the overidentifying restrictions, which is reported by default when the model is overidentified.

alpha displays a separate estimation table for the adjustment parameters, which is not displayed by default.

pi displays a separate estimation table for the parameters in $\Pi = \alpha\beta'$, which is not displayed by default.

noptable suppresses the estimation table for the elements of the Π matrix, which is displayed by default when the parameters in the cointegrating equations are not identified.

mai displays a separate estimation table for the parameters in the moving-average impact matrix, which is not displayed by default.

noetable suppresses the main estimation table that contains information about the estimated adjustment parameters and the short-run parameters, which is displayed by default.

dforce displays the estimation tables for the short-run parameters and α and β—if the latter two are requested—when the parameters in β are not identified. By default, when the specified constraints do not identify the parameters in the cointegrating equations, estimation tables are displayed only for Π and the MAI.

$\boxed{\text{Max options}}$

maximize_options: <u>iter</u>ate(*#*), <u>nolog</u>, <u>trace</u>, <u>toltr</u>ace, <u>tol</u>erance(*#*), <u>af</u>rom(*matrix$_a$*), <u>bf</u>rom(*matrix$_b$*); see [R] **maximize**.

toltrace displays the relative differences for the log likelihood and the coefficient vector at every iteration. This option cannot be specified if no constraints are defined or if nolog is specified.

afrom(*matrix$_a$*) specifies a $1 \times (K*r)$ row vector with starting values for the adjustment parameters, where K is the number of endogenous variables and r is the number of cointegrating equations specified in the rank() option. The starting values should be ordered as they are reported in e(alpha). This option cannot be specified if no constraints are defined.

bfrom(*matrix$_b$*) specifies a $1 \times (m_1 * r)$ row vector with starting values for the parameters of the cointegrating equations, where m_1 is the number of variables in the trend-augmented system and r is the number of cointegrating equations specified in the rank() option. (See *Methods and Formulas* for more details about m_1.) The starting values should be ordered as they are reported in e(betavec). As discussed in *Methods and Formulas*, for some trend specifications, e(beta) contains parameter estimates that are not obtained directly from the optimization algorithm. bfrom() should only specify starting values for the parameters reported in e(betavec). This option cannot be specified if no constraints are defined.

Remarks

Remarks are presented under the headings

> *Introduction*
> *Specification of constants and trends*
> *Collinearity*

Introduction

VECMs are used to model the stationary relationships between multiple time series that contain unit roots. vec implements Johansen's methodology for estimating the parameters of a VECM.

[TS] **vec intro** reviews the basics of integration and cointegration and highlights why we need special methods for modeling the relationships between processes that contain unit roots. This manual entry assumes familiarity with the material in [TS] **vec intro** and provides examples illustrating how to use the vec command. See Johansen (1995) and Hamilton (1994) for more in-depth introductions to cointegration analysis.

▷ Example 1

This example uses annual data on the average per-capita disposable personal income in the eight U.S. Bureau of Economic Analysis (BEA) regions of the United States. We use data from 1948–2002 in logarithms. Unit-root tests on these series fail to reject the null hypothesis that per-capita disposable income in each region contains a unit root. Since capital and labor can move easily between the different regions of the U.S., we would expect that no one series will diverge from all the remaining series and that cointegrating relationships exist.

Below we graph the natural logs of average disposal income in the New England and the Southeast regions.

```
. use http://www.stata-press.com/data/r9/rdinc
. line ln_ne ln_se year
```

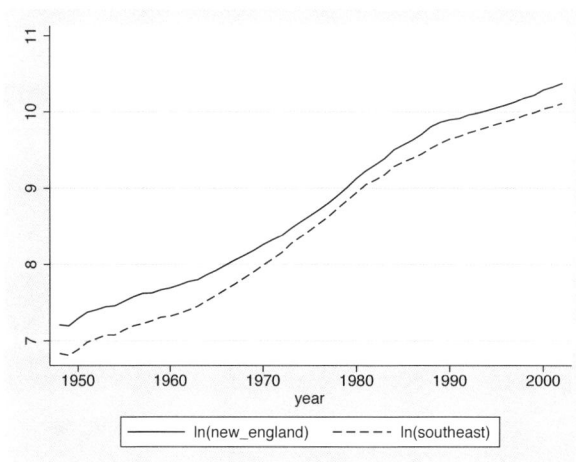

The graph indicates a differential between the two series that shrinks between 1960 and about 1980 and then grows until it stabilizes around 1990. We next estimate the parameters of a bivariate VECM with one cointegrating relationship.

```
. vec ln_ne ln_se

Vector error-correction model

Sample:    1950    2002                    No. of obs    =        53
                                           AIC           = -11.00462
Log likelihood =   300.6224                HQIC          = -10.87595
Det(Sigma_ml)  =   4.06e-08                SBIC          = -10.67004
Equation           Parms     RMSE     R-sq      chi2     P>chi2

D_ln_ne               4    .017896   0.9313   664.4668   0.0000
D_ln_se               4    .018723   0.9292   642.7179   0.0000
```

	Coef.	Std. Err.	z	P>|z|	[95% Conf. Interval]	
D_ln_ne						
_ce1						
L1.	-.4337524	.0721365	-6.01	0.000	-.5751373	-.2923675
ln_ne						
LD.	.7168658	.1889085	3.79	0.000	.3466119	1.08712
ln_se						
LD.	-.6748754	.2117975	-3.19	0.001	-1.089991	-.2597599
_cons	-.0019846	.0080291	-0.25	0.805	-.0177214	.0137521
D_ln_se						
_ce1						
L1.	-.3543935	.0754725	-4.70	0.000	-.5023168	-.2064701
ln_ne						
LD.	.3366786	.1976448	1.70	0.088	-.050698	.7240553
ln_se						
LD.	-.1605811	.2215922	-0.72	0.469	-.5948939	.2737317
_cons	.002429	.0084004	0.29	0.772	-.0140355	.0188936

```
Cointegrating equations

Equation           Parms     chi2    P>chi2

_ce1                  1    29805.02   0.0000
```

```
Identification:  beta is exactly identified
                 Johansen normalization restriction imposed
```

beta	Coef.	Std. Err.	z	P>|z|	[95% Conf. Interval]	
_ce1						
ln_ne	1	.	.	.	.	.
ln_se	-.9433708	.0054643	-172.64	0.000	-.9540807	-.9326609
_cons	-.8964065	.	.	.	.	.

The default output has three parts. The header provides information about the sample, the model fit, and the identification of the parameters in the cointegrating equation. The main estimation table contains the estimates of the short-run parameters, along with their standard errors and confidence intervals. The second estimation table reports the estimates of the parameters in the cointegrating equation, along with their standard errors and confidence intervals.

The results indicate strong support for a cointegrating equation such that

$$\text{ln\_ne} - .943\,\text{ln\_se} - .896$$

should be a stationary series. Identification of the parameters in the cointegrating equation is achieved by constraining some of them to be fixed, and fixed parameters do not have standard errors. In this example, the coefficient on ln_ne has been normalized to 1, so its standard error is missing. As discussed in *Methods and Formulas*, the constant term in the cointegrating equation is not directly estimated in this trend specification but rather is backed out from other estimates. Not all the elements of the VCE that correspond to this parameter are readily available, so the standard error for the _cons parameter is missing.

To get a better idea of how our model fits, we predict the cointegrating equation and graph it over time:

```
. predict ce, ce
. line ce year
```

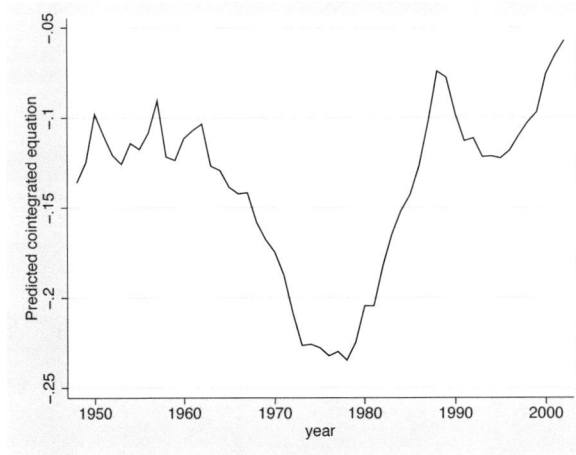

While the predicted cointegrating equation has the right appearance for the time before the mid-1960s, afterward the predicted cointegrating equation does not look like a stationary series. A better model would account for the trends in the size of the differential.

◁

As discussed in [TS] **vec intro**, simply normalizing one of the coefficients to be one is sufficient to identify the parameters of the single cointegrating vector. When there is more than one cointegrating equation, additional restrictions are required.

▷ Example 2

We have data on monthly unemployment rates in Indiana, Illinois, Kentucky, and Missouri from January 1978 through December 2003. We suspect that factor mobility will keep the unemployment rates in equilibrium. The following graph plots the data.

```
. use http://www.stata-press.com/data/r9/urates, clear
. line missouri indiana kentucky illinois t
```

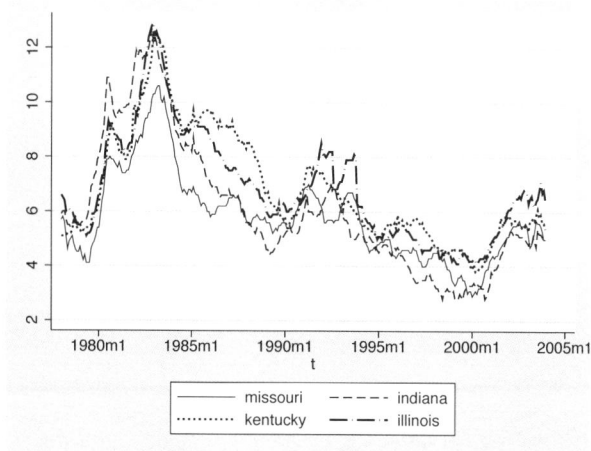

The graph shows that while the series do appear to move together, the relationship is not as clear as in the previous example. There are periods when Indiana has the highest rate and others when Indiana has the lowest rate. While the Kentucky rate moves closely with the other series for most of the sample, there is a period in the mid-1980s when the unemployment rate in Kentucky does not fall at the same rate as the other series.

We will model the series with two cointegrating equations and no linear or quadratic time trends in the original series. Since we are focusing on the cointegrating vectors, we use the `noetable` option to suppress displaying the short-run estimation table.

(*Continued on next page*)

```
. vec missouri indiana kentucky illinois, trend(rconstant) rank(2) lags(4)
> noetable
```

Vector error-correction model

Sample: 1978m5 2003m12

	No. of obs	=	308
	AIC	=	-2.306048
Log likelihood = 417.1314	HQIC	=	-2.005818
Det(Sigma_ml) = 7.83e-07	SBIC	=	-1.555184

Cointegrating equations

Equation	Parms	chi2	P>chi2
_ce1	2	133.3885	0.0000
_ce2	2	195.6324	0.0000

Identification: beta is exactly identified
                 Johansen normalization restrictions imposed

beta	Coef.	Std. Err.	z	P>|z|	[95% Conf. Interval]	
_ce1						
missouri	1	.	.	.	.	.
indiana	-2.19e-18	.	.	.	.	.
kentucky	.3493902	.2005537	1.74	0.081	-.0436879	.7424683
illinois	-1.135152	.2069063	-5.49	0.000	-1.540681	-.7296235
_cons	-.3880707	.4974323	-0.78	0.435	-1.36302	.5868787
_ce2						
missouri	1.09e-16	.	.	.	.	.
indiana	1	.	.	.	.	.
kentucky	.2059473	.2718678	0.76	0.449	-.3269038	.7387985
illinois	-1.51962	.2804792	-5.42	0.000	-2.069349	-.9698907
_cons	2.92857	.6743122	4.34	0.000	1.606942	4.250197

With the exceptions of the coefficients on kentucky in the two cointegrating equations and the constant term in the first, all the parameters are significant at the 5% level. We can re-fit the model with the Johansen normalization and the overidentifying constraint that the coefficient on kentucky in the second cointegrating equation is zero.

```
. constraint define 1 [_ce1]missouri = 1
. constraint define 2 [_ce1]indiana  = 0
. constraint define 3 [_ce2]missouri = 0
. constraint define 4 [_ce2]indiana  = 1
. constraint define 5 [_ce2]kentucky = 0
```

```
. vec missouri indiana kentucky illinois, trend(rconstant) rank(2) lags(4)
> noetable bconstraints(1/5)

Iteration 1:     log likelihood =  416.97177
 (output omitted )
Iteration 20:    log likelihood =   416.9744

Vector error-correction model

Sample:  1978m5   2003m12                        No. of obs     =         308
                                                AIC            =   -2.311522
                                                HQIC           =   -2.016134
Log likelihood =   416.9744                     SBIC           =   -1.572769
Det(Sigma_ml)  =   7.84e-07

Cointegrating equations

Equation            Parms      chi2      P>chi2

_ce1                   2      145.233    0.0000
_ce2                   1      209.9344   0.0000


Identification:  beta is overidentified
Identifying constraints:
 ( 1)   [_ce1]missouri = 1
 ( 2)   [_ce1]indiana = 0
 ( 3)   [_ce2]missouri = 0
 ( 4)   [_ce2]indiana = 1
 ( 5)   [_ce2]kentucky = 0
```

beta	Coef.	Std. Err.	z	P>|z|	[95% Conf. Interval]	
_ce1						
missouri	1	.	.	.	.	.
indiana	(dropped)					
kentucky	.2521685	.1649653	1.53	0.126	-.0711576	.5754946
illinois	-1.037453	.1734165	-5.98	0.000	-1.377343	-.6975626
_cons	-.3891102	.4726968	-0.82	0.410	-1.315579	.5373586
_ce2						
missouri	(dropped)					
indiana	1	.	.	.	.	.
kentucky	(dropped)					
illinois	-1.314265	.0907071	-14.49	0.000	-1.492048	-1.136483
_cons	2.937016	.6448924	4.55	0.000	1.67305	4.200982

```
LR test of identifying restrictions:  chi2( 1) =   .3139  Prob > chi2 = 0.575
```

The test of the overidentifying restriction does not reject the null hypothesis that the restriction is valid, and the *p*-value on the coefficient on kentucky in the first cointegrating equation indicates that it is not significant. We will leave the variable in the model and attribute the lack of significance to whatever caused the kentucky series to temporarily rise above the others from 1985 until 1990, though we could instead consider removing kentucky from the model.

Next, we look at the estimates of the adjustment parameters. In the output below, we replay the previous results. We specify the alpha option so that vec will display an estimation table for the estimates of the adjustment parameters, and we specify nobtable to suppress the table for the parameters of the cointegrating equations because we have already looked at those.

```
. vec, alpha nobtable noetable
Vector error-correction model
Sample:  1978m5   2003m12                      No. of obs     =        308
                                               AIC            = -2.311522
Log likelihood =   416.9744                    HQIC           = -2.016134
Det(Sigma_ml)  =   7.84e-07                    SBIC           = -1.572769
Adjustment parameters
Equation          Parms      chi2     P>chi2

D_missouri           2    19.39607    0.0001
D_indiana            2     6.426086   0.0402
D_kentucky           2     8.524901   0.0141
D_illinois           2    22.32893    0.0000
```

	Coef.	Std. Err.	z	P>|z|	[95% Conf. Interval]	
D_missouri						
_ce1						
L1.	-.0683152	.0185763	-3.68	0.000	-.1047242	-.0319063
_ce2						
L1.	.0405613	.0112417	3.61	0.000	.018528	.0625946
D_indiana						
_ce1						
L1.	-.0342096	.0220955	-1.55	0.122	-.0775159	.0090967
_ce2						
L1.	.0325804	.0133713	2.44	0.015	.0063732	.0587877
D_kentucky						
_ce1						
L1.	-.0482012	.0231633	-2.08	0.037	-.0936004	-.0028021
_ce2						
L1.	.0374395	.0140175	2.67	0.008	.0099657	.0649133
D_illinois						
_ce1						
L1.	.0138224	.0227041	0.61	0.543	-.0306768	.0583215
_ce2						
L1.	.0567664	.0137396	4.13	0.000	.0298373	.0836955

```
LR test of identifying restrictions:  chi2( 1) =    .3139  Prob > chi2 = 0.575
```

All the coefficients are significant at the 5% level, except those on Indiana and Illinois in the first cointegrating equation. From an economic perspective, the issue is whether the unemployment rates in Indiana and Illinois adjust when the first cointegrating equation is out of equilibrium. We could impose restrictions on one or both of those parameters and refit the model, or we could just decide to use the current results.

◁

❑ Technical Note

vec can be used to fit models in which the parameters in β are not identified, in which case only the parameters in Π and the moving-average impact matrix C are identified. When the parameters in β are not identified, the values of $\widehat{\beta}$ and $\widehat{\alpha}$ can vary depending on the starting values. However, the estimates of Π and C are identified and have known asymptotic distributions. This method is valid since these additional normalization restrictions do not impose any restriction on Π or C.

❑

Specification of constants and trends

As discussed in [TS] **vec intro**, allowing for a constant term and linear time trend allow us to write the VECM as

$$\Delta\mathbf{y}_t = \boldsymbol{\alpha}(\boldsymbol{\beta}\mathbf{y}_{t-1} + \boldsymbol{\mu} + \boldsymbol{\rho}t) + \sum_{i=1}^{p-1}\boldsymbol{\Gamma}_i\Delta\mathbf{y}_{t-i} + \boldsymbol{\gamma} + \boldsymbol{\tau}t + \boldsymbol{\epsilon}_t$$

Five different trend specifications are available:

Option in `trend()`	Parameter restrictions	Johansen (1995) notation
trend	none	$H(r)$
rtrend	$\tau = 0$	$H^*(r)$
constant	$\rho = 0$, and $\tau = 0$	$H_1(r)$
rconstant	$\rho = 0$, $\gamma = 0$, and $\tau = 0$	$H_1^*(r)$
none	$\mu = 0$ $\rho = 0$, $\gamma = 0$, and $\tau = 0$	$H_2(r)$

`trend(trend)` allows for a linear trend in the cointegrating equations and a quadratic trend in the undifferenced data. A linear trend in the cointegrating equations implies that the cointegrating equations are assumed to be trend stationary.

`trend(rtrend)` defines a restricted trend model that excludes linear trends in the differenced data but allows for linear trends in the cointegrating equations. As in the previous case, a linear trend in a cointegrating equation implies that the cointegrating equation is trend stationary.

`trend(constant)` defines a model with an unrestricted constant. This allows for a linear trend in the undifferenced data and cointegrating equations that are stationary around a nonzero mean. This is the default.

`trend(rconstant)` defines a model with a restricted constant in which there is no linear or quadratic trend in the undifferenced data. A nonzero μ allows for the cointegrating equations to be stationary around nonzero means, which provide the only intercepts for differenced data. Seasonal indicators are not allowed with this specification.

`trend(none)` defines a model that does not include a trend or a constant. When there is no trend or constant, the cointegrating equations are restricted to being stationary with zero means. Also, after adjusting for the effects of lagged endogenous variables, the differenced data is modeled as having mean zero. Seasonal indicators are not allowed with this specification.

❏ Technical Note

`vec` uses a switching algorithm developed by Boswijk (1995) to maximize the log-likelihood function when constraints are placed on the parameters. The starting values affect both the ability of the algorithm to find a maximum and its speed in finding that maximum. By default, `vec` uses the parameter estimates that correspond to Johansen's normalization. Sometimes, other starting values will cause the algorithm to find a maximum faster.

To specify starting values for the parameters in $\boldsymbol{\alpha}$, we specify a $1 \times (Kr)$ matrix in the `afrom()` option. Specifying starting values for the parameters in $\boldsymbol{\beta}$ is slightly more complicated. As explained in *Methods and Formulas*, specifying `trend(constant)`, `trend(rtrend)`, or `trend(trend)` causes some of the estimates of the trend parameters appearing in $\widehat{\boldsymbol{\beta}}$ to be "backed out". The switching algorithm only estimates the parameters of the cointegrating equations whose estimates are saved in `e(betavec)`. For this reason, only the parameters saved in `e(betavec)` can have their initial values set via `bfrom()`.

The table below describes which trend parameters in the cointegrating equations are estimated by the switching algorithm for each of the five specifications.

Trend specification	Trend parameters in cointegrating equations	Trend parameters estimated via switching algorithm
none	none	none
rconstant	_cons	_cons
constant	_cons	none
rtrend	_cons, _trend	_trend
trend	_cons, _trend	none

❑

Collinearity

As is to be expected, collinearity among variables causes some parameters to be unidentified numerically. If vec encounters perfect collinearity among the dependent variables, it exits with an error.

In contrast, if vec encounters perfect collinearity that appears to be due to too many lags in the model, vec displays a warning message and reduces the maximum lag included in the model in an effort to find a model with fewer lags in which all the parameters are identified by the data. Specifying the noreduce option causes vec to skip over these additional checks and corrections for collinearity. Thus the noreduce option can be used to force the estimation to proceed when not all the parameters are identified by the data. When some parameters are not identified due to collinearity, the results cannot be interpreted but can be used to find the source of the collinearity.

(Continued on next page)

Saved Results

vec saves in e():

Scalars

e(N)	number of observations
e(k_rank)	number of unconstrained parameters
e(k_eq)	number of equations
e(k_dv)	number of dependent variables
e(beta_iden)	1 if the parameters in β are identified and 0 otherwise
e(beta_icnt)	number of independent restrictions placed on β
e(df_lr)	degrees of freedom of the test of overidentifying restrictions
e(k_ce)	number of cointegrating equations
e(n_lags)	number of lags
e(df_m)	model degrees of freedom
e(ll)	log likelihood
e(chi2_res)	value of test of overidentifying restrictions
e(tmax)	maximum time
e(tmin)	minimum time
e(sbic)	value of SBIC
e(aic)	value of AIC
e(hqic)	value of HQIC
e(detsig_ml)	determinant of the estimated covariance matrix
e(k_#)	number of variables in equation #
e(rmse_#)	RMSE of equation #
e(r2_#)	R-squared of equation #
e(df_m#)	model degrees of freedom in equation #
e(chi2_#)	$\chi 2$ statistic for equation #
e(converge)	1 if the switching algorithm converged, 0 if it did not converge

Macros

e(cmd)	vec
e(trend)	trend specified
e(tsfmt)	format of the time variable
e(tvar)	name of the time variable
e(endog)	endogenous variables
e(eqnames)	equation names
e(reduce_opt)	noreduce, if noreduce is specified
e(reduce_lags)	list of maximum lags to which the model has been reduced
e(title)	title in estimation output
e(aconstraints)	constraints placed on α
e(bconstraints)	constraints placed on β
e(sindicators)	sindicators, if specified
e(properties)	b V
e(predict)	program used to implement predict

Matrices

e(b)	estimates of short-run parameters
e(V)	VCE of short-run parameter estimates
e(beta)	estimates of β
e(V_beta)	VCE of $\widehat{\beta}$
e(betavec)	directly obtained estimates of β
e(pi)	estimates of $\widehat{\Pi}$
e(V_pi)	VCE of $\widehat{\Pi}$
e(alpha)	estimates of α
e(V_alpha)	VCE of $\widehat{\alpha}$
e(omega)	estimates of $\widehat{\Omega}$
e(mai)	estimates of $\mathbf{C}$
e(V_mai)	VCE of $\widehat{\mathbf{C}}$

Functions

e(sample)	marks estimation sample

Methods and Formulas

vec is implemented as an ado-file. Remarks are presented under the headings

> *The general specification of the VECM*
> *The log-likelihood function*
> > *Unrestricted trend*
> > *Restricted trend*
> > *Unrestricted constant*
> > *Restricted constant*
> > *No trend*
> *Estimation with Johansen identification*
> *Estimation with constraints:* β *identified*
> *Estimation with constraints:* β *not identified*
> *Formulas for the information criteria*
> *Formulas for predict*

The general specification of the VECM

vec estimates the parameters of a VECM that can be written as

$$\Delta\mathbf{y}_t = \alpha\beta'\mathbf{y}_{t-1} + \sum_{i=1}^{p-1}\Gamma_i\Delta\mathbf{y}_{t-i} + \mathbf{v} + \delta t + \mathbf{w}_1 s_1 + \cdots + \mathbf{w}_m s_m + \epsilon_t \tag{1}$$

where

$\mathbf{y}_t$ is a $K \times 1$ vector of endogenous variables,

α is a $K \times r$ matrix of parameters,

β is a $K \times r$ matrix of parameters,

$\Gamma_1, \ldots, \Gamma_{p-1}$ are $K \times K$ matrices of parameters,

$\mathbf{v}$ is a $K \times 1$ vector of parameters,

δ is a $K \times 1$ vector of trend coefficients,

t is a linear time trend,

$s_1, \ldots, s_m$ are orthogonalized seasonal indicators specified in the `sindicators()` option, and $\mathbf{w}_1, \ldots, \mathbf{w}_m$ are $K \times 1$ vectors of coefficients on the orthogonalized seasonal indicators.

There are two types of deterministic elements in (1): the trend, $\mathbf{v} + \delta t$, and the orthogonalized seasonal terms, $\mathbf{w}_1 s_1 + \cdots + \mathbf{w}_m s_m$. Johansen (1995, chapter 11) shows that inference about the number of cointegrating equations is based on nonstandard distributions and that the addition of any term that generalizes the deterministic specification in (1) changes the asymptotic distributions of the statistics used for inference on the number of cointegrating equations and the asymptotic distribution of the ML estimator of the cointegrating equations. In fact, Johansen (1995, 84) notes that including event indicators causes the statistics used for inference on the number of cointegrating equations to have asymptotic distributions that must be computed case by case. For this reason, event indicators may not be specified in the current version of `vec`.

If seasonal indicators are included in the model, they cannot be collinear with a constant term. If they are collinear with a constant term, one of the indicator variables is dropped.

As discussed in *Specification of constants and trends*, we can reparametrize the model as

$$\Delta \mathbf{y}_t = \boldsymbol{\alpha}(\boldsymbol{\beta} \mathbf{y}_{t-1} + \boldsymbol{\mu} + \boldsymbol{\rho} t) + \sum_{i=1}^{p-1} \boldsymbol{\Gamma}_i \Delta \mathbf{y}_{t-i} + \boldsymbol{\gamma} + \boldsymbol{\tau} t + \boldsymbol{\epsilon}_t \tag{2}$$

The log-likelihood function

We can maximize the log-likelihood function much more easily by writing it in concentrated form. In fact, as discussed below, in the simple case with the Johansen normalization on $\boldsymbol{\beta}$ and no constraints on $\boldsymbol{\alpha}$, concentrating the log-likelihood function produces an analytical solution for the parameter estimates.

To concentrate the log likelihood, rewrite (2) as

$$\mathbf{Z}_{0t} = \boldsymbol{\alpha} \widetilde{\boldsymbol{\beta}}' \mathbf{Z}_{1t} + \boldsymbol{\Psi} \mathbf{Z}_{2t} + \boldsymbol{\epsilon}_t \tag{3}$$

where $\mathbf{Z}_{0t}$ is a $K \times 1$ vector of variables $\Delta \mathbf{y}_t$, $\boldsymbol{\alpha}$ is the $K \times r$ matrix of adjustment coefficients, and $\boldsymbol{\epsilon}_t$ is a $K \times 1$ vector of independently and identically distributed normal vectors with mean 0 and contemporaneous covariance matrix $\boldsymbol{\Omega}$. $\mathbf{Z}_{1t}$, $\mathbf{Z}_{2t}$, $\widetilde{\boldsymbol{\beta}}$, and $\boldsymbol{\Psi}$ depend on the trend specification and are defined below.

The log-likelihood function for the model in (3) is

$$\begin{aligned} \mathrm{L} = -\frac{1}{2} \Big\{ & TK \ln(2\pi) + T \ln(|\boldsymbol{\Omega}|) \\ & + \sum_{t=1}^{T} (\mathbf{Z}_{0t} - \boldsymbol{\alpha} \widetilde{\boldsymbol{\beta}}' \mathbf{Z}_{1t} - \boldsymbol{\Psi} \mathbf{Z}_{2t})' \boldsymbol{\Omega}^{-1} (\mathbf{Z}_{0t} - \boldsymbol{\alpha} \widetilde{\boldsymbol{\beta}}' \mathbf{Z}_{1t} - \boldsymbol{\Psi} \mathbf{Z}_{2t}) \Big\} \end{aligned} \tag{4}$$

with the constraints that $\boldsymbol{\alpha}$ and $\widetilde{\boldsymbol{\beta}}$ have rank r.

Johansen (1995, chapter 6), building on Anderson (1951), shows how the $\boldsymbol{\Psi}$ parameters can be expressed as analytic functions of $\boldsymbol{\alpha}$, $\widetilde{\boldsymbol{\beta}}$, and the data, yielding the concentrated log-likelihood function

$$\begin{aligned} L_c = -\frac{1}{2} \Big\{ & TK \ln(2\pi) + T \ln(|\boldsymbol{\Omega}|) \\ & + \sum_{t=1}^{T} (\mathbf{R}_{0t} - \boldsymbol{\alpha} \widetilde{\boldsymbol{\beta}}' \mathbf{R}_{1t})' \boldsymbol{\Omega}^{-1} (\mathbf{R}_{0t} - \boldsymbol{\alpha} \widetilde{\boldsymbol{\beta}}' \mathbf{R}_{1t}) \Big\} \end{aligned} \tag{5}$$

where

$$\mathbf{M}_{ij} = T^{-1} \sum_{t=1}^{T} \mathbf{Z}_{it} \mathbf{Z}_{jt}', \qquad i, j \in \{0, 1, 2\};$$

$$\mathbf{R}_{0t} = \mathbf{Z}_{0t} - \mathbf{M}_{02} \mathbf{M}_{22}^{-1} \mathbf{Z}_{2t}; \text{ and}$$

$$\mathbf{R}_{1t} = \mathbf{Z}_{1t} - \mathbf{M}_{12} \mathbf{M}_{22}^{-1} \mathbf{Z}_{2t}.$$

The definitions of $\mathbf{Z}_{1t}$, $\mathbf{Z}_{2t}$, $\widetilde{\beta}$, and $\boldsymbol{\Psi}$ change with the trend specifications, although some of their components stay the same.

Unrestricted trend

When the trend in the VECM is unrestricted, we can define the variables in (3) directly in terms of the variables in (1):

$\mathbf{Z}_{1t} = \mathbf{y}_{t-1}$ is $K \times 1$

$\mathbf{Z}_{2t} = (\Delta \mathbf{y}_{t-1}', \ldots, \Delta \mathbf{y}_{t-p+1}', 1, t, s_1, \ldots, s_m)'$ is $\{K(p-1) + 2 + m\} \times 1$;

$\boldsymbol{\Psi} = (\boldsymbol{\Gamma}_1, \ldots, \boldsymbol{\Gamma}_{p-1}, \mathbf{v}, \boldsymbol{\delta}, \mathbf{w}_1, \ldots, \mathbf{w}_m)$ is $K \times \{K(p-1) + 2 + m\}$

$\widetilde{\beta} = \beta$ is the $K \times r$ matrix composed of the r cointegrating vectors.

In the unrestricted trend specification, $m_1 = K$, $m_2 = K(p-1) + 2 + m$, and there are $n_{\text{parms}} = Kr + Kr + K\{K(p-1) + 2 + m\}$ parameters in (3).

Restricted trend

When there is a restricted trend in the VECM in (2), $\boldsymbol{\tau} = 0$, but the intercept $\mathbf{v} = \boldsymbol{\alpha}\boldsymbol{\mu} + \boldsymbol{\gamma}$ is unrestricted. The VECM with the restricted trend can be written as

$$\Delta \mathbf{y}_t = \boldsymbol{\alpha}(\boldsymbol{\beta}', \boldsymbol{\rho}) \begin{pmatrix} \mathbf{y}_{t-1} \\ t \end{pmatrix} + \sum_{i=1}^{p-1} \boldsymbol{\Gamma}_i \Delta \mathbf{y}_{t-i} + \mathbf{v} + \mathbf{w}_1 s_1 + \cdots + \mathbf{w}_m s_m + \epsilon_t$$

This equation can be written in the form of (3) by defining

$\mathbf{Z}_{1t} - (\mathbf{y}_{t-1}', t)'$ is $(K+1) \times 1$

$\mathbf{Z}_{2t} = (\Delta \mathbf{y}_{t-1}', \ldots, \Delta \mathbf{y}_{t-p+1}', 1, s_1, \ldots, s_m)'$ is $\{K(p-1) + 1 + m\} \times 1$

$\boldsymbol{\Psi} = (\boldsymbol{\Gamma}_1, \ldots, \boldsymbol{\Gamma}_{p-1}, \mathbf{v}, \mathbf{w}_1, \ldots, \mathbf{w}_m)$ is $K \times \{K(p-1) + 1 + m\}$

$\widetilde{\beta} = (\boldsymbol{\beta}', \boldsymbol{\rho})'$ is the $(K+1) \times r$ matrix composed of the r cointegrating vectors and the r trend coefficients $\boldsymbol{\rho}$

In the restricted trend specification, $m_1 = K + 1$, $m_2 = \{K(p-1) + 1 + m\}$, and there are $n_{\text{parms}} = Kr + (K+1)r + K\{K(p-1) + 1 + m\}$ parameters in (3).

Unrestricted constant

An unrestricted constant in the VECM in (2) is equivalent to setting $\boldsymbol{\delta} = 0$ in (1), which can be written in the form of (3) by defining

$\mathbf{Z}_{1t} = \mathbf{y}_{t-1}$ is $(K \times 1)$

$\mathbf{Z}_{2t} = (\Delta \mathbf{y}_{t-1}', \ldots, \Delta \mathbf{y}_{t-p+1}', 1, s_1, \ldots, s_m)'$ is $\{K(p-1) + 1 + m\} \times 1$;

$\boldsymbol{\Psi} = (\boldsymbol{\Gamma}_1, \ldots, \boldsymbol{\Gamma}_{p-1}, \mathbf{v}, \mathbf{w}_1, \ldots, \mathbf{w}_m)$ is $K \times \{K(p-1) + 1 + m\}$

$\widetilde{\beta} = \beta$ is the $K \times r$ matrix composed of the r cointegrating vectors

In the unrestricted constant specification, $m_1 = K$, $m_2 = \{K(p-1) + 1 + m\}$, and there are $n_{\text{parms}} = Kr + Kr + K\{K(p-1) + 1 + m\}$ parameters in (3).

Restricted constant

When there is a restricted constant in the VECM in (2), it can be written in the form of (3) by defining

$\mathbf{Z}_{1t} = \left(\mathbf{y}'_{t-1}, 1\right)'$ is $(K+1) \times 1$

$\mathbf{Z}_{2t} = (\Delta\mathbf{y}'_{t-1}, \ldots, \Delta\mathbf{y}'_{t-p+1})'$ is $K(p-1) \times 1$

$\mathbf{\Psi} = (\mathbf{\Gamma}_1, \ldots, \mathbf{\Gamma}_{p-1})$ is $K \times K(p-1)$

$\widetilde{\boldsymbol{\beta}} = \left(\boldsymbol{\beta}', \boldsymbol{\mu}\right)'$ is the $(K+1) \times r$ matrix composed of the r cointegrating vectors and the r constants in the cointegrating relations.

In the restricted trend specification, $m_1 = K + 1$, $m_2 = K(p-1)$, and there are $n_{\text{parms}} = Kr + (K+1)r + K\{K(p-1)\}$ parameters in (3).

No trend

When there is no trend in the VECM in (2), it can be written in the form of (3) by defining

$\mathbf{Z}_{1t} = \mathbf{y}_{t-1}$ is $K \times 1$

$\mathbf{Z}_{2t} = (\Delta\mathbf{y}'_{t-1}, \ldots, \Delta\mathbf{y}'_{t-p+1})'$ is $K(p-1) + m \times 1$

$\mathbf{\Psi} = (\mathbf{\Gamma}_1, \ldots, \mathbf{\Gamma}_{p-1})$ is $K \times K(p-1)$

$\widetilde{\boldsymbol{\beta}} = \boldsymbol{\beta}$ is $K \times r$ matrix of r cointegrating vectors

In the no-trend specification, $m_1 = K$, $m_2 = K(p-1)$, and there are $n_{\text{parms}} = Kr + Kr + K\{K(p-1)\}$ parameters in (3).

Estimation with Johansen identification

Not all the parameters in $\boldsymbol{\alpha}$ and $\widetilde{\boldsymbol{\beta}}$ are identified. Consider the simple case in which $\widetilde{\boldsymbol{\beta}}$ is $K \times r$ and let $\mathbf{Q}$ be a nonsingular $r \times r$ matrix. Then

$$\boldsymbol{\alpha}\widetilde{\boldsymbol{\beta}}' = \boldsymbol{\alpha}\mathbf{Q}\mathbf{Q}^{-1}\widetilde{\boldsymbol{\beta}}' = \boldsymbol{\alpha}\mathbf{Q}(\widetilde{\boldsymbol{\beta}}\mathbf{Q}^{'-1})' = \dot{\boldsymbol{\alpha}}\dot{\boldsymbol{\beta}}'$$

Substituting $\dot{\boldsymbol{\alpha}}\dot{\boldsymbol{\beta}}'$ into the log likelihood in (5) for $\boldsymbol{\alpha}\widetilde{\boldsymbol{\beta}}'$ would not change the value of the log likelihood, so some *a priori* identification restrictions must be found to identify $\boldsymbol{\alpha}$ and $\widetilde{\boldsymbol{\beta}}$. As discussed in Johansen (1995, chapters 5 and 6) and Boswijk (1995), if the restrictions exactly identify or overidentify $\widetilde{\boldsymbol{\beta}}$, the estimates of the unconstrained parameters in $\widetilde{\boldsymbol{\beta}}$ will be superconsistent, meaning that the estimates of the free parameters in $\widetilde{\boldsymbol{\beta}}$ will converge at a faster rate than estimates of the short-run parameters in $\boldsymbol{\alpha}$ and $\mathbf{\Gamma}_i$. This allows the distribution of the estimator of the short-run parameters to be derived conditional on the estimated $\widetilde{\boldsymbol{\beta}}$.

Johansen (1995, chapter 6) has proposed a normalization method for use when theory does not provide sufficient *a priori* restrictions to identify the cointegrating vector. This method has become widely adopted by researchers. Johansen's identification scheme is

$$\widetilde{\boldsymbol{\beta}}' = (\mathbf{I}_r, \breve{\boldsymbol{\beta}}') \tag{6}$$

where $\mathbf{I}_r$ is the $r \times r$ identity matrix and $\breve{\boldsymbol{\beta}}$ is a $(m_1 - r) \times r$ matrix of identified parameters.

Johansen's identification method places r^2 linearly independent constraints on the parameters in $\widetilde{\beta}$, thereby defining an exactly identified model. The total number of freely estimated parameters is $n_{\text{parms}} - r^2 = \{K + m_2 + (K + m_1 - r)r\}$, and the degrees of freedom d is calculated as the integer part of $(n_{\text{parms}} - r^2)/K$.

When only the rank and the Johansen identification restrictions are placed on the model, we can further manipulate the log likelihood in (5) to obtain analytic formulas for the parameters in $\widetilde{\beta}$, α, and Ω. For a given value of $\widetilde{\beta}$, α and Ω can be found by regressing $\mathbf{R}_{0t}$ on $\widetilde{\beta}' \mathbf{R}_{1t}$. This allows a further simplification of the problem in which

$$\alpha(\widetilde{\beta}) = \mathbf{S}_{01}\widetilde{\beta}(\widetilde{\beta}' \mathbf{S}_{11}\widetilde{\beta})^{-1}$$

$$\Omega(\widetilde{\beta}) = \mathbf{S}_{00} - \mathbf{S}_{01}\widetilde{\beta}(\widetilde{\beta}' \mathbf{S}_{11}\widetilde{\beta})^{-1}\widetilde{\beta}' \mathbf{S}_{10}$$

$$\mathbf{S}_{ij} = (1/T)\sum_{t=1}^{T} R_{it}R'_{jt} \qquad i,j \in \{0,1\}$$

Johansen (1995) shows that by inserting these solutions into equation (5), $\widehat{\beta}$ is given by the r eigenvectors $\mathbf{v}_1, \ldots, \mathbf{v}_r$ corresponding to the r largest eigenvalues $\lambda_1, \ldots, \lambda_r$ that solve the generalized eigenvalue problem

$$|\lambda_i \mathbf{S}_{11} - \mathbf{S}_{10}\mathbf{S}_{00}^{-1}\mathbf{S}_{01}| = 0 \tag{7}$$

The eigenvectors corresponding to $\lambda_1, \ldots, \lambda_r$ that solve (7) are the unidentified parameter estimates. To impose the identification restrictions in (6), we normalize the eigenvectors such that

$$\lambda_i \mathbf{S}_{11}\mathbf{v}_i = \mathbf{S}_{01}\mathbf{S}_{00}^{-1}\mathbf{S}_{01}\mathbf{v}_i \tag{8}$$

and

$$\mathbf{v}'_i \mathbf{S}_{11}\mathbf{v}_j = \begin{cases} 1 \text{ if } i = j \\ 0 \text{ otherwise} \end{cases} \tag{9}$$

At the optimum the log-likelihood function with the Johansen identification restrictions can be expressed in terms of T, K, $\mathbf{S}_{00}$, and the r largest eigenvalues

$$L_c = -\frac{1}{2}T\left\{ K\ln(2\pi) + K + \ln(|\mathbf{S}_{00}|) + \sum_{i=1}^{r}\ln(1 - \widehat{\lambda}_i) \right\}$$

where the $\widehat{\lambda}_i$ are the eigenvalues that solve (7), (8), and (9).

Using the normalized $\widehat{\beta}$, we can then obtain the estimates

$$\widehat{\alpha} = \mathbf{S}_{01}\widehat{\beta}(\widehat{\beta}' S_{11}\widehat{\beta})^{-1} \tag{10}$$

and

$$\widehat{\Omega} = \mathbf{S}_{00} - \widehat{\alpha}\widehat{\beta}' \mathbf{S}_{10}$$

Let $\widehat{\beta}_y$ be a $K \times r$ matrix that contains the estimates of the parameters in β in (1). $\widehat{\beta}_y$ differs from $\widehat{\beta}$ in that any trend parameter estimates are dropped from $\widehat{\beta}$. We can then use $\widehat{\beta}_y$ to obtain predicted values for the r nondemeaned cointegrating equations

$$\widehat{\widetilde{\mathbf{E}}}_t = \widehat{\beta}'_y \mathbf{y}_t$$

The r series in $\widetilde{\widehat{E}}_t$ are called the predicted, nondemeaned cointegrating equations because they still contain the terms μ and ρ. We want to work with the predicted, demeaned cointegrating equations. Thus we need estimates of μ and ρ. In the trend(rconstant) specification, the algorithm directly produces the estimator $\widehat{\mu}$. Similarly, in the trend(rtrend) specification, the algorithm directly produces the estimator $\widehat{\rho}$. In the remaining cases, to back out estimates of μ and ρ, we need estimates of $\mathbf{v}$ and δ, which we can obtain by estimating the parameters of the following VAR:

$$\Delta \mathbf{y}_t = \alpha \widetilde{\widehat{\mathbf{E}}}_{t-1} + \sum_{i=1}^{p-1} \Gamma_i \Delta \mathbf{y}_{t-i} + \mathbf{v} + \delta t + \mathbf{w}_1 s_1 + \cdots + \mathbf{w}_m s_m + \epsilon_t \tag{11}$$

Depending on the trend specification, we use $\widehat{\alpha}$ to back out the estimates of

$$\widehat{\mu} = (\widehat{\alpha}'\widehat{\alpha})^{-1}\widehat{\alpha}'\widehat{\mathbf{v}} \tag{12}$$

$$\widehat{\rho} = (\widehat{\alpha}'\widehat{\alpha})^{-1}\widehat{\alpha}'\widehat{\delta} \tag{13}$$

if they are not already in $\widehat{\beta}$ and are included in the trend specification.

We then augment $\widehat{\beta}_y$ to

$$\widehat{\beta}_f' = (\widehat{\beta}_y', \widehat{\mu}, \widehat{\rho})$$

where the estimates of $\widehat{\mu}$ and $\widehat{\rho}$ are either obtained from $\widehat{\beta}$ or backed out using (12) and (13). We next use $\widehat{\beta}_f$ to obtain the r predicted, demeaned cointegrating equations, $\widehat{\mathbf{E}}_t$, via

$$\widehat{\mathbf{E}}_t = \widehat{\beta}_f' (\mathbf{y}_t', 1, t)'$$

We last obtain estimates of all the short-run parameters from the VAR:

$$\Delta \mathbf{y}_t = \alpha \widehat{\mathbf{E}}_{t-1} + \sum_{i=1}^{p-1} \Gamma_i \Delta \mathbf{y}_{t-i} + \gamma + \tau t + \mathbf{w}_1 s_1 + \cdots + \mathbf{w}_m s_m + \epsilon_t \tag{14}$$

Because the estimator $\widehat{\beta}_f$ converges in probability to its true value at a rate faster than $T^{-\frac{1}{2}}$, we can take our estimated $\widehat{\mathbf{E}}_{t-1}$ as given data in (14). This allows us to estimate the variance–covariance (VCE) matrix of the estimates of the parameters in (14) using the standard VAR VCE estimator. Equation (11) can be used to obtain consistent estimates of all the parameters and of the VCE of all the parameters, except $\mathbf{v}$ and δ. The standard VAR VCE of $\widehat{\mathbf{v}}$ and $\widehat{\delta}$ is incorrect because these estimates converge at a faster rate. This is why it is important to use the predicted, demeaned cointegrating equations, $\widehat{\mathbf{E}}_{t-1}$, when estimating the short-run parameters and trend terms. In keeping with the cointegration literature, vec makes a small-sample adjustment to the VCE estimator so that the divisor is $(T - d)$ instead of T, where d represents the degrees of freedom of the model. d is calculated as the integer part of n_{parms}/K, where n_{parms} is the total number of freely estimated parameters in the model.

In the trend(rconstant) specification, the estimation procedure directly estimates μ. For trend(constant), trend(rtrend), and trend(trend), the estimates of μ are backed out using (12). In the trend(rtrend) specification, the estimation procedure directly estimates ρ. In the trend(trend) specification, the estimates of ρ are backed out using (13). Since the elements of the estimated VCE are only readily available when the estimates are obtained directly, when the trend parameter estimates are backed out, their elements in the VCE for $\widehat{\beta}_f$ are missing.

Under the Johansen identification restrictions, `vec` obtains $\widehat{\boldsymbol{\beta}}$, the estimates of the parameters in the $r \times m_1$ matrix $\widetilde{\boldsymbol{\beta}}'$ in (5). The VCE of $\mathrm{vec}(\widehat{\boldsymbol{\beta}})$ is $rm_1 \times rm_1$. Following Johansen (1995), the asymptotic distribution of $\widehat{\boldsymbol{\beta}}$ is mixed Gaussian, and its VCE is consistently estimated by

$$\left(\frac{1}{T-d} \right) (\mathbf{I}_r \otimes \mathbf{H}_J) \left\{ (\widehat{\boldsymbol{\alpha}}' \boldsymbol{\Omega}^{-1} \widehat{\boldsymbol{\alpha}}) \otimes (\mathbf{H}'_J \mathbf{S}_{11} \mathbf{H}_J) \right\}^{-1} (\mathbf{I}_r \otimes \mathbf{H}_J)' \tag{15}$$

where $\mathbf{H}_J$ is the $m_1 \times (m_1 - r)$ matrix given by $\mathbf{H}_J = (\mathbf{0}'_{r \times (m_1 - r)}, \mathbf{I}_{m_1 - r})'$. The VCE reported in `e(V_beta)` is the estimated VCE in (15) augmented with missing values to account for any backed-out estimates of $\boldsymbol{\mu}$ or $\boldsymbol{\rho}$.

The parameter estimates $\widehat{\boldsymbol{\alpha}}$ can be found either as a function of $\widehat{\boldsymbol{\beta}}$, using (10), or from the VAR in (14). The estimated VCE of $\widehat{\boldsymbol{\alpha}}$ reported in `e(V_alpha)` is given by

$$\frac{1}{(T-d)} \widehat{\boldsymbol{\Omega}} \otimes \widehat{\boldsymbol{\Sigma}}_B$$

where $\widehat{\boldsymbol{\Sigma}}_B = (\widehat{\boldsymbol{\beta}}' \mathbf{S}_{11} \widehat{\boldsymbol{\beta}})^{-1}$.

As we would expect, the estimator of $\boldsymbol{\Pi} = \boldsymbol{\alpha} \boldsymbol{\beta}'$ is

$$\widehat{\boldsymbol{\Pi}} = \widehat{\boldsymbol{\alpha}} \widehat{\boldsymbol{\beta}}'$$

and its estimated VCE is given by

$$\frac{1}{(T-d)} \widehat{\boldsymbol{\Omega}} \otimes (\widehat{\boldsymbol{\beta}} \widehat{\boldsymbol{\Sigma}}_B \widehat{\boldsymbol{\beta}}')$$

The moving-average impact matrix $\mathbf{C}$ is estimated by

$$\widehat{\mathbf{C}} = \widehat{\boldsymbol{\beta}}_\perp (\widehat{\boldsymbol{\alpha}}_\perp \widehat{\boldsymbol{\Gamma}} \widehat{\boldsymbol{\beta}}_\perp)^{-1} \widehat{\boldsymbol{\alpha}}'_\perp$$

where $\widehat{\boldsymbol{\beta}}_\perp$ is the orthogonal complement of $\widehat{\boldsymbol{\beta}}_y$, $\widehat{\boldsymbol{\alpha}}_\perp$ is the orthogonal complement of $\widehat{\boldsymbol{\alpha}}$, and $\widehat{\boldsymbol{\Gamma}} = \mathbf{I}_K - \sum_{i=1}^{p-1} \boldsymbol{\Gamma}_i$. The orthogonal complement of a $K \times r$ matrix $\mathbf{Q}$ that has rank r is a matrix $\mathbf{Q}_\perp$ of rank $K - r$, such that $\mathbf{Q}' \mathbf{Q}_\perp = \mathbf{0}$. While this operation is not uniquely defined, the results used by `vec` do not depend on the method of obtaining the orthogonal complement. `vec` uses the following method: the orthogonal complement of $\mathbf{Q}$ is given by the r eigenvectors with the highest eigenvalues from the matrix $\mathbf{Q}'(\mathbf{Q}'\mathbf{Q})^{-1}\mathbf{Q}'$.

Following Johansen (1995, chapter 13) and Drukker (2004), the VCE of $\widehat{\mathbf{C}}$ is estimated by

$$\frac{T-d}{T} \widehat{\mathbf{S}}_q \widehat{\mathbf{V}}_{\widehat{\boldsymbol{\nu}}} \widehat{\mathbf{S}}'_q \tag{16}$$

where

$$\widehat{\mathbf{S}}_q = \widehat{\mathbf{C}} \otimes \widehat{\boldsymbol{\xi}}$$

$$\widehat{\boldsymbol{\xi}} = \begin{cases} (\widehat{\boldsymbol{\xi}}_1, \widehat{\boldsymbol{\xi}}_2) & \text{if } p > 1 \\ \widehat{\boldsymbol{\xi}}_1 & \text{if } p = 1 \end{cases}$$

$$\widehat{\boldsymbol{\xi}}_1 = (\widehat{\mathbf{C}}' \widehat{\boldsymbol{\Gamma}}' - \mathbf{I}_K) \bar{\boldsymbol{\alpha}}$$

$$\bar{\alpha} = \widehat{\alpha}(\widehat{\alpha}'\widehat{\alpha})^{-1}$$

$$\widehat{\xi}_2 = \iota_{p-1} \otimes \widehat{\mathbf{C}}$$

ι_{p-1} is a $(p-1) \times 1$ vector of ones

$\widehat{\mathbf{V}}_{\widehat{\nu}}$ is the estimated VCE of $\widehat{\nu} = (\widehat{\alpha}, \widehat{\mathbf{\Gamma}}_1, \dots \widehat{\mathbf{\Gamma}}_{p-1})$

Estimation with constraints: β identified

vec can also fit models in which the adjustment parameters are subject to homogeneous linear constraints and the cointegrating vectors are subject to general linear restrictions. Mathematically, vec allows for constraints of the form

$$\mathbf{R}'_{\alpha} \mathrm{vec}(\alpha) = \mathbf{0} \tag{17}$$

where $\mathbf{R}_{\alpha}$ is a known $Kr \times n_{\alpha}$ constraint matrix, and

$$\mathbf{R}'_{\widetilde{\beta}} \mathrm{vec}(\widetilde{\beta}) = \mathbf{b} \tag{18}$$

where $\mathbf{R}_{\widetilde{\beta}}$ is a known $m_1 r \times n_{\beta}$ constraint matrix and $\mathbf{b}$ is a known $n_{\beta} \times 1$ vector of constants. While (17) and (18) are intuitive, they can be rewritten in a form to facilitate computation. Specifically, (17) can be written as

$$\mathrm{vec}(\alpha') = \mathbf{Ga} \tag{19}$$

where $\mathbf{G}$ is $Kr \times n_{\alpha}$ and $\mathbf{a}$ is $n_{\alpha} \times 1$. Equation (18) can be rewritten as

$$\mathrm{vec}(\widetilde{\beta}) = \mathbf{Hb} + \mathbf{h}_0 \tag{20}$$

where $\mathbf{H}$ is a known $n_1 r \times n_{\beta}$ matrix, $\mathbf{b}$ is an $n_{\beta} \times 1$ matrix of parameters, and $\mathbf{h}_0$ is a known $n_1 r \times 1$ matrix. See [P] **makecns** for a discussion of the different ways of specifying the constraints.

When constraints are specified via the aconstraints() and bconstraints() options, the Boswijk (1995) rank method determines whether the parameters in $\widetilde{\beta}$ are underidentified, exactly identified, or overidentified.

Boswijk (1995) uses the Rothenberg (1971) method to determine whether the parameters in $\widetilde{\beta}$ are identified. Thus the parameters in $\widetilde{\beta}$ are exactly identified if $\rho_{\beta} = r^2$, and the parameters in $\widetilde{\beta}$ are overidentified if $\rho_{\beta} > r^2$, where

$$\rho_{\beta} = \mathrm{rank}\left\{\mathbf{R}_{\widetilde{\beta}}(\mathbf{I}_r \otimes \ddot{\beta})\right\}$$

and $\ddot{\beta}$ is a full-rank matrix with the same dimensions as $\widetilde{\beta}$. The computed ρ_{β} is saved in e(beta_icnt).

Similarly, the number of freely estimated parameters in α and $\widetilde{\beta}$ is given by ρ_{jacob}, where

$$\rho_{\mathrm{jacob}} = \mathrm{rank}\left\{(\widehat{\alpha} \otimes \mathbf{I}_{m_1})\mathbf{H}, (\mathbf{I}_K \otimes \widehat{\beta})\mathbf{G}\right\}$$

Using ρ_{jacob}, we can calculate several other parameter counts of interest. In particular, the degrees of freedom of the overidentifying test are given by $(K + m_1 - r)r - \rho_{\mathrm{jacob}}$, and the number of freely estimated parameters in the model is $n_{\mathrm{parms}} = Km_2 + \rho_{\mathrm{jacob}}$.

While the problem of maximizing the log-likelihood function in (4), subject to the constraints in (17) and (18), could be handled by the algorithms in [R] **ml**, the switching algorithm of Boswijk (1995) has proven to be more convergent. For this reason, vec uses the Boswijk (1995) switching algorithm to perform the optimization.

Given starting values $(\widehat{\mathbf{b}}_0, \widehat{\mathbf{a}}_0, \widehat{\mathbf{\Omega}}_0)$, the algorithm iteratively updates the estimates until convergence is achieved, as follows:

$\widehat{\boldsymbol{\alpha}}_j$ is constructed from (19) and $\widehat{\mathbf{a}}_j$

$\widehat{\boldsymbol{\beta}}_j$ is constructed from (20) and $\widehat{\mathbf{b}}_j$

$$\widehat{\mathbf{b}}_{j+1} = \{\mathbf{H}'(\widehat{\boldsymbol{\alpha}}_j'\widehat{\mathbf{\Omega}}_j^{-1}\widehat{\boldsymbol{\alpha}}_j \otimes \mathbf{S}_{11})\mathbf{H}\}^{-1}\mathbf{H}'(\widehat{\boldsymbol{\alpha}}_j\widehat{\mathbf{\Omega}}_j^{-1} \otimes \mathbf{S}_{11})\{\text{vec}(\widehat{\mathbf{P}}) - (\widehat{\boldsymbol{\alpha}}_j \otimes \mathbf{I}_{n_{Z1}})\mathbf{h}_0\}$$

$$\widehat{\mathbf{a}}_{j+1} = \{\mathbf{G}(\widehat{\mathbf{\Omega}}_j^{-1} \otimes \widehat{\boldsymbol{\beta}}_j\mathbf{S}_{11}\widehat{\boldsymbol{\beta}}_j)\mathbf{G}\}^{-1}\mathbf{G}'(\widehat{\mathbf{\Omega}}_j^{-1} \otimes \widehat{\boldsymbol{\beta}}_j\mathbf{S}_{11})\text{vec}(\widehat{\mathbf{P}})$$

$$\widehat{\mathbf{\Omega}}_{j+1} = \mathbf{S}_{00} - \mathbf{S}_{01}\widehat{\boldsymbol{\beta}}_j\widehat{\boldsymbol{\alpha}}_j' - \widehat{\boldsymbol{\alpha}}_j\widehat{\boldsymbol{\beta}}_j'\mathbf{S}_{10} + \widehat{\boldsymbol{\alpha}}_j\widehat{\boldsymbol{\beta}}_j'\mathbf{S}_{11}\widehat{\boldsymbol{\beta}}_j\widehat{\boldsymbol{\alpha}}_j'$$

The estimated VCE of $\widehat{\boldsymbol{\beta}}$ is given by

$$\frac{1}{(T-d)}\mathbf{H}\{\mathbf{H}'(\mathbf{W} \otimes \mathbf{S}_{11})\mathbf{H}\}^{-1}\mathbf{H}'$$

where $\mathbf{W}$ is $\widehat{\boldsymbol{\alpha}}'\widehat{\mathbf{\Omega}}^{-1}\widehat{\boldsymbol{\alpha}}$. As in the case without constraints, the estimated VCE of $\widehat{\boldsymbol{\alpha}}$ can be obtained either from the VCE of the short-run parameters, as described below, or via the formula

$$\widehat{V}_{\widehat{\boldsymbol{\alpha}}} = \frac{1}{(T-d)}\mathbf{G}\left[\mathbf{G}'\left\{\widehat{\mathbf{\Omega}}^{-1} \otimes (\widehat{\boldsymbol{\beta}}'\mathbf{S}_{11}\widehat{\boldsymbol{\beta}})\mathbf{G}\right\}^{-1}\right]\mathbf{G}'$$

Boswijk (1995) notes that, as long as the parameters of the cointegrating equations are exactly identified or overidentified, the constrained ML estimator produces superconsistent estimates of $\widetilde{\boldsymbol{\beta}}$. This implies that the method of estimating the short-run parameters described above applies in the presence of constraints, as well, albeit with a caveat: when there are constraints placed on $\boldsymbol{\alpha}$, the VARs must be estimated subject to these constraints.

With these estimates and the estimated VCE of the short-run parameter matrix $\widehat{\mathbf{V}}_{\widehat{\boldsymbol{\nu}}}$, Drukker (2004) shows that the estimated VCE for $\widehat{\mathbf{\Pi}}$ is given by

$$(\widehat{\boldsymbol{\beta}} \otimes \mathbf{I}_K)\widehat{V}_{\widehat{\boldsymbol{\alpha}}}(\widehat{\boldsymbol{\beta}} \otimes \mathbf{I}_K)'$$

Drukker (2004) also shows that the estimated VCE of $\widehat{\mathbf{C}}$ can be obtained from (16) with the extension that $\widehat{V}_{\widehat{\boldsymbol{\nu}}}$ is the estimated VCE of $\widehat{\boldsymbol{\nu}}$ that takes into account any constraints on $\widehat{\boldsymbol{\alpha}}$.

Estimation with constraints: β not identified

When the parameters in $\boldsymbol{\beta}$ are not identified, only the parameters in $\mathbf{\Pi} = \boldsymbol{\alpha}\boldsymbol{\beta}$ and $\mathbf{C}$ are identified. The estimates of $\mathbf{\Pi}$ and $\mathbf{C}$ would not change if additional identification restrictions were imposed to achieve exact identification. Thus the VCE matrices for $\widehat{\mathbf{\Pi}}$ and $\widehat{\mathbf{C}}$ can be derived as if the model exactly identified $\boldsymbol{\beta}$.

Formulas for the information criteria

The AIC, SBIC, and HQIC are calculated according to their standard definitions, which include the constant term from the log likelihood; that is,

$$\text{AIC} = -2\left(\frac{\text{L}}{T}\right) + \frac{2n_{\text{parms}}}{T}$$

$$\text{SBIC} = -2\left(\frac{\text{L}}{T}\right) + \frac{\ln(T)}{T}n_{\text{parms}}$$

$$\text{HQIC} = -2\left(\frac{\text{L}}{T}\right) + \frac{2\ln\{\ln(T)\}}{T}n_{\text{parms}}$$

where n_{parms} is the total number of parameters in the model and L is the value of the log likelihood at the optimum.

Formulas for predict

xb, residuals and stdp are standard and are documented in [R] **predict**. ce causes predict to compute $\widehat{E}_t = \widehat{\beta}_f \mathbf{y}_t$ for the requested cointegrating equation.

levels causes predict to compute the predictions for the levels of the data. Let $\widehat{y}_t^d$ be the predicted value of Δy_t. Since the computations are performed for a given equation, y_t is a scalar. Using $\widehat{y}_t^d$, we can predict the level by $\widehat{y}_t = \widehat{y}_t^d + y_{t-1}$.

Since the residuals from the VECM for the differences and the residuals from the corresponding VAR in levels are identical, there is no need for an option for predicting the residuals in levels.

References

Anderson, T. W. 1951. Estimating linear restrictions on regression coefficients for multivariate normal distributions. *Annals of Mathematical Statistics* 22: 327–351.

Boswijk, H. P. 1995. Identifiability of cointegrated systems. Tinbergen Institute Discussion Paper #95–78.

Drukker, D. M. 2004. Some further results on estimation and inference in the presence of constraints on alpha in a cointegrating VECM. Working Paper, StataCorp.

Engle, R. F. and C. W. J. Granger. 1987. Cointegration and error correction: Representation, estimation and testing. *Econometrica* 55(2): 251–276.

Hamilton, J. D. 1994. *Time Series Analysis*. Princeton: Princeton University Press.

Johansen, S. 1988. Statistical analysis of cointegration vectors. *Journal of Economic Dynamics and Control* 12: 231–254.

——. 1991. Estimation and hypothesis testing of cointegration vectors in Gaussian vector autoregressive models. *Econometrica* 59: 1551–1580.

——. 1995. *Likelihood-Based Inference in Cointegrated Vector Auto-Regressive Models*. Oxford: Oxford University Press.

Maddala, G. S. and I. M. Kim. 1998. *Unit Roots, Cointegration, and Structural Change*. Cambridge, UK: Cambridge University Press.

Park, J. Y. and P. C. B. Phillips. 1988a. Statistical inference in regressions with integrated regressors: Part I. *Econometric Theory* 4: 468–497.

——. 1988b. Statistical inference in regressions with integrated regressors: Part II. *Econometric Theory* 5: 95–131.

Phillips, P. C. B. 1986. Understanding spurious regressions in econometrics. *Journal of Econometrics* 33: 311–340.

Phillips, P. C. B. and S. N. Durlauf. 1986. Multiple time series regression with integrated processes. *The Review of Economic Studies* 53(4): 473–495.

Rothenberg, T. J. 1971. Identification in parametric models. *Econometrica* 39: 577–591.

Sims, C. A., J. H. Stock, and M. W. Watson. 1990. Inference in linear time series models with some unit roots. *Econometrica* 58(1): 113–144.

Stock, J. H. 1987. Asymptotic properties of least squares estimators of cointegrating vectors. *Econometrica* 55(5): 1035–1056.

Stock, J. H. and M. W. Watson. 1988. Testing for common trends. *Journal of the American Statistical Association* 404: 1099–1107.

Watson, M. W. 1994. Vector autoregressions and cointegration. *Handbook of Econometrics*, Vol IV, Engle, R. F. and McFadden, D. L. ed. Amsterdam: Elsevier.

Also See

Complementary:	[TS] **vec postestimation**; [TS] **tsset**
Related:	[TS] **arima**, [TS] **var**, [TS] **var svar**, [R] **sureg**
Background:	[U] **11.1.10 Prefix commands**, [U] **20 Estimation and postestimation commands**, [TS] **estimation options**, [TS] **vec intro** [R] **maximize**

Title

> **vec postestimation** — Postestimation tools for vec

Description

The following postestimation commands are of special interest after vec:

command	description
fcast compute	obtain dynamic forecasts
fcast graph	graph dynamic forecasts obtained from fcast compute
irf	create and analyze IRFs and FEVDs
veclmar	LM test for autocorrelation in residuals
vecnorm	test for normally distributed residuals
vecstable	check stability condition of estimates

For information about these commands, see the corresponding entries in this manual.

In addition, the following standard postestimation commands are available:

command	description
estat	AIC, BIC, VCE, and estimation sample summary
estimates	cataloging estimation results
lincom	point estimates, standard errors, testing, and inference for linear combinations of coefficients
lrtest	likelihood-ratio test
nlcom	point estimates, standard errors, testing, and inference for nonlinear combinations of coefficients
predict	predictions, residuals, influence statistics, and other diagnostic measures
predictnl	point estimates, standard errors, testing, and inference for generalized predictions
test	Wald tests for simple and composite linear hypotheses
testnl	Wald tests of nonlinear hypotheses

See the corresponding entries in the *Stata Base Reference Manual* for details.

Syntax for predict

predict [*type*] *newvar* [*if*] [*in*] [, equation(*eqno*|*eqname*) *statistic*]

statistic	description
xb	linear prediction; the default
stdp	standard error of the linear prediction
residuals	residuals
ce	the predicted value of specified cointegrating equation
levels	one-step prediction of the level of the endogenous variable
usece(*varlist_ce*)	compute the predictions using previously predicted cointegrating equations

These statistics are available both in and out of sample; type predict ... if e(sample) ... if wanted only for the estimation sample.

Options for predict

equation(*eqno* | *eqname*) specifies to which equation you are referring.

> equation() is filled in with one *eqno* or *eqname* for options xb, residuals, stdp, ce, and levels. equation(#1) would mean that the calculation is to be made for the first equation, equation(#2) would mean the second, and so on. Alternatively, you could refer to the equation by its name. equation(D_income) would refer to the equation named D_income and equation(_ce1) to the first cointegrating equation, which is named _ce1 by vec.

> If you do not specify equation(), the results are as if you specified equation(#1).

xb, the default, calculates the linear prediction for the specified equation. The form of the VECM implies that these fitted values are the one-step predictions for the first-differenced variables.

stdp calculates the standard error of the linear prediction for the specified equation.

residuals calculates the residuals from the specified equation of the VECM.

ce calculates the predicted value of the specified cointegrating equation.

levels calculates the one-step prediction of the level of the endogenous variable in the requested equation.

usece(*varlist_ce*) specifies that previously predicted cointegrating equations saved under the names in *varlist_ce* be used to compute the predictions. The number of variables in the *varlist_ce* must equal the number of cointegrating equations specified in the model.

For more information on using predict after multiple-equation estimation commands, see [R] **predict**.

Methods and Formulas

All postestimation commands listed above are implemented as ado-files.

Also See

Complementary:	[TS] **vec**;
	[TS] **fcast compute**, [TS] **fcast graph**, [TS] **irf**, [TS] **veclmar**,
	[TS] **vecnorm**, [TS] **vecstable**,
	[R] **estimates**, [R] **lincom**, [R] **lrtest**, [R] **nlcom**, [R] **predictnl**,
	[R] **test**, [R] **testnl**
Background:	[U] **13.5 Accessing coefficients and standard errors**,
	[U] **20 Estimation and postestimation commands**,
	[R] **estat**, [R] **predict**

Title

> **veclmar** — Obtain LM statistics for residual autocorrelation after vec

Syntax

> veclmar [, *options*]

options	description
mlag(#)	use # for the maximum order of autocorrelation; default is mlag(2)
estimates(*estname*)	use previously saved results *estname*; default is to use active results
separator(#)	draw separator line after every # rows

veclmar can be used only after vec; see [TS] **vec**.

You must tsset your data before using veclmar; see [TS] **tsset**.

Description

veclmar implements a Lagrange-multiplier (LM) test for autocorrelation in the residuals of vector error-correction models (VECMs).

Options

mlag(#) specifies the maximum order of autocorrelation to be tested. The integer specified in mlag() must be greater than 0; the default is 2.

estimates(*estname*) requests that veclmar use the previously obtained set of vec estimates saved as *estname*. By default, veclmar uses the active results. See [R] **estimates** for information on saving and restoring estimation results.

separator(#) specifies how many rows should appear in the table between separator lines. By default, separator lines do not appear. For example, separator(1) would draw a line between each row, separator(2) between every other row, and so on.

Remarks

Estimation, inference, and postestimation analysis of (VECMs) is predicated on the errors not being autocorrelated. veclmar implements the LM test for autocorrelation in the residuals of a VECM discussed in Johansen (1995, 21–22). The test is performed at lags $j = 1, \ldots, \text{mlag}()$. For each j, the null hypothesis of the test is that there is no autocorrelation at lag j.

▷ Example 1

We fit a VECM using the regional income data described in [TS] **vec** and then call veclmar to test for autocorrelation.

```
. use http://www.stata-press.com/data/r9/rdinc
. vec ln_ne ln_se
 (output omitted )
. veclmar, mlag(4)
   Lagrange-multiplier test
```

lag	chi2	df	Prob > chi2
1	8.9586	4	0.06214
2	4.9809	4	0.28926
3	4.8519	4	0.30284
4	0.3270	4	0.98801

```
   H0: no autocorrelation at lag order
```

At the 5% level, we cannot reject the null hypothesis that there is no autocorrelation in the residuals for any of the orders tested. Thus this test does not find any evidence of model misspecification.

◁

Saved Results

veclmar saves in r():

Matrix

r(lm) χ^2, df, and p-values

Methods and Formulas

veclmar is implemented as an ado-file.

Consider a VECM without any trend:

$$\Delta\mathbf{y}_t = \boldsymbol{\alpha}\boldsymbol{\beta}\mathbf{y}_{t-1} + \sum_{i=1}^{p-1}\boldsymbol{\Gamma}_i\Delta\mathbf{y}_{t-i} + \boldsymbol{\epsilon}_t$$

As discussed in [TS] **vec**, as long as the parameters in the cointegrating vectors, $\boldsymbol{\beta}$, are exactly identified or overidentified, the estimates of these parameters are superconsistent. This implies that the $r \times 1$ vector of estimated cointegrating relations

$$\widehat{\mathbf{E}}_t = \widehat{\boldsymbol{\beta}}\mathbf{y}_t \tag{1}$$

can be used as data with standard estimation and inference methods. When the parameters of the cointegrating equations are not identified, (1) does not provide consistent estimates of $\widehat{\mathbf{E}}_t$; in these cases, veclmar exits with an error message.

The VECM above can be rewritten as

$$\Delta\mathbf{y}_t = \boldsymbol{\alpha}\widehat{\mathbf{E}}_t + \sum_{i=1}^{p-1}\boldsymbol{\Gamma}_i\Delta\mathbf{y}_{t-i} + \boldsymbol{\epsilon}_t$$

which is just a VAR with $p - 1$ lags where the endogenous variables have been first-differenced and is augmented with the exogenous variables $\widehat{\mathbf{E}}$. veclmar fits this VAR and then calls varlmar to compute the LM test for autocorrelation.

The above discussion assumes no trend and implicitly ignores constraints on the parameters in α. As discussed in vec, the other four trend specifications considered by Johansen (1995, section 5.7) complicate the estimation of the free parameters in β but do not alter the basic result that the $\widehat{\mathbf{E}}_t$ can be used as data in the subsequent VAR. Similarly, constraints on the parameters in α imply that the subsequent VAR must be estimated with these constraints applied, but $\widehat{\mathbf{E}}_t$ can still be used as data in the VAR.

See [TS] **varlmar** for more information on the Johansen LM test.

Reference

Johansen, S. 1995. *Likelihood-Based Inference in Cointegrated Vector Auto-Regressive Models.* Oxford: Oxford University Press.

Also See

Complementary:	[TS] **vec**
Related:	[TS] **var**, [TS] **varlmar**
Background:	[TS] **vec intro**

Title

> **vecnorm** — Test for normally distributed disturbances after vec

Syntax

vecnorm [, *options*]

options	description
jbera	report Jarque–Bera statistic; default is to report all three statistics
skewness	report skewness statistic; default is to report all three statistics
kurtosis	report kurtosis statistic; default is to report all three statistics
estimates(*estname*)	use previously saved results *estname*; default is to use active results
dfk	make small-sample adjustment when computing the estimated variance–covariance matrix of the disturbances
separator(#)	draw separator line after every # rows

vecnorm can be used only after vec; see [TS] **vec**.

Description

vecnorm computes and reports a series of statistics against the null hypothesis that the disturbances in a VECM are normally distributed.

Options

jbera requests that the Jarque–Bera statistic and any other explicitly requested statistics be reported. By default, the Jarque–Bera, skewness, and kurtosis statistics are reported.

skewness requests that the skewness statistic and any other explicitly requested statistic be reported. By default, the Jarque–Bera, skewness, and kurtosis statistics are reported.

kurtosis requests that the kurtosis statistic and any other explicitly requested statistic be reported. By default, the Jarque–Bera, skewness, and kurtosis statistics are reported.

estimates(*estname*) requests that vecnorm use the previously obtained set of vec estimates saved as *estname*. By default, vecnorm uses the active results. See [R] **estimates** for information about saving and restoring estimation results.

dfk requests that a small-sample adjustment be made when computing the estimated variance–covariance matrix of the disturbances.

separator(#) specifies how many rows should appear in the table between separator lines. By default, separator lines do not appear. For example, separator(1) would draw a line between each row, separator(2) between every other row, and so on.

Remarks

vecnorm computes a series of test statistics for the null hypothesis that the disturbances in a VECM are normally distributed. For each equation and all equations jointly, up to three statistics may be computed: a skewness statistic, a kurtosis statistic, and the Jarque–Bera statistic. By default, all three statistics are reported; if you specify only one statistic, the others are not reported. The Jarque–Bera statistic tests skewness and kurtosis jointly. The single-equation results are against the null hypothesis that the disturbance for that particular equation is normally distributed. The results for all the equations are against the null that all K disturbances have a K-dimensional multivariate normal distribution. Failure to reject the null hypothesis indicates lack of model misspecification.

As noted by Johansen (1995, 141), the log likelihood for the VECM is derived assuming the errors are independently and identically distributed (i.i.d.) normal, though many of the asymptotic properties can be derived under the weaker assumption that the errors are merely i.i.d. Many researchers still prefer to test for normality. vecnorm uses the results from vec to produce a series of statistics against the null hypothesis that the K disturbances in the VECM are normally distributed.

▷ Example 1

This example uses vecnorm to test for normality after estimating the parameters of a VECM using the regional income data.

```
. use http://www.stata-press.com/data/r9/rdinc
. vec ln_ne ln_se
(output omitted)
. vecnorm
```

Jarque-Bera test

Equation	chi2	df	Prob > chi2
D_ln_ne	0.094	2	0.95417
D_ln_se	0.586	2	0.74608
ALL	0.680	4	0.95381

Skewness test

Equation	Skewness	chi2	df	Prob > chi2
D_ln_ne	.05982	0.032	1	0.85890
D_ln_se	.243	0.522	1	0.47016
ALL		0.553	2	0.75835

Kurtosis test

Equation	Kurtosis	chi2	df	Prob > chi2
D_ln_ne	3.1679	0.062	1	0.80302
D_ln_se	2.8294	0.064	1	0.79992
ALL		0.126	2	0.93873

The Jarque–Bera results present test statistics for each equation and for all equations jointly against the null hypothesis of normality. For the individual equations, the null hypothesis is that the disturbance term in that equation has a univariate normal distribution. For all equations jointly, the null hypothesis is that the K disturbances come from a K dimensional normal distribution. In this example, the single-equation and overall Jarque–Bera statistics do not reject the null of normality.

The single-equation skewness test statistics are for the null hypotheses that the disturbance term in each equation has zero skewness, which is the skewness of a normally distributed variable. The row marked ALL shows the results for a test that the disturbances in all equations jointly have zero skewness. The skewness results shown above do not suggest non-normality.

The kurtosis of a normally distributed variable is three, and the kurtosis statistics presented in the table test the null hypothesis that the disturbance terms have kurtosis consistent with normality. The results in this example do not reject the null hypothesis.

◁

The statistics computed by vecnorm are based on the estimated variance–covariance matrix of the disturbances. vec saves the ML estimate of this matrix, which vecnorm uses by default. Specifying the dfk option instructs vecnorm to make a small-sample adjustment to the variance–covariance matrix before computing the test statistics.

Saved Results

vecnorm saves in r():

Macros
    r(dfk)   dfk, if specified
Matrices
    r(jb)   Jarque–Bera χ^2, df, and p-values
    r(skewness) skewness χ^2, df, and p-values
    r(kurtosis) kurtosis χ^2, df, and p-values

Methods and Formulas

vecnorm is implemented as an ado-file.

As discussed in *Methods and Formulas* of [TS] **vec**, a cointegrating VECM can be rewritten as a VAR in first differences that includes the predicted cointegrating equations as exogenous variables. vecnorm computes the tests discussed in [TS] **varnorm** for the corresponding augmented VAR in first differences. See *Methods and Formulas* of [TS] **veclmar** for more information on this approach.

When the parameters of the cointegrating equations are not identified, the consistent estimates of the cointegrating equations are not available, and, in these cases, vecnorm exits with an error message.

References

Hamilton, J. D. 1994. *Time Series Analysis*. Princeton: Princeton University Press.

Jarque, C. M. and A. K. Bera. 1987. A test for normality of observations and regression residuals. *International Statistical Review* 55: 163-172.

Johansen, S. 1995. *Likelihood-Based Inference in Cointegrated Vector Auto-Regressive Models*. Oxford: Oxford University Press.

Lütkepohl, H. 1993. *Introduction to Multiple Time Series Analysis*. 2nd ed. New York: Springer.

Also See

Complementary:	[TS] **vec**
Background:	[TS] **vec intro**

Title

> **vecrank** — Estimate the cointegrating rank using Johansen's framework

Syntax

vecrank *depvar* [*if*] [*in*] [, *options*]

options	description
Model	
<u>l</u>ags(*#*)	use *#* for the maximum lag in underlying VAR model
trend(<u>c</u>onstant)	include an unrestricted constant in model; the default
trend(<u>rc</u>onstant)	include a restricted constant in model
trend(<u>t</u>rend)	include a linear trend in the cointegrating equations and a quadratic trend in the undifferenced data
trend(<u>rt</u>rend)	include a restricted trend in model
trend(<u>n</u>one)	do not include a trend or a constant model
Adv. model	
<u>s</u>indicators(*varlist*$_{si}$)	use normalized seasonal indicator variables *varlist*$_{si}$
noreduce	do not perform checks and corrections for collinearity among lags of dependent variables
Reporting	
<u>not</u>race	do not report the trace statistic
<u>max</u>	report maximum-eigenvalue statistic
<u>ic</u>	report information criteria
level99	report 1% critical values instead of 5% critical values
levela	report both 1% and 5% critical values

You must tsset your data before using vecrank; see [TS] **tsset**.

depvar may contain time-series operators; see [U] **11.4.3 Time-series varlists**.

by, rolling, and statsby may be used with vecrank; see [U] **11.1.10 Prefix commands**.

vecrank does not allow gaps in the data.

Description

vecrank produces statistics used to determine the number of cointegrating equations in a vector error-correction model (VECM).

Options

> ⌐ **Model** ⌐

lags(*#*) specifies the number of lags in the VAR representation of the model. The VECM will include one fewer lag of the first-differences. The number of lags must be greater than zero but small enough so that the degrees of freedom used by the model are less than the number of observations.

trend(*trend_spec*) specifies one of five trend specifications to include in the model. See [TS] **vec intro** and [TS] **vec** for descriptions. The default is trend(constant).

$sindicators(varlist_{si})$ specifies normalized seasonal indicator variables to be included in the model. The indicator variables specified in this option must be normalized as discussed in Johansen (1995, 84). If the indicators are not properly normalized, the likelihood-ratio-based tests for the number of cointegrating equations do not converge to the asymptotic distributions derived by Johansen. For details, see *Methods and Formulas* of [TS] **vec**. sindicators() cannot be specified with trend(none) or trend(rconstant)

noreduce causes vecrank to skip the checks and corrections for collinearity among the lags of the dependent variables. By default, vecrank checks if the current lag specification causes some of the regressions performed by vecrank to contain perfectly collinear variables and reduces the maximum lag until the perfect collinearity is removed. See *Collinearity* in [TS] **vec** for more information.

notrace requests that the output for the trace statistic not be displayed. The default is to display the trace statistic.

max requests that the output for the maximum-eigenvalue statistic be displayed. The default is to not display this output.

ic causes the output for the information criteria to be displayed. The default is to not display this output.

level99 causes the 1% critical values to be displayed instead of the default 5% critical values.

levela causes both the 1% and the 5% critical values to be displayed.

Remarks

Remarks are presented under the headings

Introduction
The trace statistic
The maximum-eigenvalue statistic
Minimizing an information criterion

Introduction

Before estimating the parameters of a VECM models, you must choose the number of lags in the underlying VAR, the trend specification, and the number of cointegrating equations. vecrank offers several ways of determining the number of cointegrating vectors conditional on a trend specification and lag order.

vecrank implements three types of methods for determining r, the number of cointegrating equations in a VECM. The first is Johansen's "trace" statistic method, the second is his "maximum eigenvalue" statistic method. The third method chooses r to minimize an information criterion.

All three methods are based on Johansen's maximum likelihood (ML) estimator of the parameters of a cointegrating VECM. The basic VECM is

$$\Delta \mathbf{y}_t = \alpha\beta'\mathbf{y}_{t-1} + \sum_{t=1}^{p-1} \Gamma_i \Delta\mathbf{y}_{t-i} + \epsilon_t$$

where $\mathbf{y}$ is a $(K \times 1)$ vector of I(1) variables, $\boldsymbol{\alpha}$ and $\boldsymbol{\beta}$ are $(K \times r)$ parameter matrices with rank $r < K$, $\boldsymbol{\Gamma}_1, \ldots, \boldsymbol{\Gamma}_{p-1}$ are $(K \times K)$ matrices of parameters, and ϵ_t is a $(K \times 1)$ vector of normally distributed errors that is serially uncorrelated but has contemporaneous covariance matrix $\boldsymbol{\Omega}$.

Building on the work of Anderson (1951), Johansen (1995) derives an ML estimator for the parameters and two likelihood-ratio (LR) tests for inference on r. These LR tests are known as the trace statistic and the maximum-eigenvalue statistic because the log likelihood can be written as the log of the determinant of a matrix plus a simple function of the eigenvalues of another matrix.

Let $\lambda_1, \ldots, \lambda_K$ be the K eigenvalues used in computing the log likelihood at the optimum. Furthermore, assume that these eigenvalues are sorted from the largest λ_1 to the smallest λ_K. If there are $r < K$ cointegrating equations, $\boldsymbol{\alpha}$ and $\boldsymbol{\beta}$ have rank r and the eigenvalues $\lambda_{r+1}, \ldots, \lambda_K$ are zero.

The trace statistic

The null hypothesis of the trace statistic is that there are no more than r cointegrating relations. Restricting the number of cointegrating equations to be r or less implies that the remaining $K - r$ eigenvalues are zero. Johansen (1995, chapters 11 and 12) derives the distribution of the trace statistic

$$-T \sum_{i=r+1}^{K} \ln(1 - \widehat{\lambda}_i)$$

where T is the number of observations and the $\widehat{\lambda}_i$ are the estimated eigenvalues. For any given value of r, large values of the trace statistic are evidence against the null hypothesis that there are r or fewer cointegrating relations in the VECM.

One of the problems in determining the number of cointegrating equations is that the process involves more than one statistical test. Johansen (1995, chapters 6, 11, and 12) derives a method based on the trace statistic that has nominal coverage despite evaluating multiple tests. This method can be interpreted as being an estimator $\widehat{r}$ of the true number of cointegrating equations r_0. The method starts testing at $r = 0$ and accepts as $\widehat{r}$ the first value of r for which the trace statistic fails to reject the null.

▷ Example 1

We have quarterly data on the natural logs of aggregate consumption, investment, and GDP in the United States from the first quarter of 1959 through the fourth quarter of 1982. As discussed in King et al. (1991), the balanced-growth hypothesis in economics implies that we would expect to find two cointegrating equations among these three variables. In the output below, we use `vecrank` to determine the number of cointegrating equations using Johansen's multiple-trace test method.

```
. use http://www.stata-press.com/data/r9/balance2
(macro data for VECM/balance study)

. vecrank y i c, lags(5)
                    Johansen tests for cointegration
Trend: constant                                    Number of obs =      91
Sample:    1960q2    1982q4                                   Lags =       5
```

maximum rank	parms	LL	eigenvalue	trace statistic	5% critical value
0	39	1231.1041	.	46.1492	29.68
1	44	1245.3882	0.26943	17.5810	15.41
2	47	1252.5055	0.14480	3.3465*	3.76
3	48	1254.1787	0.03611		

The header produces information about the sample, the trend specification, and the number of lags included in the model. The main table contains a separate row for each possible value of r, the number of cointegrating equations. Recall that when $r = 3$, all three variables in this model are stationary.

In this example, because the trace statistic at $r = 0$ of 46.1492 exceeds its critical value of 29.68, we reject the null hypothesis of no cointegrating equations. Similarly, because the trace statistic at $r = 1$ of 17.581 exceeds its critical value of 15.41, we reject the null hypothesis that there are 1 or fewer cointegrating equations. In contrast, since the trace statistic at $r = 2$ of 3.3465 is less than its critical value of 3.76, we cannot reject the null hypothesis that there are 2 or fewer cointegrating equations. Since Johansen's method for estimating r is to accept as $\hat{r}$ the first r for which the null hypothesis is not rejected, we accept $r = 2$ as our estimate of the number of cointegrating equations between these three variables. The "*" by the trace statistic at $r = 2$ indicates that this is the value of r selected by Johansen's multiple-trace test procedure. The eigenvalue shown in the last line of output computes the trace statistic in the preceding line.

◁

▷ Example 2

In the previous example, we used the default 5% critical values. We can estimate r using 1% critical values instead by specifying the level99 option.

```
. vecrank y i c, lags(5) level99
                    Johansen tests for cointegration
Trend: constant                                    Number of obs =      91
Sample:    1960q2    1982q4                                   Lags =       5
```

maximum rank	parms	LL	eigenvalue	trace statistic	1% critical value
0	39	1231.1041	.	46.1492	35.65
1	44	1245.3882	0.26943	17.5810*	20.04
2	47	1252.5055	0.14480	3.3465	6.65
3	48	1254.1787	0.03611		

The output indicates that switching from the 5% to the 1% level changes the resulting estimate from $r = 2$ to $r = 1$.

◁

The maximum-eigenvalue statistic

The alternative hypothesis of the trace statistic is that the number of cointegrating equations is strictly larger than the number r assumed under the null hypothesis. Instead, we could assume a given r under the null hypothesis and test this against the alternative that there are $r + 1$ cointegrating equations. Johansen (1995, chapters 6, 11, and 12) derives an LR test of the null of r cointegrating relations against the alternative of $r + 1$ cointegrating relations. Because the part of the log likelihood that changes with r is a simple function of the eigenvalues of a $(K \times K)$ matrix, this test is known as the maximum-eigenvalue statistic. This method is used less frequently than the trace statistic method because no solution to the multiple-testing problem has yet been found.

▷ Example 3

In the output below, we re-examine the balanced-growth hypothesis. We use the `levela` option to obtain both the 5% and 1% critical values, and we use the `notrace` option to suppress the table of trace statistics.

```
. vecrank y i c, lags(5) max levela notrace
                        Johansen tests for cointegration
   Trend: constant                              Number of obs =      91
   Sample:   1960q2   1982q4                              Lags =       5

   maximum                             max     5% critical   1% critical
      rank    parms       LL    eigenvalue   statistic      value         value
         0       39   1231.1041                  28.5682      20.97         25.52
         1       44   1245.3882     0.26943       14.2346     14.07         18.63
         2       47   1252.5055     0.14480        3.3465      3.76          6.65
         3       48   1254.1787     0.03611
```

We can reject $r = 1$ in favor of $r = 2$ at the 5% level but not at the 1% level. As with the trace statistic method, whether we choose to specify one or two cointegrating equations in our VECM will depend on the significance level we use here.

◁

Minimizing an information criterion

Many multiple-testing problems in the time-series literature have been solved by defining an estimator that minimizes an information criterion with known asymptotic properties. Selecting the lag length in an autoregressive model is probably the best-known example. Gonzalo and Pitarakis (1998) and Aznar and Salvador (2002) have shown that this approach can be applied to determining the number of cointegrating equations in a VECM. As in the lag-length selection problem, choosing the number of cointegrating equations that minimizes either the Schwarz Bayesian information criterion (SBIC) or the Hannan and Quinn information criterion (HQIC) provides a consistent estimator of the number of cointegrating equations.

▷ Example 4

We use these information-criteria methods to estimate the number of cointegrating equations in our balanced-growth data.

```
. vecrank y i c, lags(5) ic notrace
                    Johansen tests for cointegration
Trend: constant                              Number of obs =      91
Sample:   1960q2   1982q4                              Lags =       5
```

maximum rank	parms	LL	eigenvalue	SBIC	HQIC	AIC
0	39	1231.1041		-25.12401	-25.76596	-26.20009
1	44	1245.3882	0.26943	-25.19009	-25.91435	-26.40414
2	47	1252.5055	0.14480	-25.19781*	-25.97144*	-26.49463
3	48	1254.1787	0.03611	-25.18501	-25.97511	-26.50942

Both the SBIC and the HQIC estimators suggest that there are two cointegrating equations in the balanced-growth data.

◁

Saved Results

vecrank saves the following in e():

Scalars

e(N)	number of observations
e(k_eq)	number of equations
e(k_dv)	number of dependent variables
e(tmin)	minimum time
e(tmax)	maximum time
e(n_lags)	number of lags
e(k_ce95)	number of cointegrating equations chosen by multiple trace tests with level(95)
e(k_ce99)	number of cointegrating equations chosen by multiple trace tests with level(99)
e(k_cesbic)	number of cointegrating equations chosen by minimizing SBIC
e(k_cehqic)	number of cointegrating equations chosen by minimizing HQIC

Macros

e(cmd)	vecrank
e(tmins)	formatted minimum time
e(tmaxs)	formatted maximum time
e(trend)	trend specified
e(reduced_lags)	list of maximum lags to which the model has been reduced
e(reduce_opt)	noreduce, if noreduce option specified
e(tsfmt)	format for current time variable

Matrices

e(max)	vector of maximum-eigenvalue statistics
e(trace)	vector of trace statistics
e(lambda)	vector of eigenvalues
e(k_rank)	vector of numbers of unconstrained parameters
e(hqic)	vector of HQIC values
e(sbic)	vector of SBIC values
e(aic)	vector of AIC values
e(ll)	vector of log-likelihood values

Methods and Formulas

vecrank is implemented as an ado-file.

As shown in *Methods and Formulas* of [TS] **vec**, given a lag, trend, and seasonal specification when there are $0 \leq r \leq K$ cointegrating equations, the log likelihood with the Johansen identification restrictions can be written as

$$L = -\frac{1}{2}T \left[K \{\ln(2\pi) + 1\} + \ln(|S_{00}|) + \sum_{i=1}^{r} \ln\left(1 - \widehat{\lambda}_i\right) \right] \qquad (1)$$

where the $(K \times K)$ matrix S_{00} and the eigenvalues $\widehat{\lambda}_i$ are defined in *Methods and Formulas* of [TS] **vec**.

The trace statistic compares the null hypothesis that there are r or fewer cointegrating relations with the alternative hypothesis that there are more than r cointegrating equations. Under the alternative hypothesis, the log likelihood is

$$L_A = -\frac{1}{2}T \left[K \{\ln(2\pi) + 1\} + \ln(|S_{00}|) + \sum_{i=1}^{K} \ln\left(1 - \widehat{\lambda}_i\right) \right] \qquad (2)$$

Thus the LR test that compares the unrestricted model in (2) with the restricted model in (1) is given by

$$LR_{\text{trace}} = -T \sum_{i=r+1}^{K} \ln\left(1 - \widehat{\lambda}_i\right)$$

As discussed by Johansen (1995), the trace statistic has a nonstandard distribution under the null hypothesis because the null hypothesis places restrictions on the coefficients on $\mathbf{y}_{t-1}$, which is assumed to have $K - r$ random-walk components. vecrank reports the Osterwald-Lenum (1992) critical values.

The maximum-eigenvalue statistic compares the null model containing r cointegrating relations with the alternative model that has $r + 1$ cointegrating relations. Thus using these two values for r in (1) and a few lines of algebra implies that the LR test of this hypothesis is

$$LR_{\max} = -T\ln\left(1 - \widehat{\lambda}_{r+1}\right)$$

As in the case of the trace statistic, because this test involves restrictions on the coefficients on a vector of I(1) variables, the test statistic's distribution will be nonstandard. vecrank reports the Osterwald-Lenum (1992) critical values.

The formulas for the AIC, SBIC, and HQIC are given in *Methods and Formulas* of [TS] **vec**.

Søren Johansen (1939–) earned degrees in mathematical statistics at the University of Copenhagen, where he is now based. In addition to making contributions to mathematical statistics, probability theory, and medical statistics, he has worked mostly in econometrics, in particular, on the theory of cointegration.

References

Anderson, T. W. 1951. Estimating linear restrictions on regression coefficients for multivariate normal distributions. *Annals of Mathematical Statistics* 22: 327–351.

Aznar, A. and M. Santos. 2002. Selecting the rank of the cointegration space and the form of the intercept using an information criterion. *Econometric Theory* 18: 926–947.

Engle, R. F. and C. W. J. Granger. 1987. Cointegration and error correction: Representation, estimation and testing. *Econometrica* 55(2): 251–276.

Gonzalo, J. and J. Y. Pitarakis. 1998. Specification via model selection in vector error-correction models. *Economics Letters* 60: 321–328.

Hamilton, J. D. 1994. *Time Series Analysis*. Princeton: Princeton University Press.

Hubrich, K., H. Lütkepohl, and P. Saikkonen. 2001. A review of systems cointegration tests. *Econometric Reviews* 20(3): 247–318.

Johansen, S. 1988. Statistical analysis of cointegration vectors. *Journal of Economic Dynamics and Control* 12: 231–254.

——. 1991. Estimation and hypothesis testing of cointegration vectors in Gaussian vector autoregressive models. *Econometrica* 59: 1551–1580.

——. 1995. *Likelihood-Based Inference in Cointegrated Vector Auto-Regressive Models*. Oxford: Oxford University Press.

King, R. G., C. I. Plosser, J. H. Stock, and M. W. Watson. 1991. Stochastic trends and economic fluctuations. *The American Economic Review* 81(4): 819–840.

Lütkepohl, H. 1993. *Introduction to Multiple Time Series Analysis*. 2nd ed. New York: Springer.

Maddala, G. S. and I. M. Kim. 1998. *Unit Roots, Cointegration, and Structural Change*. Cambridge, UK: Cambridge University Press.

Osterwald-Lenum, M. 1992. A note with quantiles of the asymptotic distribution of the maximum likelihood cointegration rank test statistics. *Oxford Bulletin of Economics and Statistics* 54(3): 461–472.

Park, J. Y. and P. C. B. Phillips. 1988a. Statistical inference in regressions with integrated regressors: Part I. *Econometric Theory* 4: 468–497.

——. 1988b. Statistical inference in regressions with integrated regressors: Part II. *Econometric Theory* 5: 95–131.

Phillips, P. C. B. 1986. Understanding spurious regressions in econometrics. *Journal of Econometrics* 33: 311–340.

Phillips, P. C. B. and S. N. Durlauf. 1986. Multiple time series regression with integrated processes. *The Review of Economic Studies* 53(4): 473–495.

Sims, C. A., J. H. Stock, and M. W. Watson. 1990. Inference in linear time series models with some unit roots. *Econometrica* 58(1): 113–144.

Stock, J. H. 1987. Asymptotic properties of least squares estimators of cointegrating vectors. *Econometrica* 55(5): 1035–1056.

Stock, J. H. and M. W. Watson. 1988. Testing for common trends. *Journal of the American Statistical Association* 404: 1099–1107.

Watson, M. W. 1994. Vector autoregressions and cointegration. *Handbook of Econometrics*, Vol IV, Engle, R. F. and McFadden, D. L. eds, Amsterdam: Elsevier.

Also See

Complementary:	[TS] **tsset**, [TS] **vec**
Background:	[U] **11.1.10 Prefix commands**,
	[TS] **vec intro**

Title

> **vecstable** — Check the stability condition of VECM estimates

Syntax

vecstable $\left[\,,\ options\,\right]$

options	description
Main	
estimates(*estname*)	use previously saved results *estname*; default is to use active results
amat(*matrix_name*)	save the companion matrix as *matrix_name*
graph	graph eigenvalues of the companion matrix
dlabel	label eigenvalues with the distance from the unit circle
modlabel	label eigenvalues with the modulus
rlopts(*cline_options*)	affect rendition of reference unit circle
nogrid	suppress polar grid circles
pgrid($\left[\,\ldots\,\right]$) $\left[\,\ldots\,\right]$	specify radii and appearance of polar grid circles; see *Options* for details
Add plot	
addplot(*plot*)	add other plots to the generated graph
Y-Axis, X-Axis, Title, Caption, Legend, Overall	
twoway_options	any options other than by() documented in [G] ***twoway_options***

vecstable can be used only after vec; see [TS] **vec**.

Description

vecstable checks the eigenvalue stability condition in a vector error-correction model (VECM) fitted using vec.

Options

> **Main**

estimates(*estname*) requests that vecstable use the previously obtained set of [TS] **vec** estimates saved as *estname* By default, vecstable uses the active results. See [R] **estimates** for information about saving and restoring estimation results.

amat(*matrix_name*) specifies a valid Stata matrix name by which the companion matrix can be saved. The companion matrix is referred to as the **A** matrix in Lütkepohl (1993) and [TS] **varstable**. The default is not to save the companion matrix.

graph causes vecstable to draw a graph of the eigenvalues of the companion matrix.

dlabel labels the eigenvalues with their distances from the unit circle. dlabel can only be specified with graph and cannot be specified with modlabel.

modlabel labels the eigenvalues with their moduli. modlabel can only be specified with graph and cannot be specified with dlabel.

rlopts(*cline_options*) affect the rendition of the reference unit circle; see [G] *cline_options*.

nogrid suppresses the polar grid circles. nogrid can only be specified with graph.

pgrid([*numlist*][, *line_options*]) [pgrid([*numlist*][, *line_options*]) ...

pgrid([*numlist*][, *line_options*])] determines the radii and appearance of the polar grid circles. By default, the graph includes nine polar grid circles with radii .1, .2, ..., .9 that have the grid linestyle. The *numlist* specifies the radii for the polar grid circles. The *line_options* determine the appearance of the polar grid circles; see [G] *line_options*. Since the pgrid() option can be repeated, circles with different radii can have distinct appearances. pgrid() can only be specified with graph.

    Add plot

addplot(*plot*) adds specified plots to the generated graph; see [G] *addplot_option*.

    Y-Axis, X-Axis, Title, Caption, Legend, Overall

twoway_options are any of the options documented in [G] *twoway_options* excluding by(). These include options for titling the graph (see [G] *title_options*) and options for saving the graph to disk (see [G] *saving_option*).

Remarks

Inference after vec requires that the cointegrating equations be stationary and that the number of cointegrating equations be correctly specified. While the methods implemented in vecrank identify the number of stationary cointegrating equations, they assume that the individual variables are I(1). vecstable provides indicators of whether the number of cointegrating equations is misspecified or whether the cointegrating equations, which are assumed to be stationary, are not stationary.

vecstable is analogous to varstable. vecstable uses the coefficient estimates from the previously fitted VECM to back out estimates of the coefficients of the corresponding VAR and then compute the eigenvalues of the companion matrix. See [TS] **varstable** for details about how the companion matrix is formed and about how to interpret the resulting eigenvalues in the case of covariance-stationary VAR models.

If a VECM has K endogenous variables and r cointegrating vectors, there will be $K - r$ unit moduli in the companion matrix. If any of the remaining moduli computed by vecrank are too close to one, either the cointegrating equations are not stationary or there is another common trend and the rank() specified in the vec command is too high. Unfortunately, there is no general distribution theory that allows you to determine if an estimated root is too close to one for all the cases that commonly arise in practice.

▷ Example 1

In [TS] **vec**, we estimated the parameters of a bivariate VECM of the natural logs of the average disposable incomes in two of the economic regions created by the U.S. Bureau of Economic Analysis. In that example, we concluded that the predicted cointegrating equation was probably not stationary. Here we continue that example by refitting that model and using vecstable to analyze the eigenvalues of the companion matrix of the corresponding VAR.

```
. use http://www.stata-press.com/data/r9/rdinc
. vec ln_ne ln_se
  (output omitted)
```

```
. vecstable
Eigenvalue stability condition
```

Eigenvalue		Modulus
1		1
.9477854		.947785
.2545357 +	.2312756i	.343914
.2545357 −	.2312756i	.343914

```
The VECM specification imposes a unit modulus
```

The output contains a table showing the eigenvalues of the companion matrix and their associated moduli. The table shows that one of the roots is 1. The table footer reminds us that the specified VECM imposes one unit modulus on the companion matrix.

The output indicates that there is a real root at about .95. Although there is no distribution theory to measure how close this root is to one, following other discussions in the literature (e.g., Johansen [1995, 137–138]), we conclude that the root of .95 supports our earlier analysis, in which we concluded that the predicted cointegrating equation is probably not stationary.

If we had included the `graph` option with `vecstable`, the following graph would have been displayed:

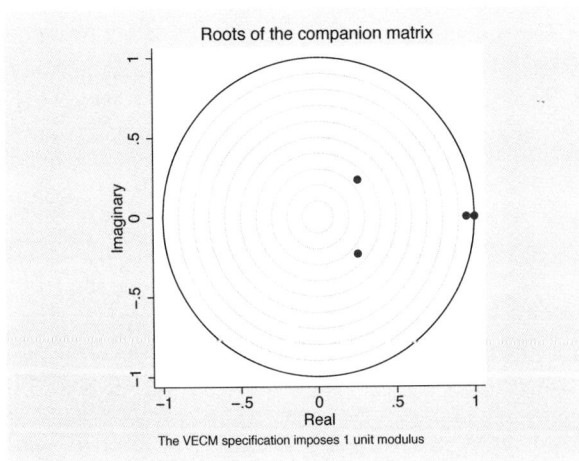

The graph plots the eigenvalues of the companion matrix with the real component on the x-axis and the imaginary component on the y-axis. Although the information is the same as in the table, the graph shows visually how close the root with modulus .95 is to the unit circle.

◁

Saved Results

vecstable saves in r():

Scalars
  r(unitmod)   number of unit moduli imposed on the companion matrix
Matrices
  r(Re)   real part of the eigenvalues of **A**
  r(Im)   imaginary part of the eigenvalues of **A**
  r(Modulus)   moduli of the eigenvalues of **A**

where **A** is the companion matrix of the VAR that corresponds to the VECM.

Methods and Formulas

vecstable is implemented as an ado-file.

vecstable uses the formulas given in *Methods and Formulas* of [TS] **irf create** to obtain estimates of the parameters in the corresponding VAR from the vec estimates. With these estimates, the calculations are identical to those discussed in [TS] **varstable**. In particular, the derivation of the companion matrix, **A**, from the VAR point estimates is given in [TS] **varstable**.

References

Hamilton, J. D. 1994. *Time Series Analysis.* Princeton: Princeton University Press.

Johansen, S. 1995. *Likelihood-Based Inference in Cointegrated Vector Auto-Regressive Models.* Oxford: Oxford University Press.

Lütkepohl, H. 1993. *Introduction to Multiple Time Series Analysis.* 2nd ed. New York: Springer.

Also See

Complementary: [TS] **vec**

Background: [TS] **vec intro**

Title

> **wntestb** — Bartlett's periodogram-based test for white noise

Syntax

> wntestb *varname* [*if*] [*in*] [, *options*]

options	description
Main	
<u>table</u>	display a table instead of graphical output
<u>level</u>(#)	set confidence level; default is level(95)
Plot	
marker_options	change look of markers (color, size, etc.)
marker_label_options	add marker labels; change look or position
cline_options	add connecting lines; change look
Add plot	
addplot(*plot*)	add other plots to the generated graph
Y-Axis, X-Axis, Title, Caption, Legend, Overall	
twoway_options	any options other than by() documented in [G] ***twoway_options***

You must tsset your data before using wntestb; see [TS] **tsset**. In addition, the time series must be dense (nonmissing with no gaps in the time variable) in the specified sample.

varname may contain time-series operators; see [U] **11.4.3 Time-series varlists**.

Description

wntestb performs Bartlett's periodogram-based test for white noise. The result is presented graphically by default but, optionally, may be presented as text in a table.

Options

<u>Main</u>

table displays the test results as a table instead of as the default graph.

level(#) specifies the confidence level, as a percentage, for the confidence bands included on the graph. The default is level(95) or as set by set level; see [U] **20.6 Specifying the width of confidence intervals**.

<u>Plot</u>

marker_options specify the look of markers plotted at the correlation points. This look includes the marker symbol, the marker size, its color and outline; see [G] ***marker_options***.

marker_label_options specify if and how the markers are to be labeled; see [G] ***marker_label_options***.

418

cline_options specify if the points are to be connected with lines and the rendition of those lines; see [G] *cline_options*.

⌐ Add plot ⌐

addplot(*plot*) adds specified plots to the generated graph; see [G] *addplot_option*.

⌐ Y-Axis, X-Axis, Title, Caption, Legend, Overall ⌐

twoway_options are any of the options documented in [G] *twoway_options*, excluding by(). These include options for titling the graph (see [G] *title_options*) and saving the graph to disk (see [G] *saving_option*).

Remarks

Bartlett's test is a test of the null hypothesis that the data come from a white-noise process of uncorrelated random variables having a constant mean and a constant variance.

For a discussion of this test, see Bartlett (1955, 92–94), Newton (1988, 172), or Newton (1996).

▷ Example 1

In this example, we generate two time series and show the graphical and statistical tests that can be obtained from this command. The first time series is a white-noise process, and the second is a white-noise process with an embedded deterministic cosine curve.

```
. drop _all
. set seed 12393
. set obs 100
obs was 0, now 100
. gen x1 = invnormal(uniform())
. gen x2 = invnormal(uniform()) + cos(2*_pi*(_n-1)/10)
. gen time = _n
. tsset time
        time variable:  time, 1 to 100
```

We can then submit the white-noise data to the wntestb command by typing

```
. wntestb x1
```

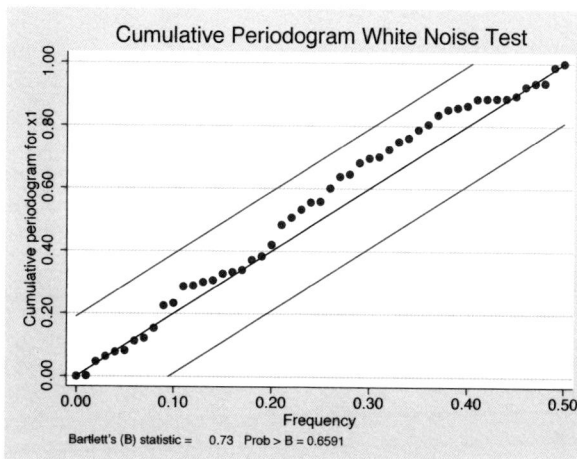

We can see in the graph that the values never appear outside of the confidence bands. We also note that the test statistic has a p-value of .66, so we conclude that the process is not different from white noise. If we had only wanted the statistic without the plot, we could have used the `table` option.

Turning our attention to the other series (x2), we type

`. wntestb x2`

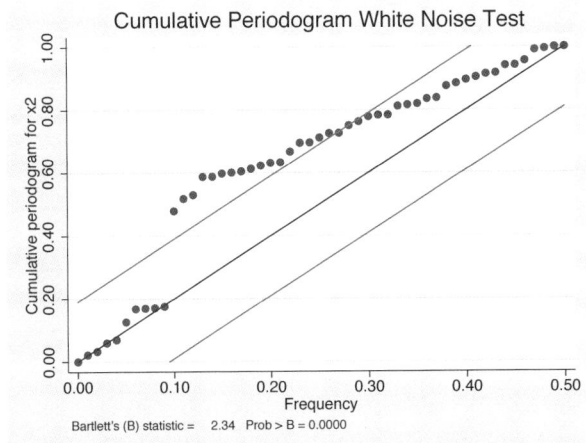

Cumulative Periodogram White Noise Test

Bartlett's (B) statistic = 2.34   Prob > B = 0.0000

Here the process does appear outside of the bands. In fact, it steps out of the bands at a frequency of .1 (exactly as we synthesized this process). We also have confirmation from the test statistic, at a p-value less than .0001, that the process is significantly different from white noise.

◁

Saved Results

`wntestb` saves in `r()`:

Scalars

`r(stat)`	Bartlett's statistic	`r(p)`	probability value

Methods and Formulas

`wntestb` is implemented as an ado-file.

If $x(1), \ldots, x(T)$ is a realization from a white-noise process with variance σ^2, the spectral distribution would be given by $F(\omega) = \sigma^2 \omega$ for $\omega \in [0, 1]$, and we would expect the cumulative periodogram (see [TS] **cumsp**) of the data to be close to the points $S_k = k/q$ for $q = [n/2] + 1, k = 1, \ldots, q$. $[n/2]$ is the maximum integer less than or equal to $n/2$.

Except for $\omega = 0$ and $\omega = .5$, the random variables $2\widehat{f}(\omega_k)/\sigma^2$ are asymptotically independently and identically distributed as χ_2^2. Since χ_2^2 is the same as twice a random variable distributed exponentially with mean 1, the cumulative periodogram has approximately the same distribution as the ordered values from a uniform (on the unit interval) distribution. Feller (1948) shows that this results in

$$\lim_{q \to \infty} \Pr\left(\max_{1 \le k \le q} \sqrt{q}\left|U_k - \frac{k}{q}\right| \le a\right) = \sum_{j=-\infty}^{\infty} (-1)^j e^{-2a^2 j^2} = G(a)$$

where U_k is the ordered uniform quantile. The Bartlett statistic is computed as

$$B = \max_{1 \leq k \leq q} \sqrt{\frac{n}{2}} \left| \widehat{F}_k - \frac{k}{q} \right|$$

where $\widehat{F}_k$ is the cumulative periodogram defined in terms of the sample spectral density $\widehat{f}$ (see [TS] **pergram**) as

$$\widehat{F}_k = \frac{\sum_{j=1}^{k} \widehat{f}(\omega_j)}{\sum_{j=1}^{q} \widehat{f}(\omega_j)}$$

The associated p-value for the Bartlett statistic and the confidence bands on the graph are computed as $1 - G(B)$ using Feller's result.

Maurice Stevenson Bartlett (1910–2002) was a British statistician. Apart from a short period in industry, he spent his career teaching and researching at the universities of Cambridge, Manchester, London (University College), and Oxford. His many contributions include work on the statistical analysis of multivariate data (especially factor analysis) and time series and on stochastic models of population growth, epidemics, and spatial processes.

Acknowledgment

wntestb is based on the **wntestf** command by H. Joseph Newton (1996), Department of Statistics, Texas A&M University.

References

Bartlett, M. S. 1955. *An Introduction to Stochastic Processes with Special Reference to Methods and Applications.* Cambridge: Cambridge University Press.

Feller, W. 1948. On the Kolmogorov–Smirnov theorems for empirical distributions. *Annals of Mathematical Statistics* 19: 177–189.

Gani, J. 2002. Professor M. S. Barlett FRS, 1910–2002. *The Statistician* 51: 399–402.

Newton, H. J. 1988. *TIMESLAB: A Time Series Laboratory*. Pacific Grove, CA: Wadsworth & Brooks/Cole.

———. 1996. sts12: A periodogram-based test for white noise. *Stata Technical Bulletin* 34: 36–39. Reprinted in *Stata Technical Bulletin Reprints*, vol. 6, pp. 203–207.

Olkin, I. 1989. A conversation with Maurice Bartlett. *Statistical Science* 4: 151–163.

Also See

Complementary:	[TS] **tsset**
Related:	[TS] **corrgram**, [TS] **cumsp**, [TS] **pergram**, [TS] **wntestq**
Background:	*Stata Graphics Reference Manual*

Title

wntestq — Portmanteau (Q) test for white noise

Syntax

wntestq *varname* $\big[$ *if* $\big]$ $\big[$ *in* $\big]$ $\big[$, lags(#) $\big]$

You must tsset your data before using wntestq; see [TS] **tsset**. In addition, the time series must be dense (nonmissing with no gaps in the time variable) in the specified sample.

varname may contain time-series operators; see [U] **11.4.3 Time-series varlists**.

Description

wntestq performs the portmanteau (or Q) test for white noise.

Option

lags(#) specifies the number of autocorrelations to calculate. The default is to use $\min([n/2]-2, 40)$ where $[n/2]$ is the greatest integer less than or equal to $n/2$.

Remarks

Box and Pierce (1970) developed a portmanteau test of white noise that was refined by Ljung and Box (1978). See also Diggle (1990, section 2.5).

▷ Example 1

In the example shown in [TS] **wntestb**, we generated two time series. One (x1) was a white-noise process, and the other (x2) was a white noise process with an embedded cosine curve. Here we compare the output of the two tests.

```
. set seed 12393
. set obs 100
obs was 0, now 100
. gen x1 = invnormal(uniform())
. gen x2 = invnormal(uniform()) + cos(2*_pi*(_n-1)/10)
. gen time = _n
. tsset time
        time variable:  time, 1 to 100
. wntestb x1, table
Cumulative periodogram white-noise test
```

Bartlett's (B) statistic	=	0.7311
Prob > B	=	0.6591

```
. wntestq x1

Portmanteau test for white noise
─────────────────────────────────────────────
 Portmanteau (Q) statistic  =    27.2378
 Prob > chi2(40)            =     0.9380

. wntestb x2, table

Cumulative periodogram white-noise test
─────────────────────────────────────────────
 Bartlett's (B) statistic  =     2.3364
 Prob > B                  =     0.0000

. wntestq x2

Portmanteau test for white noise
─────────────────────────────────────────────
 Portmanteau (Q) statistic  =   182.2446
 Prob > chi2(40)            =     0.0000
```

This example shows that both tests agree. For the first process, the Bartlett and portmanteau test results in nonsignificant test statistics: a p-value of 0.6591 for `wntestb` and one of 0.9380 for `wntestq`.

For the second process, each of the tests has a significant result to less than 0.0001.

◁

Saved Results

`wntestq` saves in `r()`:

Scalars

r(stat)	Q statistic	r(p)	Probability value
r(df)	degrees of freedom		

Methods and Formulas

`wntestq` is implemented as an ado-file.

The portmanteau test relies on the fact that if $x(1), \ldots, x(n)$ is a realization from a white-noise process, then

$$Q = n(n+2) \sum_{j=1}^{m} \frac{1}{n-j} \, \widehat{\rho}^{\,2}(j) \longrightarrow \chi_m^2$$

where m is the number of autocorrelations calculated (equal to the number of lags specified) and $\longrightarrow$ indicates convergence in distribution to a χ^2 distribution with m degrees of freedom. $\widehat{\rho}_j$ is the estimated autocorrelation for lag j; see [TS] **corrgram** for details.

References

Box, G. E. P. and D. A. Pierce. 1970. Distribution of residual autocorrelations in autoregressive-integrated moving average time series models. *Journal of the American Statistical Association* 65: 1509–1526.

Diggle, P. J. 1990. *Time Series: A Biostatistical Introduction.* Oxford: Oxford University Press.

Ljung, G. M. and G. E. P. Box. 1978. On a measure of lack of fit in time series models. *Biometrika* 65: 297–303.

Sperling, R. and C. F. Baum. 2001. sts19: Multivariate portmanteau (Q) test for white noise. *Stata Technical Bulletin* 60: 39–41. Reprinted in *Stata Technical Bulletin Reprints*, vol. 10, pp. 373–375.

Also See

Complementary: [TS] **tsset**

Related: [TS] **corrgram**, [TS] **cumsp**, [TS] **wntestb**

Title

> **xcorr** — Cross-correlogram for bivariate time series

Syntax

> xcorr *varname*$_1$ *varname*$_2$ $\begin{bmatrix} if \end{bmatrix}$ $\begin{bmatrix} in \end{bmatrix}$ $\begin{bmatrix} , options \end{bmatrix}$

options	description
Main	
<u>gene</u>rate(*newvar*)	create *newvar* containing cross-correlation values
<u>table</u>	display a table instead of graphical output
<u>nopl</u>ot	do not include the character-based plot in tabular output
<u>lags</u>(#)	include # lags and leads in graph
Plot	
<u>base</u>(#)	value to drop to; default is 0
marker_options	change look of markers (color, size, etc.)
marker_label_options	add marker labels; change look or position
line_options	change look of dropped lines
Add plot	
<u>addp</u>lot(*plot*)	add other plots to the generated graph
Y-Axis, X-Axis, Title, Caption, Legend, Overall	
twoway_options	any options other than by() documented in [G] ***twoway_options***

You must tsset your data before using xcorr; see [TS] **tsset**.
varname$_1$ and *varname*$_2$ may contain time-series operators; see [U] **11.4.3 Time-series varlists**.

Description

xcorr plots the sample cross-correlation function.

Options

> **Main**

generate(*newvar*) specifies a new variable to contain the cross-correlation values.

table requests that the results be presented as a table rather than the default graph.

noplot requests that the tabular output not include the character-based plot of the cross-correlations.

lags(#) indicates the number of lags and leads to include in the graph. The default is to use $\min(\lfloor n/2 \rfloor - 2, 20)$.

> **Plot**

base(#) specifies the value from which the lines should extend. The default is base(0).

marker_options, *marker_label_options*, and *line_options* affect the rendition of the plotted cross-correlations.

   marker_options specify the look of markers plotted at the correlation points. This look includes the marker symbol, the marker size, its color and outline; see [G] *marker_options*.

   marker_label_options specify if and how the markers are to be labeled; see [G] *marker_label_options*.

   line_options specify the look of the dropped lines, including pattern, width, and color; see [G] *line_options*.

⌐ Add plot ⌐_____

addplot(*plot*) provides a way to add other plots to the generated graph. See [G] *addplot_option*.

⌐ Y-Axis, X-Axis, Title, Caption, Legend, Overall ⌐_____

twoway_options are any of the options documented in [G] *twoway_options*, excluding by(). These include options for titling the graph (see [G] *title_options*) and options for saving the graph to disk (see [G] *saving_option*).

Remarks

▷ Example 1

We have a bivariate time series (Box, Jenkins, and Reinsel 1994, Series J) on the input and output of a gas furnace, where 296 paired observations on the input (gas rate) and output (percent CO_2) were recorded every 9 seconds. The cross-correlation function is given by

```
. use http://www.stata-press.com/data/r9/furnace
(TIMESLAB: Gas furnace)

. xcorr input output, xline(5) lags(40)
```

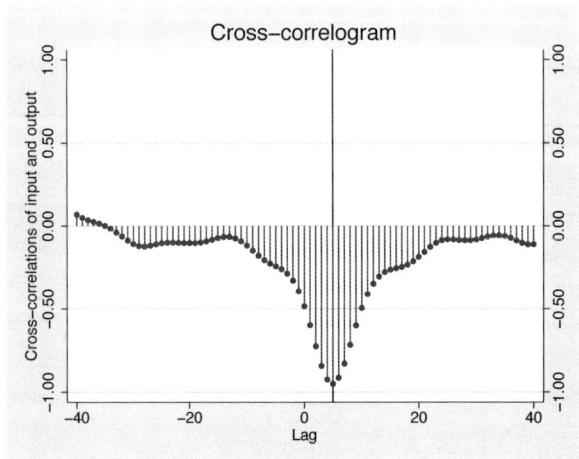

Note that we included a vertical line at lag 5, as there is a well-defined peak at this value. This peak indicates that the output lags the input by 5 time periods. Further, the fact that the correlations are negative indicates that as input (coded gas rate) is increased, output ($\%CO_2$) decreases.

We may obtain the table of autocorrelations and the character-based plot of the cross-correlations (analogous to the univariate time-series command `corrgram`) by specifying the `table` option.

```
. xcorr input output, table lags(20)
                      -1        0         1
   LAG      CORR    [Cross-correlation]

   -20    -0.1033                 |
   -19    -0.1027                 |
   -18    -0.0998                 |
   -17    -0.0932                 |
   -16    -0.0832                 |
   -15    -0.0727                 |
   -14    -0.0660                 |
   -13    -0.0662                 |
   -12    -0.0751                 |
   -11    -0.0927                 |
   -10    -0.1180                 |
    -9    -0.1484               --|
    -8    -0.1793               --|
    -7    -0.2059               --|
    -6    -0.2266               --|
    -5    -0.2429               --|
    -4    -0.2604              ---|
    -3    -0.2865              ---|
    -2    -0.3287              ---|
    -1    -0.3936             ----|
     0    -0.4845             ----|
     1    -0.5985            -----|
     2    -0.7251           ------|
     3    -0.8429          -------|
     4    -0.9246         --------|
     5    -0.9503         --------|
     6    -0.9146         --------|
     7    -0.8294          -------|
     8    -0.7166           ------|
     9    -0.5998            -----|
    10    -0.4952             ----|
    11    -0.4107             ----|
    12    -0.3479              ---|
    13    -0.3049              ---|
    14    -0.2779              ---|
    15    -0.2632              ---|
    16    -0.2548              ---|
    17    -0.2463               --|
    18    -0.2332               --|
    19    -0.2135               --|
    20    -0.1869               --|
```

Once again, the well-defined peak is apparent in the plot.

◁

Methods and Formulas

`xcorr` is implemented as an ado-file.

The cross-covariance function of lag k for time series x_1 and x_2 is given by

$$\text{Cov}\Big\{x_1(t), x_2(t+k)\Big\} = R_{12}(k)$$

Note that this function is not symmetric about lag zero; that is,

$$R_{12}(k) \neq R_{12}(-k)$$

We define the cross-correlation function as

$$\rho_{ij}(k) = \text{Corr}\Big\{x_i(t), x_j(t+k)\Big\} = \frac{R_{ij}(k)}{\sqrt{R_{ii}(0)R_{jj}(0)}}$$

where ρ_{11} and ρ_{22} are the autocorrelation functions for x_1 and x_2, respectively. The sequence $\rho_{12}(k)$ is the cross-correlation function and is drawn for lags $k \in (-Q, -Q+1, \ldots, -1, 0, 1, \ldots, Q-1, Q)$.

Note that if $\rho_{12}(k) = 0$ for all lags, x_1 and x_2 are not cross-correlated.

References

Box, G. E. P., G. M. Jenkins, and G. C. Reinsel. 1994. *Time Series Analysis: Forecasting and Control*. 3rd ed. Englewood Cliffs, NJ: Prentice Hall.

Hamilton, J. D. 1994. *Time Series Analysis*. Princeton: Princeton University Press.

Newton, H. J. 1988. *TIMESLAB: A Time Series Laboratory*. Pacific Grove, CA: Wadsworth & Brooks/Cole.

Also See

Complementary:	[TS] **tsset**
Related:	[TS] **corrgram**, [TS] **pergram**

Glossary

ARCH model. An autoregressive conditional heteroskedasticity (ARCH) model is a regression model in which the conditional variance is modeled as an autoregressive (AR) process. The ARCH(m) model is

$$y_t = \mathbf{x}_t\boldsymbol{\beta} + \epsilon_t$$
$$\mathrm{E}(\epsilon_t^2|\epsilon_{t-1}^2, \epsilon_{t-2}^2, \ldots) = \alpha_0 + \alpha_1\epsilon_{t-1}^2 + \cdots + \alpha_m\epsilon_{t-m}^2$$

where ϵ_t is a white-noise error term. The equation for y_t represents the conditional mean of the process, and the equation for $\mathrm{E}(\epsilon_t^2|\epsilon_{t-1}^2, \epsilon_{t-2}^2, \ldots)$ specifies the conditional variance as an autoregressive function of its past realizations. Although the conditional variance changes over time, the unconditional variance is time invariant because y_t is a stationary process. Modeling the conditional variance as an AR process raises the implied unconditional variance, making this model particularly appealing to researchers modeling fat-tailed data, such as financial data.

ARIMA model. An autoregressive integrated moving-average (ARIMA) model is a time-series model suitable for use with integrated processes. In an ARIMA(p, d, q) model, the data is differenced d times to obtain a stationary series, and then an ARMA(p, q) model is fit to this differenced data. ARIMA models that include exogenous explanatory variables are known as ARMAX models.

ARMA model. An autoregressive moving-average (ARMA) model is a time-series model in which the current period's realization is the sum of an autoregressive (AR) process and a moving-average (MA) process. An ARMA(p, q) model includes p AR terms and q MA terms. ARMA models with just a few lags are often able to fit data, as well as pure AR or MA models with many more lags.

ARMAX model. An ARMAX model is a time-series model in which the current period's realization is an ARMA process plus a linear function of a set of a exogenous variables. Equivalently, an ARMAX model is a linear regression model in which the error term is specified to follow an ARMA process.

autocorrelation function. The autocorrelation function (ACF) expresses the correlation between periods t and $t - k$ of a time series as function of the time t and the lag k. For a stationary time series, the ACF does not depend on t and is symmetric about $k = 0$, meaning that the correlation between periods t and $t - k$ is equal to the correlation between periods t and $t + k$.

autoregressive process. An autoregressive process is a time-series model in which the current value of a variable is a linear function of its own past values and a white-noise error term. A first-order autoregressive process, denoted as an AR(1) process, is $y_t = \rho y_{t-1} + \epsilon_t$. An AR($p$) model contains p lagged values of the dependent variable.

Cochrane–Orcutt estimator. This estimation is a linear regression estimator that can be used when the error term exhibits first-order autocorrelation. An initial estimate of the autocorrelation parameter ρ is obtained from OLS residuals, and then OLS is performed on the transformed data $\widetilde{y}_t = y_t - \rho y_{t-1}$ and $\widetilde{\mathbf{x}}_t = \mathbf{x}_t - \rho\mathbf{x}_{t-1}$.

cointegrating vector. A cointegrating vector specifies a stationary linear combination of nonstationary variables. Specifically, if each of the variables $x_1, x_2, \ldots, x_k$ is integrated of order one and there exists a set of parameters $\beta_1, \beta_2, \ldots, \beta_k$ such that $z_t = \beta_1 x_1 + \beta_2 x_2 + \cdots + \beta_k x_k$ is a stationary process, the variables $x_1, x_2, \ldots, x_k$ are said to be cointegrated, and the vector $\boldsymbol{\beta}$ is known as a cointegrating vector.

conditional variance. While the conditional variance is simply the variance of a conditional distribution, in time-series analysis the conditional variance is frequently modeled as autoregressive process, giving rise to ARCH models.

correlogram. A correlogram is a table or graph showing the sample autocorrelations or partial autocorrelations of a time series.

covariance stationarity. A process is covariance stationary if (1) the mean of the process is finite and independent of t, (2) the unconditional variance of the process is finite and independent of t, and (3) the covariance between periods t and $t - s$ is finite and depends on $t - s$ but not on t or s themselves. Covariance-stationary processes are also known as weakly stationary processes.

cross-correlation function. The cross-correlation function expresses the correlation between one series at time t and another series at time $t - k$ as a function of the time t and lag k. If both series are stationary, the function does not depend on t. The function is not symmetric about $k = 0$: $\rho_{12}(k) \neq \rho_{12}(-k)$.

difference operator. The difference operator Δ denotes the change in the value of a variable from period $t - 1$ to period t. Formally, $\Delta y_t = y_t - y_{t-1}$, and $\Delta^2 y_t = \Delta(y_t - y_{t-1}) = (y_t - y_{t-1}) - (y_{t-1} - y_{t-2}) = y_t - 2y_{t-1} + y_{t-2}$.

dynamic forecast. A dynamic forecast is one in which the current period's forecast is calculated using forecasted values for prior periods.

exponential smoothing. Exponential smoothing is a method of smoothing a time series in which the smoothed value at period t is equal to a fraction α of the series value at time t plus a fraction $1 - \alpha$ of the previous period's smoothed value. The fraction α is known as the smoothing parameter.

forecast-error variance decomposition. Forecast-error variance decompositions measure the fraction of the error in forecasting variable i after h periods that is attributable to the orthogonalized shocks to variable j.

forward operator. The forward operator F denotes the value of a variable at time $t + 1$. Formally, $Fy_t = y_{t+1}$, and $F^2 y_t = Fy_{t+1} = y_{t+2}$.

frequency-domain analysis. Frequency-domain analysis is analysis of time-series data by considering its frequency properties. The spectral density and distribution functions are key components of frequency-domain analysis, so it is often called spectral analysis. In Stata, the `cumsp` and `pergram` commands are used to analyze the sample spectral distribution and density functions, respectively.

GARCH model. A generalized autoregressive conditional heteroskedasticity (GARCH) model is a regression model in which the conditional variance is modeled as an ARMA process. The GARCH(m, k) model is

$$y_t = \mathbf{x}_t \boldsymbol{\beta} + \epsilon_t$$
$$\sigma_t^2 = \gamma_0 + \gamma_1 \epsilon_{t-1}^2 + \cdots + \gamma_m \epsilon_{t-m}^2 + \delta_1 \sigma_{t-1}^2 + \cdots + \delta_k \sigma_{t-k}^2$$

where the equation for y_t represents the conditional mean of the process and σ_t represents the conditional variance. See [TS] **arch** or Hamilton (1994, chapter 21) for details on how the conditional variance equation can be viewed as an ARMA process. GARCH models are often used because the ARMA specification often allows the conditional variance to be modeled with fewer parameters than are required by a pure ARCH model. Many extensions to the basic GARCH model exist; see [TS] **arch** for those that are implemented in Stata. See also *ARCH model*.

generalized least-squares estimator. A generalized least-squares (GLS) estimator is used to estimate the parameters of a regression function when the error term is heteroskedastic or autocorrelated. In the linear case, GLS is sometimes described as "OLS on transformed data" because the GLS estimator can be implemented by applying an appropriate transformation to the dataset and then using OLS.

Holt–Winters smoothing. A set of methods for smoothing time-series data that assume the value of a time series at time t can be approximated as the sum of a mean term that drifts over time, as well as a time trend whose strength also drifts over time. Variations of the basic method allow for seasonal patterns in data, as well.

impulse–response function. An impulse–response function (IRF) measures the effect of a shock to an endogenous variable on itself or another endogenous variable. The kth impulse–response function of variable i on variable j measures the effect on variable j in period $t + k$ in response to a one-unit shock to variable i in period t, holding everything else constant.

independent and identically distributed. A series of observations is independently and identically distributed (i.i.d.) if each observation is an independent realization from the same underlying distribution. In some contexts, the definition is relaxed to mean only that the observations are independent and have identical means and variances; see Davidson and MacKinnon (1993, 42).

integrated process. A nonstationary process is integrated of order d, written I(d), if the process must be differenced d times to produce a stationary series. An I(1) process y_t is one in which Δy_t is stationary.

lag operator. The lag operator L denotes the value of a variable at time $t - 1$. Formally, $Ly_t = y_{t-1}$, and $L^2 y_t = Ly_{t-1} = y_{t-2}$.

moving-average process. A moving-average process is a time-series process in which the current value of a variable is modeled as a weighted average of current and past realizations of a white-noise process and, optionally, a time-invariant constant. By convention, the weight on the current realization of the white-noise process is equal to one, and the weights on the past realizations are known as the moving-average (MA) coefficients. A first-order moving-average process, denoted as an MA(1) process, is $y_t = \theta \epsilon_{t-1} + \epsilon_t$.

Newey–West covariance matrix. The Newey–West covariance matrix is a member of the class of heteroskedasticity-and-autocorrelation-consistent (HAC) covariance matrix estimators used with time-series data that produces covariance estimates that are robust to both arbitrary heteroskedasticity and autocorrelation up to a prespecified lag.

orthogonalized impulse–response function. An orthogonalized impulse–response function (OIRF) measures the effect of an orthogonalized shock to an endogenous variable on itself or another endogenous variable. An orthogonalized shock is one that affects a single variable at time t but no other variables. See [TS] **irf create** for a discussion of the difference between IRFs and OIRFs.

partial autocorrelation function. The partial autocorrelation function (PACF) expresses the correlation between periods t and $t - k$ of a time series as a function of the time t and lag k, after controlling for the effects of intervening lags. For a stationary time series, the PACF does not depend on t. The PACF is not symmetric about $k = 0$: the partial autocorrelation between y_t and y_{t-k} is not equal to the partial autocorrelation between y_t and y_{t+k}.

periodogram. A periodogram is a graph of the spectral density function of a time series as a function of frequency. Note that the `pergram` command first standardizes the amplitude of the density by the sample variance of the time series, then plots the logarithm of that standardized density. Peaks in the periodogram represent cyclical behavior in the data.

Prais–Winsten estimator. A Prais–Winsten estimator is a linear regression estimator that is used when the error term exhibits first-order autocorrelation; see also *Cochrane–Orcutt estimator*. Here the first observation in the dataset is transformed as $\widetilde{y}_1 = \sqrt{1 - \rho^2}\, y_1$ and $\widetilde{\mathbf{x}}_1 = \sqrt{1 - \rho^2}\, \mathbf{x}_1$, so that the first observation is not lost. The Prais–Winsten estimator is a generalized least-squares estimator.

random walk. A random walk is a time-series process in which the current period's realization is equal to the previous period's realization plus a white-noise error term: $y_t = y_{t-1} + \epsilon_t$. A *random walk with drift* also contains a nonzero time-invariant constant: $y_t = \delta + y_{t-1} + \epsilon_t$. The constant term δ is known as the drift parameter. An important property of random-walk processes is that the best predictor of the value at time $t + 1$ is the value at time t plus the value of the drift parameter.

seasonal difference operator. The period-s seasonal difference operator Δ_s denotes the difference in the value of a variable at time t and time $t - s$. Formally, $\Delta_s y_t = y_t - y_{t-s}$, and $\Delta_s^2 y_t = \Delta_s(y_t - y_{t-s}) = (y_t - y_{t-s}) - (y_{t-s} - y_{t-2s}) = y_t - 2y_{t-s} + y_{t-2s}$.

serial correlation. Serial correlation refers to regression errors that are correlated over time. If a regression model does not contained lagged dependent variables as regressors, the OLS estimates are consistent in the presence of mild serial correlation, but the covariance matrix is incorrect. When the model includes lagged dependent variables and the residuals are serially correlated, the OLS estimates are biased and inconsistent. See, for example, Davidson and MacKinnon (1993, chapter 10) for more information.

serial correlation tests. Because OLS estimates are at least inefficient and potentially biased in the presence of serial correlation, econometricians have developed many tests to detect it. Popular ones include the Durbin–Watson (1950, 1951, 1971) test, the Breusch–Pagan (1980) test, and Durbin's (1970) alternative test. See [R] **regress postestimation time series**.

smoothing. Smoothing a time series refers to the process of extracting an overall trend in the data. The motivation behind smoothing is the belief that a time series exhibits a trend component as well as an irregular component and that the analyst is interested only in the trend component. Some smoothers also account for seasonal or other cyclical patterns as well.

spectral analysis. See *frequency-domain analysis*.

spectral density function. The spectral density function is the derivative of the spectral distribution function. Intuitively, the spectral density function $f(\omega)$ indicates the amount of variance in a time series that is attributable to sinusoidal components with frequency ω. See also *spectral distribution function*. The spectral density function is sometimes called the *spectrum*.

spectral distribution function. The (normalized) spectral distribution function $F(\omega)$ of a process describes the proportion of variance that can be explained by sinusoids with frequencies in the range $(0, \omega)$, where $0 \le \omega \le \pi$. The spectral distribution and density functions used in frequency-domain analysis are closely related to the autocorrelation function used in time-domain analysis; see, for example, Chatfield (2004, chapter 6).

spectrum. See *spectral density function*.

strict stationarity. A process is strictly stationary if the joint distribution of $y_1, \ldots, y_k$ is the same as the joint distribution of $y_{1+\tau}, \ldots, y_{k+\tau}$ for all k and τ. Intuitively, shifting the origin of the series by τ units has no effect on the joint distributions.

SVAR. A structural vector autoregression (SVAR) is a type of VAR in which short- or long-run constraints are placed on the resulting impulse–response functions. The constraints are usually motivated by economic theory and therefore allow causal interpretations of the IRFs to be made.

time-domain analysis. Time-domain analysis is analysis of data viewed as a sequence of observations observed over time. The autocorrelation function, linear regression, ARCH models, and ARIMA models are common tools used in time-domain analysis.

unit-root process. A unit-root process is one which is integrated of order one, meaning that the process is nonstationary but that first-differencing the process produces a stationary series. The simplest example of a unit-root process is the random walk. See Hamilton (1994, chapter 15) for a discussion of when general ARMA processes may contain a unit root.

unit-root tests. Whether a process has a unit root has both important statistical and economic ramifications, so a variety of tests have been developed to test for them. Among the earliest tests proposed is the one by Dickey and Fuller (1979), though most researchers now use an improved variant called the augmented Dickey–Fuller test instead of the original version. Other common unit-root tests implemented in Stata include the DF–GLS test of Elliot, Rothenberg, and Stock (1996) and the Phillips–Perron (1988) test. See [TS] **dfuller**, [TS] **dfgls**, and [TS] **pperron**.

VAR. A vector autoregression (VAR) is a multivariate regression technique in which each dependent variable is regressed on lags of itself and on lags of all the other dependent variables in the model. Occasionally, exogenous variables are included in the model as well.

VECM. A vector error-correction model (VECM) is a type of VAR that is used with variables that are cointegrated. Although first-differencing variables that are integrated of order one makes them stationary, fitting a VAR to such first-differenced variables results in misspecification error if the variables are cointegrated. See *The multivariate VECM specification* in [TS] **vec intro** for more on this point.

white noise. A variable u_t represents a white-noise process if the mean of u_t is zero, the variance of u_t is σ^2, and the covariance between u_t and u_s is zero for all $s \neq t$. Gaussian white noise refers to white noise in which u_t is normally distributed.

References

Breusch, T. S. and A. R. Pagan. 1980. The Lagrange multiplier test and its applications to model specification in econometrics. *Review of Economic Studies* 47: 334–355.

Chatfield, C. 2004. *The Analysis of Time Series: An Introduction*. 6th ed. London: Chapman & Hall.

Davidson, R. and J. G. MacKinnon. 1993. *Estimation and Inference in Econometrics*. Oxford: Oxford University Press.

Dickey, D. A. and W. A. Fuller. 1979. Distribution of the estimators for autoregressive time series with a unit root. *Journal of the American Statistical Association* 74: 427–431.

Durbin, J. 1970. Testing for serial correlation in least squares regression when some of the regressors are lagged dependent variables. *Econometrica* 38: 410–421.

Durbin, J. and G. S. Watson. 1950. Testing for serial correlation in least squares regression I. *Biometrika* 37: 409–428.

——. 1951. Testing for serial correlation in least squares regression II. *Biometrika* 38: 159–178.

——. 1971. Testing for serial correlation in least squares regression III. *Biometrika* 58: 1–19.

Elliot, G., T. Rothenberg, and J. H. Stock. 1996. Efficient tests for an autoregressive unit root. *Econometrica* 64: 813–836.

Hamilton, J. D. 1994. *Time Series Analysis*. Princeton: Princeton University Press.

Phillips, P. C. B. and P. Perron. 1988. Testing for a unit root in time series regression. *Biometrika* 75: 335–346.

Subject and author index

This is the subject and author index for the *Stata Time-Series Reference Manual*. Readers interested in topics other than time series should see the combined subject index in the *Stata Quick Reference and Index*, which indexes
the *Getting Started with Stata for Macintosh Manual*, the *Getting Started with Stata for Unix Manual*, the *Getting Started with Stata for Windows Manual*, the *Stata User's Guide*, the *Stata Base Reference Manual*, the *Stata Data Management Reference Manual*, the *Stata Graphics Reference Manual*, the *Stata Programming Reference Manual*, the *Stata Longitudinal/Panel Data Reference Manual*, the *Stata Multivariate Statistics Reference Manual*, the *Stata Survey Data Reference Manual*, the *Stata Survival Analysis & Epidemiological Tables Reference Manual*, and this manual.
Readers interested in Mata topics should see the index at the end of the *Mata Reference Manual*.

Semicolons set off the most important entries from the rest. Sometimes no entry will be set off with semicolons, meaning that all entries are equally important.

A

Abraham, B., [TS] **tssmooth**, [TS] **tssmooth dexponential**, [TS] **tssmooth exponential**, [TS] **tssmooth hwinters**, [TS] **tssmooth shwinters**
ac command, [TS] **corrgram**
Akaike, H., [TS] **varsoc**
Amemiya, T., [TS] **varsoc**
Amisano, G., [TS] **irf create**, [TS] **var intro**, [TS] **var svar**, [TS] **vargranger**, [TS] **varwle**
Anderson, T. W., [TS] **vec**, [TS] **vecrank**
Ansley, C. F., [TS] **arima**
A-PARCH, [TS] **arch**
ARCH,
    postestimation, [TS] **arch postestimation**
    regression, [TS] **arch**
arch command, [TS] **arch**
ARCH effects, estimation, [TS] **arch**
ARIMA,
    postestimation, [TS] **arima postestimation**
    regression, [TS] **arima**
arima command, [TS] **arima**
ARMA, [TS] **arch**, [TS] **arima**
ARMAX model, [TS] **arima**
autocorrelation, [TS] **arch**, [TS] **arima**, [TS] **corrgram**, [TS] **newey**, [TS] **prais**, [TS] **var**, [TS] **varlmar**
autoregression, *see* VAR
autoregressive
    conditional heteroskedasticity, [TS] **arch**
    integrated moving average, [TS] **arima**; [TS] **arch**
    moving average, [TS] **arima**; [TS] **arch**
Aznar, A., [TS] **vecrank**

B

Bartlett, M. S., [TS] **wntestb**
Bartlett's
    bands, [TS] **corrgram**
    periodogram test, [TS] **wntestb**
Baum, C. F., [TS] **arch**, [TS] **arima**, [TS] **dfgls**, [TS] **rolling**, [TS] **time series**, [TS] **tsset**, [TS] **var**, [TS] **wntestq**
Becketti, S., [TS] **corrgram**
Bera, A. K., [TS] **arch**, [TS] **varnorm**, [TS] **vecnorm**
Berndt, E. K., [TS] **arch**, [TS] **arima**
Black, F., [TS] **arch**
Bollerslev, T., [TS] **arch**, [TS] **arima**
Boswijk, H. P., [TS] **vec**
Bowerman, B., [TS] **tssmooth**, [TS] **tssmooth dexponential**, [TS] **tssmooth exponential**, [TS] **tssmooth hwinters**, [TS] **tssmooth shwinters**
Box, G. E. P., [TS] **arima**, [TS] **corrgram**, [TS] **cumsp**, [TS] **dfuller**, [TS] **pergram**, [TS] **pperron**, [TS] **wntestq**, [TS] **xcorr**
Brockwell, P. J., [TS] **corrgram**

C

Chatfield, C., [TS] **arima**, [TS] **corrgram**, [TS] **pergram**, [TS] **tssmooth**, [TS] **tssmooth dexponential**, [TS] **tssmooth exponential**, [TS] **tssmooth hwinters**, [TS] **tssmooth ma**, [TS] **tssmooth shwinters**
Chatterjee, S., [TS] **prais**
Cheung, Y., [TS] **dfgls**
Chou, R. Y., [TS] **arch**
Christiano, L. J., [TS] **irf create**, [TS] **var svar**
Cochrane, D., [TS] **prais**
Cochrane–Orcutt regression, [TS] **prais**
cointegration, [TS] **fcast compute**, [TS] **fcast graph**, [TS] **vec intro**, [TS] **vec**, [TS] **veclmar**, [TS] **vecnorm**, [TS] **vecrank**, [TS] **vecstable**
conditional variance, [TS] **arch**
correlogram, [TS] **corrgram**
corrgram command, [TS] **corrgram**
Cox, N. J., [TS] **tssmooth hwinters**, [TS] **tssmooth shwinters**
cross-correlogram, [TS] **xcorr**
cumsp command, [TS] **cumsp**
cumulative spectral distribution, empirical, [TS] **cumsp**

D

data manipulation, [TS] **tsappend**, [TS] **tsfill**, [TS] **tsreport**, [TS] **tsrevar**, [TS] **tsset**
David, J. S., [TS] **arima**
Davidson, R., [TS] **arch**, [TS] **arima**, [TS] **varlmar**
Davis, R. A., [TS] **corrgram**
DeGroot, M. H., [TS] **arima**
dfgls command, [TS] **dfgls**

X

Y

Z